Lecture Notes in Electrical Engineering

Volume 876

The book series *Lecture Notes in Electrical Engineering* (LNEE) publishes the latest developments in Electrical Engineering - quickly, informally and in high quality. While original research reported in proceedings and monographs has traditionally formed the core of LNEE, we also encourage authors to submit books devoted to supporting student education and professional training in the various fields and applications areas of electrical engineering. The series cover classical and emerging topics concerning:

- Communication Engineering, Information Theory and Networks
- Electronics Engineering and Microelectronics
- Signal, Image and Speech Processing
- Wireless and Mobile Communication
- Circuits and Systems
- Energy Systems, Power Electronics and Electrical Machines
- Electro-optical Engineering
- Instrumentation Engineering
- Avionics Engineering
- Control Systems
- Internet-of-Things and Cybersecurity
- Biomedical Devices, MEMS and NEMS

For general information about this book series, comments or suggestions, please contact leontina.dicecco@springer.com.

To submit a proposal or request further information, please contact the Publishing Editor in your country:

China

Jasmine Dou, Editor (jasmine.dou@springer.com)

India, Japan, Rest of Asia

Swati Meherishi, Editorial Director (Swati.Meherishi@springer.com)

Southeast Asia, Australia, New Zealand

Ramesh Nath Premnath, Editor (ramesh.premnath@springernature.com)

USA, Canada:

Michael Luby, Senior Editor (michael.luby@springer.com)

All other Countries:

Leontina Di Cecco, Senior Editor (leontina.dicecco@springer.com)

**** This series is indexed by EI Compendex and Scopus databases. ****

More information about this series at https://link.springer.com/bookseries/7818

Mohan Kolhe · Aziz Muhammad ·
Abdel El Kharbachi · Tri Yogi Yuwono
Editors

Recent Advances in Renewable Energy Systems

Select Proceedings of ICOME 2021

Springer

Editors
Mohan Kolhe
Faculty of Engineering and Science
University of Agder
Grimstad, Norway

Abdel El Kharbachi
Electrochemical Energy Storage
Helmholtz-Institute Ulm
Ulm, Germany

Aziz Muhammad
Department of Mechanical
and Biofunctional Systems
The University of Tokyo
Tokyo, Japan

Tri Yogi Yuwono
Department of Mechanical Engineering
Sepuluh Nopember Institute of Technology
Surabaya, Indonesia

ISSN 1876-1100 ISSN 1876-1119 (electronic)
Lecture Notes in Electrical Engineering
ISBN 978-981-19-1580-2 ISBN 978-981-19-1581-9 (eBook)
https://doi.org/10.1007/978-981-19-1581-9

This Springer imprint is published by the registered company Springer Nature Singapore Pte Ltd.
The registered company address is: 152 Beach Road, #21-01/04 Gateway East, Singapore 189721, Singapore

Preface

In the World Energy Outlook 2021, the International Energy Agency (IEA) states that there will be an increase in global average temperature past 1.5 °C in around 2030 and will continue to increase until it reaches 2.6 °C in 2100, although demand for fossil fuels tends to slow down until 2030 and will decrease slightly around 2050. Unfortunately, not all agendas at the 26th Conference of the Parties (COP26) in November 2021 in Glasgow can be agreed. In particular regarding maintaining the temperature increase constantly at 1.5 degrees, the elimination/reduction of subsidies for the use of coal and fossil fuels must be implemented, as well as efforts to produce a balanced decision text (balance text) between the obligation to increase ambition and targets (mitigation) by state parties and the obligation to fulfil funding commitments from developed countries to developing countries (www.menlhk.go.id). It may be due to the diversity of economic conditions and the needs of each country. The economic crisis in various countries, especially in developing countries, due to the prolonged COVID-19 pandemic has hampered efforts to increase access to clean energy, such as electricity and cooking fuel from clean energy. The state budget is more concentrated on overcoming the pandemic and economic recovery due to the pandemic.

The IEA states four actions to contain the 1.5 °C temperature increase: a massive push for clean electrification; a new focus on realizing the full potential of energy efficiency; concerted efforts to prevent leakage from fossil fuel operations; and the drive for clean energy innovation. We researchers or academics from universities must have a contribution to participate in creating a clean and sustainable environment, which we can do through our innovations, especially related to clean energy. For this reason, the Department of Mechanical Engineering, Institut Teknologi Sepuluh Nopember-Surabaya-Indonesia (ITS), held the International Conference on Mechanical Engineering for the fifth time. Researchers from various universities presented their innovative works at the conference, particularly related to energy efficiency and technology.

This book is a collection of promising innovations selected from the Fifth International Conference on Mechanical Engineering (ICOME-2021) proceedings. The variety of innovations in this book, covering aerodynamics, fluid mechanics, cooling and air conditioning systems, turbo machineries, thermodynamics, heat transfer, combustion systems, alternative fuels, etc., are intended to address future challenges. These innovations are developed with technology that leads to sustainable development, especially discussing energy efficiency in renewable energy resources and technology. Finally, we hope that the publication of this book can inspire every reader to be aware of the importance of clean energy for the future of the world as a noble legacy for future generations.

Surabaya, Indonesia
February 2022

Tri Yogi Yuwono

Contents

Computational Fluid Dynamics Study of a Steam Reformer Unit Performance to Produce Hydrogen Fuel for PEM Fuel Cell Applications 1
Hussein A. Z. AL-bonsrulah, Dhinakaran Veeman, and M. V. Reddy

Fuel Cell Power Pack with Integrated Metal Hydride Hydrogen Storage for Powering Electric Forklift 19
Ivan Tolj, Mykhaylo Lototskyy, Adrian Parsons, and Sivakumar Pasupathi

Testing an In-House CFD Code for Solving Gas–Solid Flow with Different Simulation Parameters 29
Is Bunyamin Suryo, Tri Yogi Yuwono, and Uwe Schnell

A Numerical Study for the Prediction of Unmanned Aerial Vehicle Aerodynamic Performance Based on Dihedral and Tip-Twist Angles of the Wing 37
Adi Susanto and Arif Wahjudi

A Numerical Study for Prediction of Unmanned Aerial Vehicle Aerodynamic Performance Based on Chord Tip and Offset of the Wing 45
Firiana Firdaus, Arif Wahjudi, and Wawan Aries Widodo

Potential of a Grid-Tied PV System: A Field Study in Hot and Sunny Climate Region 53
I Nyoman Suamir, I Wayan Temaja, and I Nengah Ardita

Simulation and Dynamic System Modeling in an Elastically Supported Rigid Cylinder for Vibration Energy Harvesting 61
Subekti, Harus Laksana Guntur, Vivien S. Djanali, and Achmad Syaifudin

Parameters Analysis of Vortex Formation on Conical Basin of Gravitational Water Vortex Power Plant (GWVPP) ... 69
Erna Septyaningrum, Ridho Hantoro, Sarwono, and Ester Carolina

Solar Canopy with IoT-Based Single-Axis Solar Tracking System as a Solution for Utilizing Urban Open Parking Area ... 79
Radix Kautsar Ramadhan, Hafiz Rayhan Gunawan, and Galang Adi Saputro

Investigation of the Four Runner Blade Arrangement Against the Power of Kaplan Turbine ... 87
Sirojuddin, Nadia Sari Dewi, and Ragil Sukarno

Investigation of the Runner Blade Arrangements on a 3-Blade Kaplan Turbine Against Turbine Power ... 97
Sirojuddin, Alya Awanis Zahara, and Ragil Sukarno

Analysis Study of Performance and Reliability Impact in Boiler Through Differential Coal Calorific Value (Case Study: Pelabuhan Ratu Coal-Fired Power Plant) ... 105
Hendra Yudisaputro, M. Nur Yuniarto, Yohanes, and Agus Wibawa

Effect of the Oxide Scale on Tube Boiler Remaining Life of a 600 MW Coal Power Plant ... 113
Diki Purwadi, Suwarno, and Vivien S. Djanali

Numerical Study of the Generator Lubricant Cooler Air-Side Flow to Increase the Reliability of GTG#1.3 PLTGU Muara Karang ... 121
Aris Kurniawan and Sutardi

Optimization of Coal Blending with Backpropagation Neural Networks (BPNN) and Genetic Algorithms (GA) in Tangential In-Furnace Blending Boilers ... 131
Mohamad Kurnadi, Sutikno, and M. Khoirul Effendi

Exergy Analysis in Gas Turbine Power Plant with Different Offline Compressor Washing Methods ... 145
Arif Budianto and D. Bambang Arip

Swappable Battery Innovation as a Drone Frame Structure with Purpose to Increasing the Flight Time Duration ... 153
Muhammad Haekal Shafi, Valiant Tirta Amarta, Ferdina Ramadhansyah, Puguh Pambudi, and Alief Wikarta

Changes the Governor Valve Operation Mode to Improve Efficiency HP Turbine ... 161
Aripin Gandi Marbun

Energy Absorption Analysis on Crash-Module Shape and Configuration of Medium-Speed Train 171
Achmad Syaifudin, Agus Windharto, Andri Setiawan, and Abdul Rochman Farid

Numerical Study Air Preheat Coil as Porous Medium to Analyse Flow Characteristics and Improve Productivity in Muara Karang Unit 4 Steam Power Plant 181
Rizal Mahendra Pratama and Tri Yogi Yuwono

Numerical Study for the Modified Cooler of SAF Motor at Power Plant 189
Khoirul Huda, Prabowo, Bambang Arip Dwiyantoro, and Teguh Widjayanto

Numerical Study of the Effect of Cooling Air on Low-Pressure Air Cooler in a Two-Stage Reciprocating Compressor 197
Verry Mardiananta Arsana and Sutardi

Numerical Study of Fin Height Effects with the Staggered Arrangement in Annular-finned Tube Heat Exchangers 205
Deluxe La, Prabowo, and Tri Vicca Kusumadewi

Study of the Effect of Changes in Fill Grid Cooling Tower Unit 2 Salak Geothermal Power Plant 213
Yuansah, Wawan Aries Widodo, and Wahyu Somantri

Performance Assessment of Drying Machine Using CFD 223
Mokhammad Fahmi Izdiharrudin, Ridho Hantoro, and Erna Septyaningrum

Cold Isostatic Pressing Treatment in the Preparation of Al and Y-Doped LLZO ($Li_{6.15}La_3Zr_{1.75}Al_{0.2}Y_{0.25}O_{12-\delta}$) Solid Electrolyte 231
Septia Kurniawati Arifah, Fitria Rahmawati, and Yuniawan Hidayat

The Effect of Inlet Air Cooling to Power Output Enhancement of Gas Turbine 241
Iman Firmansyah and Prabowo

Numerical Study for Evaluating Effect of Mass Flow Rate Toward Particle Circulation Rate on Seal Pot in Circulating Fluidized Bed Boiler 249
Nur Ikhwan and Denny Oktavianto

Study of Effect of Intake Air Temperature of Compressor Gas Turbine on Exergy Destruction in Tambak Lorok Combined Cycle Power Plants 257
Marzuki, Ary Bachtiar Krishna Putra, and Christiono Utomo

Experimental Study Combustion Air Optimization to Increase Efficiency Due to Changes in Coal Quality with Variation of Load in Labuan Steam Power Plant 271
Adriska Simon Prayoga

Numerical Study Effect Using Low Rank Coal on Flow Characteristics, Combustion, and Furnace Exit Gas Temperature on Tangentially Fired Pulverized Coal Boiler 350 MWe 281
Arief Laga Putra, Wawan Aries Widodo, and Ardi Nugroho

Numerical Study of Emissions on DDF Engine with 20% CNG with Variation on Compression Ratio 289
Betty Ariani, I. Made Ariana, and Aguk Zuhdi M. Fathallah

Numerical Study of Gas Mixing Effect on Block 3 and Block 4 Muara Tawar's Gas Turbine Combustion Stability 295
Danan Tri Yulianto and Bambang Sudarmanta

Numerical Study of the Effects of Burner Tilt and Coal Optimization on Combustion Characteristics of 350 MWe Tangentially Fired Pulverized Coal Boiler 303
Halim and Ary Bachtiar Krishna Putra

Thermogravimetric and Kinetic Analysis on Peat Combustion Through Coats-Redfern Fitting Model 311
Ardianto Prasetiyo and Sukarni Sukarni

Co-Combustion of Water Hyacinth (*Eichhornia crassipes*) and Coal: Thermal Behavior and Kinetics Analysis Under the Coats-Redfern Method 321
Sukarni Sukarni, Nandang Mufti, Avita Ayu Permanasari, Ardianto Prasetiyo, Poppy Puspitasari, and Anwar Johari

Effect of Stored Dexlite and Palm Oil Biodiesel on Fuel Properties, Performance, and Emission of Single-Cylinder Diesel Engines 333
Atok Setiyawan, Kuntang Winangun, and Vernanda Sania

Production of Bioethanol from Corn Straw by Co-immobilization of *Saccharomyces cerevisiae* and *Aspergillus niger* in Na-Alginate: Review and Potential Study 341
Karenina Anisya Pratiwi, Petra Arde Septia Graha, Shinta Dewi Surya Pertiwi, Yuliana Dewi Puspitasari, Muhammad Dimas Hafani, Afan Hamzah, Arief Widjaja, and Soeprijanto

Analysis of the Characteristic and Performance Development of Coconut Biodiesel 353
Tri Vicca Kusumadewi and Digdo Listyadi Setyawan

The Analysis of Biogas Production from Mixed Cow Dung with Watermelon Rind as an Alternative Fuel 361
Faisal Manta, Doddy Suanggana, Putra Dilto Tondok, and Firman Ali Nuryanto

Tar Reduction in a Three-Stage Refuse-Derived Fuel Gasification System by Adjusting Intake Air Ratio 369
Bambang Sudarmanta, Atok Setiawan, Is Bunyamin Suryo, Harsono, and Sigit Mujiarto

The Effect of Corncob-Active Carbon Adsorbent Mass on Methane and Carbon Dioxide Content on Biogas Purification 377
Slamet Wahyudi and Lia Fitriya

Author Index 385

Editors and Contributors

About the Editors

Prof. Mohan Kolhe is with the University of Agder (Norway) as full professor in electrical power engineering with focus in smart grid and renewable energy in the Faculty of Engineering and Science. He has also received the offer of full professorship in smart grid from the Norwegian University of Science and Technology (NTNU). He is a leading renewable energy technologist and has previously held academic positions at the world's prestigious universities e.g. University College London (UK/Australia), University of Dundee (UK); University of Jyvaskyla (Finland); Hydrogen Research Institute, QC (Canada); etc. He has been successful in winning competitive research funding from the prestigious research councils (e.g. Norwegian Research Council, EU, EPSRC, BBSRC, NRP, etc.) for his work on sustainable energy systems. He has published extensively in the energy systems engineering.

Dr. Aziz Muhammad is currently an associate professor at Institute of Innovative Research, Tokyo Institute of Technology, Tokyo, Japan. He obtained his bachelor's degree from Department of Mechanical and Aerospace Engineering, Faculty of Engineering, Kyushu University (2004), master's degree and Doctor's degree from Engineering, Department of Intelligent Machinery and Systems Engineering, Faculty of Engineering, Kyushu University. His major areas of research interests include Energy systems, Process design, Power generation, Carbon capture and storage, Hydrogen production, Renewable Energy, Energy conservation, Energy and exergy analysis, Exergy recovery, Electric vehicle, Batteries, and Smart grid.

Dr. Abdel El Kharbachi DEA, Ph.D. is currently an associate professor at the Electrochemical Energy Storage of Helmholtz Institute Ulm (HIU), Germany. He obtained his M.Sc. (Chemistry) and DEA (Chemistry) from the University of Liège, and also his Ph.D. From Institut National Polytechnique de Grenoble, France. Abdel El Kharbachi, DEA, Ph.D. Has served as a research associate in France at both the

CEA-Saclay (for ITER project) and LRCS Lab-Amiens (Laboratoire de Réactivité et Chimie des Solides). He has a panoply of research experiences in materials characterization and thermodynamics, and inorganic analytical (electro)chemistry. In 2015–2018, he worked at IFE-Norway as a researcher and project leader for battery-related activities based on hydrides. Late 2018, Abdel joined the Fichtner Group HIU/KIT in the framework of CELEST (Center for Electrochemical Energy Storage Ulm & Karlsruhe) as a senior scientist working and supervising the research on novel oxy-fluoride materials for Li/Na ion and solid-state batteries. Since 2019, he serves as Guest Editor of the International Journal of Hydrogen Energy and Editorial Board member of MDPI journals (Sustainability, Crystals etc.).

Prof. Tri Yogi Yuwono is currently at the Department of Mechanical Engineering, Sepuluh Nopember Institute of Technology-Indonesia, the university where he obtained his bachelor's degree and was the president (2011–2015). He obtained his master's (DEA) and Ph.D. degrees in 1990 and 1993 respectively from the Institute National Polytechnique de Grenoble (INPG), France. He has published more than 55 international papers. He is a reviewer in several respected international journals. His main research interests are fluid flow, wind turbines, and hydrokinetic turbines.

Contributors

Hussein A. Z. AL-bonsrulah Iraq Ministry of Oil, Midland Refineries Company, Najaf Refinery, Najaf, Iraq

Valiant Tirta Amarta Department of Mechanical Engineering, Institut Teknologi Sepuluh Nopember, Surabaya, Indonesia

I Nengah Ardita Mechanical Engineering Department, Bali, Indonesia

Betty Ariani Naval Architecture, Universitas Muhammadiyah Surabaya, Surabaya, Indonesia

Septia Kurniawati Arifah Research Group of Solid-State Chemistry & Catalysis, Chemistry Department, Sebelas Maret University, Surakarta, Indonesia

Verry Mardiananta Arsana The Department of Mechanical Engineering, FTIRS-ITS Collaboration Program With PT Perusahaan Listrik Negara, Surabaya, Indonesia

D. Bambang Arip Department of Mechanical Engineering, Institut Teknologi Sepuluh Nopember, Surabaya, Indonesia

Arif Budianto Department of Mechanical Engineering, Institut Teknologi Sepuluh Nopember, Surabaya, Indonesia

Ester Carolina Sepuluh Nopember Institute of Technology, Surabaya, Indonesia

Nadia Sari Dewi State University of Jakarta, Jakarta, Indonesia

Vivien S. Djanali Department of Mechanical Engineering, Institut Teknologi Sepuluh Nopember, Kampus Keputih-Sukolilo, Surabaya, Indonesia

Bambang Arip Dwiyantoro Departement of Mechanical Engineering, Institut Teknologi Sepuluh Nopember, Surabaya, Indonesia

Abdul Rochman Farid Division of Product Supports, Indonesian Railway Company Ltd. (PT. INKA), Surabaya, Indonesia

Aguk Zuhdi M. Fathallah Department of Marine Engineering, Institut Teknologi Sepuluh Nopember, Surabaya, Indonesia

Firiana Firdaus Institut Teknologi Sepuluh Nopember, Surabaya, JawaTimur, Indonesia

Iman Firmansyah Department of Mechanical Engineering, Institut Teknologi Sepuluh Nopember, Surabaya, Indonesia

Lia Fitriya Brawijaya University, Malang, Indonesia

Petra Arde Septia Graha Departement of Industrial Chemical Engineering, Faculty of Vocational Studies, Institut Teknologi Sepuluh Nopember, Surabaya, Indonesia

Hafiz Rayhan Gunawan Institut Teknologi Sepuluh Nopember, Surabaya, Indonesia

Harus Laksana Guntur Mechanical Engineering Department, Institut Teknologi Sepuluh Nopember, Surabaya, Indonesia

Muhammad Dimas Hafani Departement of Industrial Chemical Engineering, Faculty of Vocational Studies, Institut Teknologi Sepuluh Nopember, Surabaya, Indonesia

Halim PT. Indonesia Power, Jakarta, Indonesia;
Department of Mechanical Engineering, Sepuluh Nopember Institute of Technology, Surabaya, Indonesia

Afan Hamzah Departement of Industrial Chemical Engineering, Faculty of Vocational Studies, Institut Teknologi Sepuluh Nopember, Surabaya, Indonesia

Ridho Hantoro Sepuluh Nopember Institute of Technology, Surabaya, Indonesia; Department of Engineering Physics, Faculty of Industrial Technology, Intitut Teknologi Sepuluh Nopember, Surabaya, Indonesia

Harsono Institut Teknologi Sepuluh Nopember, Sukolilo, Surabaya, Indonesia

Yuniawan Hidayat Research Group of Solid-State Chemistry & Catalysis, Chemistry Department, Sebelas Maret University, Surakarta, Indonesia

Khoirul Huda Departement of Mechanical Engineering, Institut Teknologi Sepuluh Nopember, Surabaya, Indonesia

Nur Ikhwan Mechanical Engineering Department, Institut Teknologi Sepuluh Nopember, Surabaya, Indonesia

Mokhammad Fahmi Izdiharrudin Department of Engineering Physics, Faculty of Industrial Technology, Intitut Teknologi Sepuluh Nopember, Surabaya, Indonesia

Anwar Johari Centre of Hydrogen Energy, Institute of Future Energy, Universiti Teknologi Malaysia, Johor Bahru, Malaysia

M. Khoirul Effendi Department of Mechanical Engineering, Institut Teknologi Sepuluh Nopember, Surabaya, Indonesia

Mohamad Kurnadi Department of Mechanical Engineering, Institut Teknologi Sepuluh Nopember, Surabaya, Indonesia

Aris Kurniawan Department of Mechanical Engineering, FTIRS-ITS Collaboration Program With PT Perusahaan Listrik Negara, Jakarta, Indonesia

Tri Vicca Kusumadewi Department of Mechanical Engineering, Sepuluh Nopember Institute of Technology (ITS), Surabaya, Indonesia

Deluxe La Department of Mechanical Engineering, Institut Teknologi Sepuluh Nopember, Surabaya, Indonesia

Mykhaylo Lototskyy HySA Systems Competence Center, South African Institute for Advanced Materials Chemistry (SAIAMC), University of the Western Cape, Bellville, South Africa

I. Made Ariana Department of Marine Engineering, Institut Teknologi Sepuluh Nopember, Surabaya, Indonesia

Faisal Manta Department of Mechanical Engineering, Kalimantan Institute of Technology, Balikpapan, Indonesia

Aripin Gandi Marbun Department of Mechanical Engineering, Institut Teknologi Sepuluh Nopember (ITS), Surabaya, Indonesia

Marzuki Department of Mechanical Engineering, Institut Teknologi Sepuluh Nopember, Surabaya, Indonesia

Nandang Mufti Center of Advanced Materials for Renewable Energy (CAMRY), Universitas Negeri Malang, Malang, Indonesia;
Department of Physics, Universitas Negeri Malang, Malang, Indonesia

Sigit Mujiarto Institut Teknologi Sepuluh Nopember, Sukolilo, Surabaya, Indonesia

Ardi Nugroho Electricity Technology Development, PT. Pembangkitan Jawa-Bali, Surabaya, Indonesia

M. Nur Yuniarto Department of Mechanical Engineering, Institut Teknologi Sepuluh Nopember, Surabaya, Indonesia

Firman Ali Nuryanto Department of Mechanical Engineering, Kalimantan Institute of Technology, Balikpapan, Indonesia

Denny Oktavianto Mechanical Engineering Department, Institut Teknologi Sepuluh Nopember, Surabaya, Indonesia

Puguh Pambudi Department of Mechanical Engineering, Institut Teknologi Sepuluh Nopember, Surabaya, Indonesia

Adrian Parsons HySA Systems Competence Center, South African Institute for Advanced Materials Chemistry (SAIAMC), University of the Western Cape, Bellville, South Africa

Sivakumar Pasupathi HySA Systems Competence Center, South African Institute for Advanced Materials Chemistry (SAIAMC), University of the Western Cape, Bellville, South Africa

Avita Ayu Permanasari Center for Renewable and Sustainable Energy Engineering (CRSEE), Department of Mechanical Engineering, Universitas Negeri Malang, Malang, Indonesia;
Center of Advanced Materials for Renewable Energy (CAMRY), Universitas Negeri Malang, Malang, Indonesia

Shinta Dewi Surya Pertiwi Departement of Industrial Chemical Engineering, Faculty of Vocational Studies, Institut Teknologi Sepuluh Nopember, Surabaya, Indonesia

Prabowo Departement of Mechanical Engineering, Institut Teknologi Sepuluh Nopember, Surabaya, Indonesia

Ardianto Prasetiyo Center for Renewable and Sustainable Energy Engineering (CRSEE), Department of Mechanical Engineering, Universitas Negeri Malang, Malang, Indonesia

Rizal Mahendra Pratama Institut Teknologi Sepuluh Nopember, Surabaya, Indonesia

Karenina Anisya Pratiwi Departement of Industrial Chemical Engineering, Faculty of Vocational Studies, Institut Teknologi Sepuluh Nopember, Surabaya, Indonesia

Diki Purwadi Department of Mechanical Engineering, Institut Teknologi Sepuluh Nopember, Kampus Keputih-Sukolilo, Surabaya, Indonesia

Poppy Puspitasari Center for Renewable and Sustainable Energy Engineering (CRSEE), Department of Mechanical Engineering, Universitas Negeri Malang, Malang, Indonesia;

Center of Advanced Materials for Renewable Energy (CAMRY), Universitas Negeri Malang, Malang, Indonesia

Yuliana Dewi Puspitasari Departement of Industrial Chemical Engineering, Faculty of Vocational Studies, Institut Teknologi Sepuluh Nopember, Surabaya, Indonesia

Arief Laga Putra Tanjung Awar-Awar CFPP, PT. Pembangkitan Jawa-Bali, Tuban, East Java, Indonesia

Ary Bachtiar Krishna Putra Department of Mechanical Engineering, Sepuluh Nopember Institute of Technology, Surabaya, Indonesia

Fitria Rahmawati Research Group of Solid-State Chemistry & Catalysis, Chemistry Department, Sebelas Maret University, Surakarta, Indonesia

Radix Kautsar Ramadhan Institut Teknologi Sepuluh Nopember, Surabaya, Indonesia

Ferdina Ramadhansyah Department of Industrial Mechanical Engineering, Institut Teknologi Sepuluh Nopember, Surabaya, Indonesia

M. V. Reddy Centre of Excellence in Transportation Electrification and Energy Storage (CETEES), Institute of Research Hydro-Québec, Varennes, QC, Canada

Vernanda Sania Mechanical Engineering Department, Institut Teknologi Sepuluh Nopember, Surabaya, Indonesia

Galang Adi Saputro Institut Teknologi Sepuluh Nopember, Surabaya, Indonesia

Sarwono Sepuluh Nopember Institute of Technology, Surabaya, Indonesia

Uwe Schnell Institut Für Feuerungs- Und Kraftwerkstechnik (IFK), Universitaet Stuttgart, Stuttgart, Germany

Erna Septyaningrum Department of Engineering Physics, Faculty of Industrial Technology, Intitut Teknologi Sepuluh Nopember, Surabaya, Indonesia

Andri Setiawan Graduate Program of Supply Chain Management, Institut Teknologi Sepuluh Nopember, Surabaya, Indonesia

Atok Setiawan Institut Teknologi Sepuluh Nopember, Sukolilo, Surabaya, Indonesia

Atok Setiyawan Mechanical Engineering Department, Institut Teknologi Sepuluh Nopember, Surabaya, Indonesia

Digdo Listyadi Setyawan Department of Mechanical Engineering, University of Jember, Jember, Indonesia

Muhammad Haekal Shafi Department of Mechanical Engineering, Institut Teknologi Sepuluh Nopember, Surabaya, Indonesia

Adriska Simon Prayoga Department of Mechanical Engineering, Institut Teknologi Sepuluh Nopember, Surabaya, Indonesia

Sirojuddin State University of Jakarta, Jakarta, Indonesia

Soeprijanto Departement of Industrial Chemical Engineering, Faculty of Vocational Studies, Institut Teknologi Sepuluh Nopember, Surabaya, Indonesia

Wahyu Somantri PT. Indonesia Power, Surabaya, Indonesia

I Nyoman Suamir Mechanical Engineering Department, Bali, Indonesia

Doddy Suanggana Department of Mechanical Engineering, Kalimantan Institute of Technology, Balikpapan, Indonesia

Subekti Graduate Program of Engineering Department, Institut Teknologi Sepuluh Nopember, Surabaya, Indonesia;
Mechanical Engineering Department, Universitas Mercu Buana, West Jakarta, Indonesia

Bambang Sudarmanta Department of Mechanical Engineering, Institut Teknologi Sepuluh Nopember, Surabaya, Indonesia

Sukarni Sukarni Center for Renewable and Sustainable Energy Engineering (CRSEE), Department of Mechanical Engineering, Universitas Negeri Malang, Malang, Indonesia;
Center of Advanced Materials for Renewable Energy (CAMRY), Universitas Negeri Malang, Malang, Indonesia

Ragil Sukarno State University of Jakarta, Jakarta, Indonesia

Is Bunyamin Suryo Department of Mechanical Engineering, Institut Tekhnologi Sepuluh Nopember, Sukolilo, Surabaya, Indonesia

Adi Susanto Department of Mechanical Engineering, Institut Teknologi Sepuluh Nopember, Surabaya, Indonesia

Sutardi Department of Mechanical Engineering, FTIRS-ITS, Surabaya, Indonesia;
Department of Mechanical Engineering, Institut Teknologi Sepuluh Nopember, Surabaya, Indonesia

Suwarno Department of Mechanical Engineering, Institut Teknologi Sepuluh Nopember, Kampus Keputih-Sukolilo, Surabaya, Indonesia

Achmad Syaifudin Department of Mechanical Engineering, Institut Teknologi Sepuluh Nopember, Surabaya, Indonesia

I Wayan Temaja Mechanical Engineering Department, Bali, Indonesia

Ivan Tolj Department of Thermodynamics, Faculty of Mechanical Engineering and Naval Architecture, University of Split, Split, Croatia

Putra Dilto Tondok Department of Mechanical Engineering, Kalimantan Institute of Technology, Balikpapan, Indonesia

Christiono Utomo Department of Mechanical Engineering, Institut Teknologi Sepuluh Nopember, Surabaya, Indonesia

Dhinakaran Veeman Centre for Computational Mechanics, Chennai Institute of Technology, Chennai, India

Arif Wahjudi Department of Mechanical Engineering, Institut Teknologi Sepuluh Nopember, Surabaya, JawaTimur, Indonesia

Slamet Wahyudi Brawijaya University, Malang, Indonesia

Agus Wibawa PT. Pembangkitan Jawa Bali, Surabaya, Indonesia

Arief Widjaja Departement of Chemical Engineering. Industrial and System Technology Faculty, Institut Teknologi Sepuluh Nopember, Surabaya, Indonesia

Teguh Widjayanto PT PJB Services, Surabaya, Indonesia

Wawan Aries Widodo Department of Mechanical Engineering, Institut Teknologi Sepuluh Nopember, Surabaya, JawaTimur, Indonesia

Alief Wikarta Department of Mechanical Engineering, Institut Teknologi Sepuluh Nopember, Surabaya, Indonesia

Kuntang Winangun Mechanical Engineering Department, Institut Teknologi Sepuluh Nopember, Surabaya, Indonesia

Agus Windharto Department of Industrial Design, Institut Teknologi Sepuluh Nopember, Surabaya, Indonesia

Yohanes Department of Mechanical Engineering, Institut Teknologi Sepuluh Nopember, Surabaya, Indonesia

Yuansah Department of Mechanical Engineering, Institut Teknologi Sepuluh Nopember, Surabaya, Indonesia

Hendra Yudisaputro PT. Indonesia Power, Jakarta, Indonesia;
Department of Mechanical Engineering, Institut Teknologi Sepuluh Nopember, Surabaya, Indonesia

Danan Tri Yulianto Department of Mechanical Engineering, Institut Teknologi Sepuluh Nopember, Surabaya, Indonesia

Tri Yogi Yuwono Department of Mechanical Engineering, Institut Tekhnologi Sepuluh Nopember, Sukolilo, Surabaya, Indonesia

Alya Awanis Zahara State University of Jakarta, Jakarta, Indonesia

Computational Fluid Dynamics Study of a Steam Reformer Unit Performance to Produce Hydrogen Fuel for PEM Fuel Cell Applications

Hussein A. Z. AL-bonsrulah[1](✉), Dhinakaran Veeman[2], and M. V. Reddy[3]

[1] Iraq Ministry of Oil, Midland Refineries Company, Najaf Refinery, Najaf 54001, Iraq
huseenabd541@gmail.com

[2] Centre for Computational Mechanics, Chennai Institute of Technology, Chennai 600069, India

[3] Centre of Excellence in Transportation Electrification and Energy Storage (CETEES), Institute of Research Hydro-Québec, 1806, Lionel-Boulet Blvd., Varennes, QC J3X 1S1, Canada

1 Introduction

Natural gas is a fossil fuel that is formed under the earth's surface because of anaerobic digestion, including plants of different forms of fatal matter. It is placed deep inside the earth and boiled as oil. Natural gas is a healthier substitute, because it is used to replace other forms of fossil fuel by running vehicles and by producing electricity. Figure 1 depicts the world chart of gas output in cubic meters a year in the nations. Figure 2 depicts the world map of the countries with proven natural gas deposits.

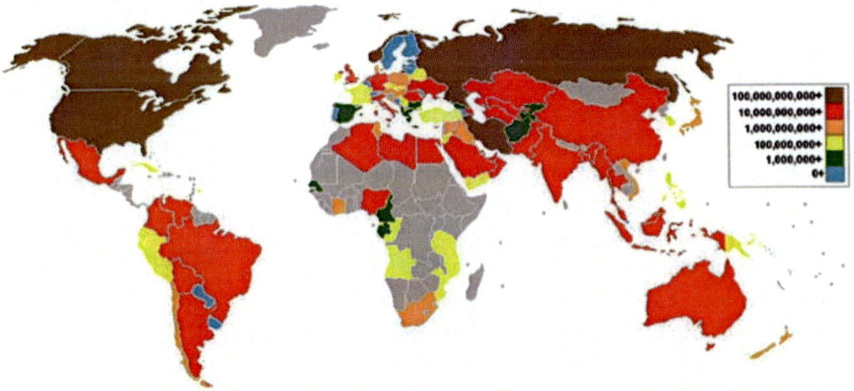

Fig. 1 Countries produce natural gas in cubic meters per annum [1]

However, natural gas releases some quantities of greenhouse gas. The main downside for natural gas is the introduction of carbon emissions into the environment due to climate change and global warming. Even as the most carbon dioxide-emitting countries have stepped up attempts to mitigate emissions, clear steps do have to be taken. While natural

M. Kolhe et al. (eds.), *Recent Advances in Renewable Energy Systems*, Lecture Notes in Electrical Engineering 876,
https://doi.org/10.1007/978-981-19-1581-9_1

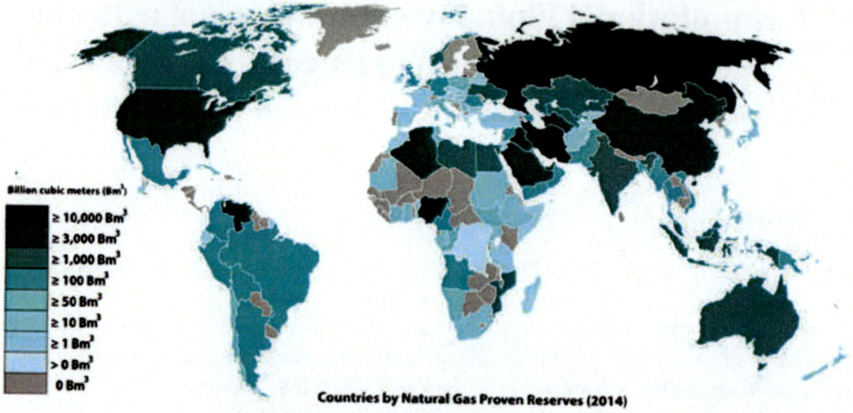

Fig. 2 Countries by natural gas proven reserves [1]

gas has been seen as a stronger source of energy than coal and oil in the past, several analysts today argue that if you look at the lifelong effect of gas extraction on end-of-life use, it may well be even less harmful to Mother Nature than oil [1, 2]. The reform of natural gas fuels would provide hydrogen for fuel cell electric vehicles today. Water vapor is the only product for the fuel cell electric vehicle tailpipes, with the hydrocarbon upstream phase, such as natural gas, as well as sources and storage for usage in electric fuel cell cars, and the gross greenhouse gas emissions have been lowered by half, although oil has been cut by more than 90% compared with today's internal fuel cells. A steam reformer extracts hydrocarbon hydrogen from natural gas, methane, and propane oils. Steam reform is the most popular commercial method of H_2 processing and is, therefore, the easiest, environmental, and protection system.

The machine has zero sulfides (H_2S) and carbon monoxide (CO) emissions, which are harmful intermediate chemicals in the manufacturing phase. The by-product of this manufacturing method is CO_2, which is a greenhouse gas (GHG). Much CO_2 is recovered and transformed into dry ice during the process stream, and the rest is released from the unit [3, 4]. Steam reform of hydrocarbon fuel offers a reliable, cost-effective, and widely used process for hydrogen production which provides energy protection that benefits the environment in the near and medium term. Fuel-reforming hydrocarbons will provide electric PEM fuel cell hydrogen. The majority of the hydrogen now manufactured in the USA is generated by the reform of steamed methane, a mature process in which high-temperature steam (700–1000 °C) is utilized to produce methane hydrogen such as natural gas [5]. In steam-methane reforming, in the presence of a catalyst to create hydrogen, carbon monoxide, and a fairly tiny quantity of carbon dioxide, methane interacts with steam under 3–15 bar pressure (1 bar = 14.5 psi) [6]. Steam reform is endothermic, i.e., heat must be provided for the reaction process. Then, carbon monoxide and steam are reacted using a catalytic converter to create additional carbon dioxide and hydrogen in what is called the "water gas shift reaction." Carbon dioxide and other contaminants are removed from the gas stream and remain mainly hydrogen in a final stage termed the "pressure swing adsorption." Steam reforming may also be used to generate hydrogen from more fuels, like ethanol, propane, or even petrol [7]. The methane and other hydrocarbons in natural gas react in partial oxidation with a limited oxygen content (usually air) that is insufficient to oxidize hydrocarbons entirely into carbon dioxide and water.

The reaction products comprise mostly hydrogen and carbon monoxide (and nitrogen if the process is performed with air rather than purity), and a relatively tiny amount of carbon dioxide and other molecules, as well as less than the stoichiometric oxygen available [8]. The carbon monoxide then interacts with water to make up carbon dioxide and additional hydrogen in a water–gas shift process. Partial oxidation is a process that gives off heat. Typically, the procedure is significantly more rapid than reforming steam and requires a smaller reactor vessel. Initially, this process generates less hydrogen per unit of the input fuel, as can be demonstrated with chemical reactions in partial oxidation, than with the steam reform of the same fuel. Low-cost reform of natural gas can offer today hydrogen for electric cell automobiles and other uses [9]. In the end, the DOE predicts that natural gas hydrogen production will increase alongside renewable power, nuclear power, coal (carbon capture and storage) production, and other low-carbon household energy supplies. New methods have been created due to high hydrogen generation by steam reform and some disadvantages arising from the complexity of processes, such as high stages, thermodynamic limitations, and low energy efficiency. In ammonia factories and during the refining process, major industrial uses of hydrogen are made and sulfur is extracted from petroleum. One that utilizes more than 1.5 million cubic meters a day is the wide use of hydrogen (50 million standard cubic feet per day) [10]. Hydrogen is generally generated in these circumstances at the place in which it is used.

2 Experimental Setup

2.1 PEM Fuel Cells and Hydrogen Fuel

As a chemical, PEM fuel cells use hydrogen. The marketing of PEM fuel cells are accompanied by the marketing of hydro-energy technology—e.g., hydrogen development, supply, and storage technologies—particularly for the transport and stationary electricity generation markets. This implies that hydrogen must be readily accessible (not as a scientific gas, but as a carrier) until fuel cells can be completely traded [3–11]. The fuel cells, on the other side, may very well be the guiding factor to advance hydrogen power technology. PEM fuel cells own a lot of special characteristics, for instance, a rise in power consumption, without pollution, sound, alternation, and possible cost-efficient, rendering them desirable in a large number of applications although a small hydrogen provision. The hydrogen power plants are designed for hydrogen-powered cars, but may also be used to power small machines.

Hydrogen has several special characteristics that render it a perfect energy carrier, specifically [2, 3]:

- It can be extracted from relatively high efficiency and turned into fuel.

- Water abundantly accessible is the raw material for the processing of hydrogen. Hydrogen is a completely sustainable resource, and the utilization of hydrogen through electrochemical conversion helps in safe water.
- It can be found in gaseous form (for large storage), in liquid form (for air and space transport), or metal or chemical hydrides (convenient for surface vehicles and other relatively limited storage requirements).
- Can be transported over long distances via pipelines or tankers (in some cases more efficiently and economically than electricity).
- It can be transferred to other energy sources more easily than the other gasoline.
- As a transport carrier, hydrogen is environmentally friendly and does not induce waste, greenhouse gas, or any adverse environmental consequences through its packaging, transport, or final application. Hydrogen is not harmful in itself.
- When correctly treated, hydrogen is a very healthy fuel.

The processing of hydrogen needs raw materials. Technologies for the development of hydrogen from fossil fuels used to generate industrial hydrogen have been produced. That involves reforming the vapor of natural gas, partial oxidation of hydrocarbons, and coal gasification. The schematic of the steam reformer device with an ambient pressure PEM fuel cell pile working by hydrogen results from the combustion of natural gas, propane, or LPG [4] as shown in Fig. 3. A steam reformer (SR), a water gas shift (WGS), and heat exchangers compose the fuel conversion complex. Oxidation of the residual fuel gas heats the endothermic reaction region (RFG). This component is classified as post oxidation (PO). The PEM fuel cell stack works from a WGS reactor via hydrogen. Before joining the SR, fuel is combined with water and heated. At the exit, the hot syngas is cooled. Following WGS, water is diluted to raise the amount of hydrogen. The cold syngas is reheated easily in the stack. The fuel cell uses part of the hydrogen, and RFG is sent to PO [3–12]. Such developments, however, would not reduce fossil fuel dependency and CO_2 output. Direct thermal (and catalytic) hydrocarbon cracking is the only process that can generate hydrogen without CO_2 production from fossil fuels. This approach was used for biomass processing, but it is in an early phase of growth for cost-effective hydrogen production, showing no technical viability in the labs [13]. Water electrolysis is a completely evolved method for the production of hydrogen capacity from several to thousands of /hr. Water electrolysis is an advanced technique. It is reasonably effective, but since high energy is required (electricity), hydrogen produced by water is highly expensive [4–14].

2.2 Security Measures of Hydrogen as Fuel

The unusual physical properties of hydrogen differ considerably from conventional fuels. In certain instances, some of these properties could theoretically render hydrogen less harmful, while others could make hydrogen more dangerous. Table 1 displays the essential properties of hydrogen and Table 2 compares the properties of hydrogen with other fuels and summarizes their effect on defense [2, 3]. Hydrogen has the smallest molecule and is more likely to exit into tiny holes than other liquid or gas fuels. If there is a leak, for whatever reason, hydrogen can spread much quicker than any other gasoline, reducing the risk of threat. Hydrogen is far more rugged and transparent than tar, propane, or

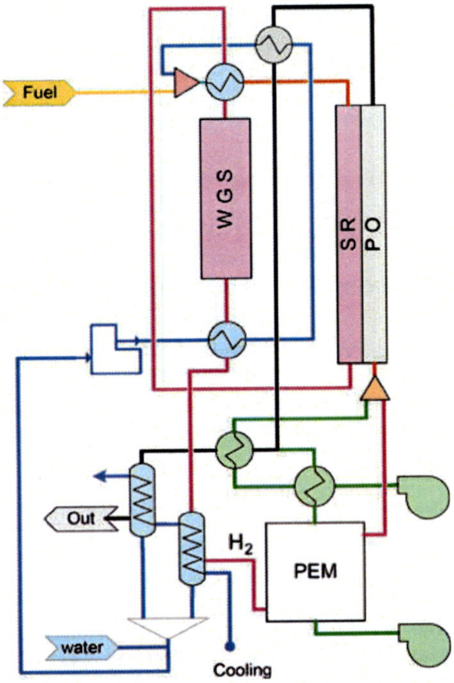

Fig. 3 PEM fuel cell and steam reformer system

natural gas. In very high-volume concentrations, the hydrogen/air combination can burn between 4 and 75% of the hydrogen in the air.

Certain fuels have lower flammability, i.e., 5.3–15% of natural gas, 2.1–10% of propane, and 1–7.8% of petrol. This difference, though, has no real significance. In a variety of actual spills, the key factor for evaluating when a leak ignites is the lower flammability level. The lower hydrogen threshold is 4 times that of fuel, 1.9 times that of propane and much lower than that of natural gas. Hydrogen has very low combustion energy (0.02 MJ), around one order fewer than other fuels. Ignore energy is the fuel/air ratio factor, and the hydrogen-air content is at least 30% higher. Hydrogen combustion energy is equivalent to natural gas at the lower flammability limit (LFL). Hydrogen blaze strength is seven times that of fuel or natural gas. Hydrogen fires are much more prone to causing deflagration or even detonation than most liquids. However, the detonation capacity depends on the exact fuel/air ratio in a complicated manner, in particular the temperature and the small space geometry. Hydrogen detonation is highly controversial under open circumstances. The lower amounts of hydrogen fuel to air detonation are 13–18%, double the ratio of natural gas and twelve times the ratio of gas. Since the lower flammability is 4%, an explosion may happen only in the most unlikely scenarios; for example, hydrogen should first build up in closed space to achieve a concentration of 13% without combustion, and at that stage, it may cause an ignition source. Hydrogen provides the lowest explosive energy of any substance in a blast, and the hydrogen content is 22 times larger than that of gasoline vapor.

Table 1 Hydrogen properties

Property	Value
Molecular weight	2.016 [kg/kmol]
Density	0.0838 [kg/m^3]
Higher heating value	141.9 [MJ/kg], 11.89 [MJ/m^3]
Lower heating value	119.9 [MJ/kg], 10.05 [MJ/m^3]
Boiling temperature	20.3 [K]
Density as liquid	70.8 [kg/m^3]
Critical temperature	32.94 [K]
Critical pressure	12.84 [bar]
Critical density	31.40 [kg/m^3]
Self-ignition temperature	858 [K]
Ignition limits in air	4–75 [vol.%]
Stoichiometric mixture in air	29.53 [vol.%]
Flame temperature in air	2318 [K]
Diffusion coefficient	0.61 [m2/s]
Specific heat	14.89 [kJ/kg K]

Table 2 List of properties related to hydrogen safety as opposed to other fuels

Property (compared with other fuels)		Risk (compared with other fuels)
Leak probability	Higher	Dangerous
Volume of fuel released in leak	Higher	Same
Energy of fuel released in leak	Lower	Safe
Diffusivity and buoyancy	Higher	Safe
Lower flammability limit in air	Higher	Same
Minimum ignition energy	Lower	Same
Ignition energy at LFL	Approximately same	Same
Flame velocity	Higher	Dangerous
Lower detonability fuel/air ratio	Higher	Safe
Explosive energy per energy stored	Lower	Safe
Flame visibility	Lower	Dangerous
Flame emissivity	Lower	Safe
Flame fumes toxicity	Lower	Safe
Fuel toxicity	Lower	Safe

The hydrogen flame is almost invisible and can be dangerous, so people cannot realize that a hydrogen flame is burning. This can be remedied by applying the requisite luminosity to the chemicals. The low emissivity of hydrogen flames ensures that radiant heat transmission is far less likely to cause inflammation or destroy artifacts and those around them. Those who inhale dust threaten petroleum gas and soot, but hydrogen fires just produce water vapor (unless secondary materials start burning). Liquid hydrogen poses a broad variety of protective risks, such as cold burns and intensified fuel leaching. The massive release of liquid hydrogen has similar properties to the release of gasoline, but can dissipate much quicker. Another possible danger in the case of a breakdown of the pressure release valve is the catastrophic eruption of a molten liquid–vapor. In situations where the car is inoperative, or in the event of a collision, certain risks can be handled. Fire, explosions, or contamination usually faces possible threats. As in fire, neither hydrogen nor gas is toxic. The latter may be ignored. Hydrogen may arrive from fuel storage, fuel production chains, or from the fuel cell itself as a fire or explosion. The fuel cell is the least risk since the thin polymer membrane absorbs oxygen and hydrogen from the fuel cell. In the event of membrane failure, hydrogen and oxygen would be combined and the fuel cell would immediately lose its power, which is easily understood by the control system. In this case, the transmission lines are automatically severed. The fuel cell operating temperature (60–90 °C) is too low to cause thermal inflammation, yet hydrogen and oxygen must be mixed to determine the combustion conditions at the catalytic surface. However, the theoretical danger would be negligible owing to the limited volume of hydrogen in fuel cells and supply chains. Hydrogen tends to present equal risks to other fuels. In reality, contrary to general perception, hydrogen is in several respects better than gas and natural gas [15–17].

2.3 Steam Reformer Unit to Produce Hydrogen Fuel

Steam reforming is one of the most common and, at the same time, least costly methods for the manufacture of hydrogen [3–11] today. Its benefits derive from its high operating performance and low maintenance and manufacturing costs. Methane is the most commonly used raw material, followed by ethane, propane, butane, and sulfur-free natural gas. However, hydrocarbon gas fuel such as propane is converted into hydrogen at a much lower temperature than methane, also in natural gas mixtures [3–18]. Therefore, propane fuel is a good choice for hydrogen production for fuel cell vehicles.

Naturally, propane is used in conjunction with other hydrocarbons manufactured during the production of natural gas and oil refining. It also extracts propane from heated crude oil using a distillation device. Then, it is isolated, liquefied by pressurization, and deposited in pressure vessels.

The basic reaction is:

$$C_nH_m + 2nH_2O \rightarrow (2n + 0.5m)H_2 + nCO_2$$

where n and m are the numbers of carbon and hydrogen atoms in the hydrocarbon fuel, respectively.

3 Mathematical Model of Steam Reformer and Numerical Method

Mathematical modeling is a suitable way to convert the physical system into a mathematical system in the form of governing equations. A greater care has to be taken to convert the physical system into a mathematical system [19, 20]. Since problem of steam reformer system is a transport phenomenon which involves heat, fluid flow, and chemical reactions [21, 22], selection of suitable numerical method [23, 24] and the formulation of physical model to acceptable level of mathematical formulation is important [25].

3.1 Model Definition

The purpose of this study is to do a parametric analysis utilizing a steam reformer model using 3-dimensional multi-physics computational fluid dynamics (CFD). The reformer effects must be carefully analyzed and checked in numerous operating conditions. This study helps to identify critical parameters and offers insight into the physical processes that contribute to steam reformer performance under various operating conditions. CFD simulation presents computational methods for continuity, dynamics, energy, and transport equations for organisms, as well as calculation of the catalyst bed reaction rate. The CFD model provides a much more efficient and lower-cost computer-aided approach for modeling and optimizing the steam reformer. The entire electronics realm comprises insulating coats, heating pipes, and brittle catalytic pads for the reformer system. Figure 4 displays such a device's geometry. The chemical reformation takes place in a porous catalytic bed, where the endothermal reaction process is supplied with energy by heated tubes.

The reactor is in a sweater. In stoichiometric numbers, hydrocarbon fuels, including natural gas and steam, are combined and linked through reactor inlet. Heat gasses in the opposite direction are passed across a series of tubes perforating the reactor bed for heating purposes. A complete three-dimensional, strongly related framework of mass, energy, and dynamic equations must be used in this work to explain the steam reformer system. COMSOL® is a general-purpose multi-physics platform program for the modeling of engineering systems for computational fluid dynamics (CFD). The COMSOL® multi-physics program offers predefined physics interfaces to model a broad variety of physics phenomena, including many popular, multiple physics links. The physics interfaces are basic user interfaces for a specific area of science or engineering, with all facets for controlling the phenomenon concerned—from specifying model parameters to discrediting the effects of the solution [26]. Therefore, in the current research, the COMSOL® multi-physics model of computational fluid dynamics will be used to examine the steam reformer machine in depth (Table 3).

Reformer water and propane contribute to hydrogen and carbon dioxide:

$$C_3H_8 + 6H_2O \xrightarrow{k} 10H_2 + 3CO_2$$

An overall cinematic model was constructed from experiments [4], where the first reaction rate [SI unit: mol/(m^3·s)] was observed in propane Concentrate:

$$r = k c_{C3H8} \tag{1}$$

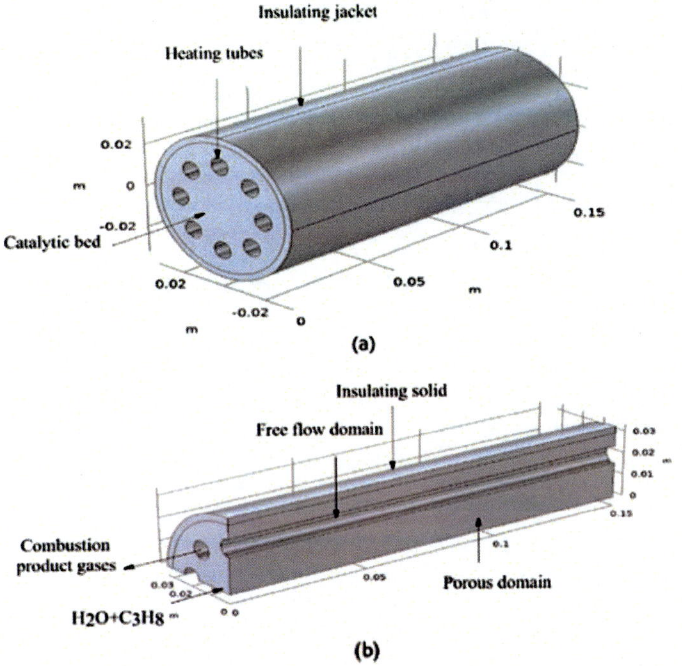

Fig. 4 Three-dimensional computational model geometry of the steam reformer unit

The constant rate depends on the temperature:

$$k = A\exp\left(-\frac{E}{R_g T}\right)$$

In A 7 × 10^5 (SI: 1/s) and E 83.14 (SI: kJ/mol).

3.2 Fluid Flow—Reformers

In this research work, Darcy's law has been utilized to model the fluid flow behavior. Darcy's law explains gaseous species movement across the reformer bed:

$$\nabla \cdot \left(\rho\left(-\frac{k}{\eta}\nabla p_{sr}\right)\right) = 0$$

The permeability of the porous material (SI unit: kg/m^3), the viscosity (η) (SI unit: Pa·s), the permeability of the porous medium (SI unit: m^2), and the pressure in the reformer bed is p_{sr} (SI unit: Pa). In this scenario, Darcy's law guide settles Darcy's equation. Inlet and outlet limits define a reduction of 75 Pa over bed.

The other constraints are impermeable and are the same as:

$$-\frac{k}{\eta}\nabla p_{sr} \cdot n = 0$$

Table 3 Steam reformer parameters

Property	Value
Binary diffusion coefficient (CO_2_C_3H_8)	5.1×10^{-6} [m^2/s]
Binary diffusion coefficient (C_3H_8_H_2O)	8.4×10^{-6} [m^2/s]
Binary diffusion coefficient (CO_2_H_2)	3.6×10^{-5} [m^2/s]
Binary diffusion coefficient (H_2_H_2O)	4.9×10^{-5}[m^2/s]
Binary diffusion coefficient (CO_2_H_2O)	1.1×10^{-5} [m^2/s]
Molar mass (C_3H_8)	44.1×10^{-3} [kg/mol]
Molar mass (H_2)	2.016×10^{-3} [kg/mol]
Molar mass (CO_2)	44.01×10^{-3} [kg/mol]
Molar mass (H_2O)	18.016×10^{-3} [kg/mol]
Initial weight fraction (C_3H_8)	0.28 [1]
Initial weight fraction (H_2)	0.01 [1]
Initial weight fraction (CO_2)	0.01 [1]
Enthalpy of reaction	410×10^3 [J/mol]
Density, insulating foam	24 [kg/m^3]
Heat capacity, insulating foam	1.9 [J/kg/K]
Thermal conductivity, insulating foam	0.027 [W/m/K]
Heat capacity, reformer bed	2800 [J/kg/K]
Thermal conductivity, reformer bed	0.1 [W/m/K]
Viscosity, reformer bed	2.7×10^{-5}[Pa s]

3.3 Energy Transport—Reformer Bed

The average temperature distribution in porous beds is assessed using a one-equation approach:

$$(\rho C_P)_t \frac{\partial T_{Sr}}{\partial T} + \nabla \cdot (-k_{sr} \nabla T_{sr}) + (\rho C_P)_f \mathbf{u} \cdot \nabla T_{sr} = \mathbf{Q}$$

The bed's volumetric heat power is:

$$(\rho C_P)_t = \varepsilon(\rho C_P)_f + ((1-\varepsilon)\rho C_P)_s$$

.

In the aforementioned equations, the indices "*f*" and "s" signify fluid and solid phases, respectively, and ε is the fluid phase volume fraction. In addition, T_{sr} is the reformer bed temperature (SI unit: K) and ksr [SI unit: W/(m·K)]. Q reflects a heat supply (SI unit: W/m^3) and u fluid speed (SI unit: m/s). The equation is modeled using Fluid Heat Transfer interface.

If the porous medium is homogeneous and isotropic, it becomes:

$$\nabla \cdot (-k_{sr} \nabla T_{sr}) + (\rho C_P)_f \mathrm{u} \cdot \nabla T_{sr} = \mathrm{Q} \tag{2}$$

The heat source due to reaction is

$$Q = \Delta H_r \cdot \mathrm{r}$$

where r is Eq. 1. Propane steam reformation is endothermic, with a reaction enthalpy of $\Delta H_r = 4.1 \times 10^5$ J/mol.

Equation 2. Often allows for heat transfer in the insulating vest. With no reactions in this domain, the definition reduces to:

$$\nabla \cdot (-k_i \nabla T_{sr}) = 0$$

k_i

The boundary conditions are used to impose the real conditions during the problem solving. Inlet gas temperature is 700 K. Convective heat transfer predominates at the outlet:

$$\mathrm{n} \cdot (-k_{sr} \nabla T_{sr}) = 0$$

The Newton's law of cooling has been used to define the exchange of heat between tubes and bed using the following expression.

$$q = h_t(T_{\mathrm{sr}} - T) \quad (3)$$

where h_t is the heat transfer coefficient [SI unit: W/(m^2·K)] and T is the heating tube temperature (K). A related term explains the heat flux from the surrounding isolating jacket:

$$q = -h_j(T_{\mathrm{sr}} - T_{\mathrm{amb}})$$

where h_j is jacket heat transfer coefficient [SI unit: W/(m^2·K)], and T_{amb} is ambient temperature (K).

3.4 Mass Transport—Reformer Bed

The mass balance equations for the model are Maxwell–Stefan diffusion and convection equations in a constant state:

$$\nabla \cdot \left(\rho \omega_i u - \rho \omega_i \sum_{j=1}^{n} D_{ij} (\nabla x_j + (x_j - \omega_j) \frac{\nabla p}{p} \right) - D_i^T \frac{\nabla T}{T}) = R_i$$

In the above equations, ρ denotes density (SI unit: kg/m^3), ω_i is the mass fraction of species i, x_j is the molar fraction of species, and j is the multicomponent D_{ij} part Fick diffusivity (SI unit: m^2/s). D_i^T denotes the generalized coefficient of thermal diffusion [SI unit: kg/(m s)], T (SI unit: K), and R_i [SI unit: kg/(m^3·s)]. Mass balances are set up and addressed using the Condensed Species Transfer interface.

Propane inlet weight fraction is 0.28. Convective flux state at the outlet:

$$n \cdot \left(\left(-\rho\omega_i \sum_{j=1}^{n} D_{ij} \left(\nabla x_j + (x_j - \omega_j) \frac{\nabla p}{p} \right) \right) - D^T \frac{\nabla T}{T} \right) = 0$$

All other boundaries use separation or symmetry.

3.5 Fluid Flow—Heating Tubes

The low-compressible Navier–Stokes equations describe the heating gas flow in tubes:

$$\rho u \cdot \nabla u = \nabla \cdot \left[-pI + \eta(\nabla u + (\nabla u)^T) - \left(\frac{2\eta}{3}\right)(\nabla \cdot u)I \right]$$

$$\nabla \cdot (\rho u) = 0$$

where ρ denotes density (SI unit: kg/m^3), u represents velocity (SI unit: m/s), η denotes viscosity [SI unit: kg/ (m·s)], and p equals tube strain (SI unit: Pa). The Laminar Flow interface sets the Navier–Stokes equations and is used to model the gas flow in the tubes.

The boundary conditions are

$$u \cdot n = v_0 \quad \text{inlet}$$

$$u = 0 \quad \text{walls}$$

$$p = p_{\text{ref}} \quad \text{outlet}$$

Viscous pressures are neglected at outlet and pressure is adjusted to one atmosphere.

3.6 Energy Transport—Heating Tubes

Energy transport in tubes is defined by:

$$\nabla \cdot (-k_{ht} \nabla T) + \rho C_p \mathbf{u} \cdot \nabla T = 0$$

k_{ht}

Convective heat transfer predominates at the outlet:

$$n \cdot (-k_{ht} \nabla T) = 0$$

Heat exchange between bed and tubes is given by:

$$q = h_t(T_{\text{sr}} - T)$$

This is the same heat flux as Eq. 3, but with inverted sign.

4 Results and Discussion

A computational model has been developed to provide insight into the physical processes that contribute to the performance of a steam reformer under varied operating conditions. PEM fuel cell and steam reformer system has been used to validate the computational model. A commercial finite element package COMSOL has been utilized to create and analyze the computational model. Conservation equation such has mass, momentum, and energy equations has been solved through essential and natural boundary condition in order to reflect the physics process into a computational process. Figure 5 shows a propane weight fraction in the reformer bed. The inlet weight is 0.28 while the exit fraction is about zero.

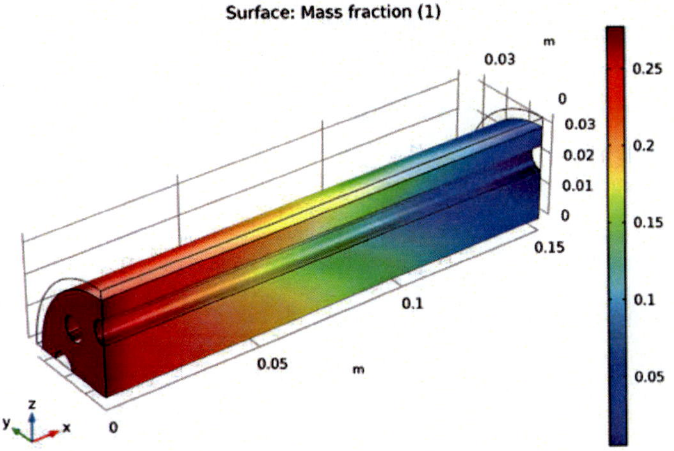

Fig. 5 Weight proportion allocation of propane in the reformer bed

The cross-sectional plot through the reformer center (Fig. 6) shows the distribution of the concentration in the bed. Because local reactivity is primarily affected by temperature, the results indicate that the heat in the tubes is adequate to allow efficient use of the entire catalytic bed.

The weight fractions of all responding organisms in the bed measured in the reactor centerline are shown in Fig. 7. The plot reveals that the whole length of the bed is engaged in propane conversion.

Figure 8 clearly shows the energy exchange between the heating tubes and the refurbishing bed. The heating tube gas enters at 900 K and leaves at around 674 K. The gas temperature in the reformer bed exceeds 700 K at the inlet, is marginal, and gradually leaves at an average temperature of 795 K.

A plot around the middle of the reactor (Fig. 9) illustrates how the temperature initially decreases due to reactions of endothermal reform. If the reaction rate is decreased based on the lower temperature and propane material, the energy provided by heating tubes decides the evolution of the temperature.

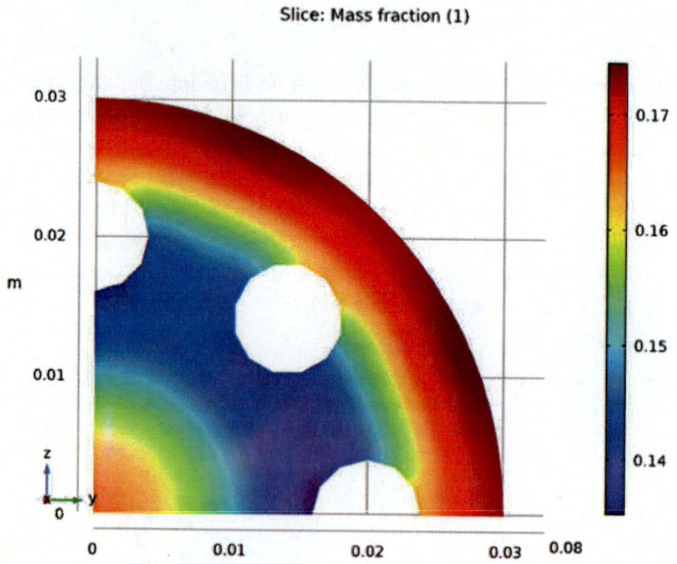

Fig. 6 Weight fraction distribution of propane across the center of the reformer bed in a cross section

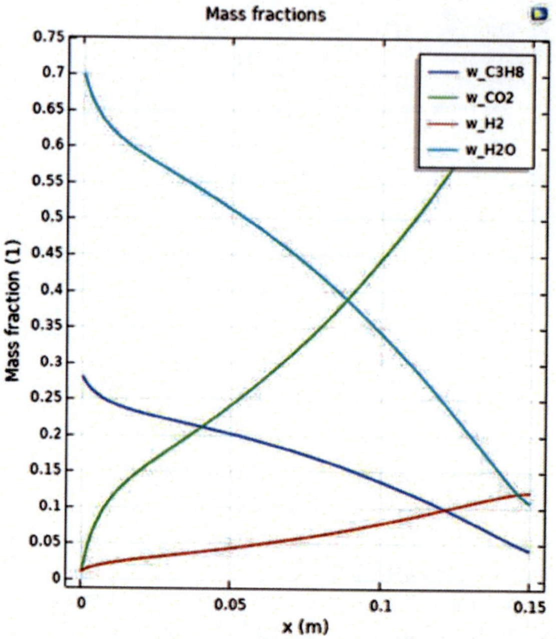

Fig. 7 Weight of the species as a feature of the reactor location, traced in the middle line of the reactor

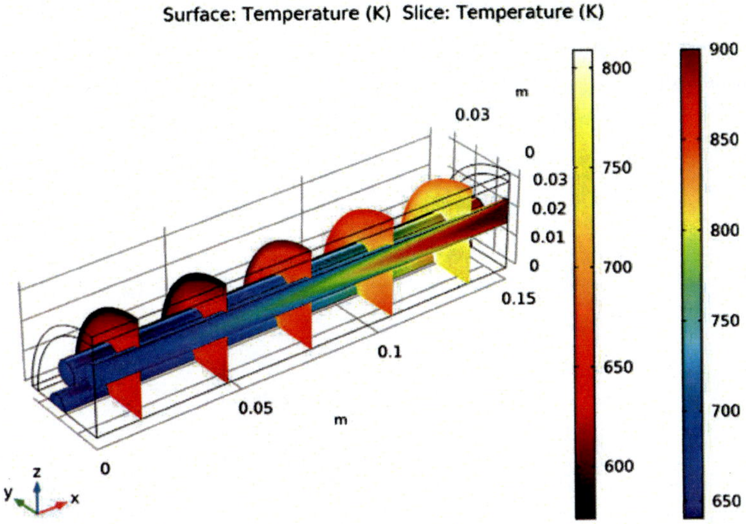

Fig. 8 Reformer device temperature distribution, including the reformer bed, heating tubing and wall isolation

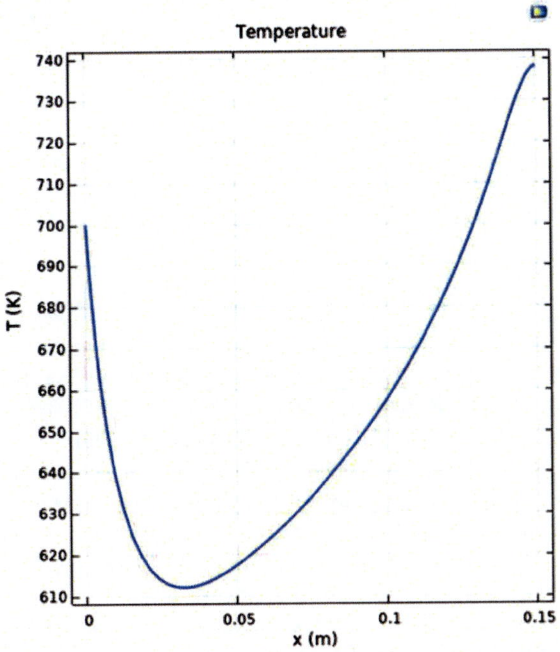

Fig. 9 Bed temperature as a function of position, plotted along the reactor centerline

Figure 10 demonstrates the speed fields of the heating gas in the tubes as well as the reacting gas in the bed. The flow is laminar in the heating tubes, and the distribution of the parabolic velocity is clearly seen. The gas velocity in the porous bed increases dramatically through the reactor, and the gas velocity at the outlet is roughly twice that at the inlet.

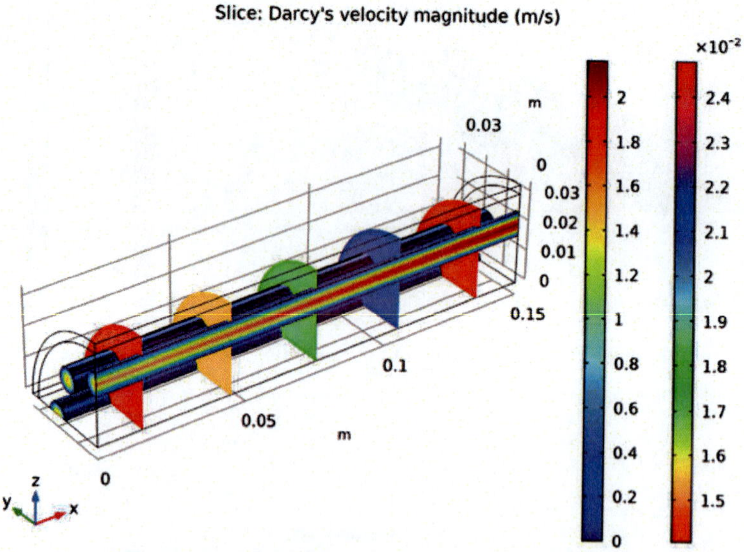

Fig. 10 Heating tube speed fields and the reformer bed

The increased velocity is largely due to expansion of the gas caused by chemical reaction and, to a lesser degree, the rise in temperature. Figure 11 shows the related differences in density in the reformer bed, accounting for the effects of composition as well as temperature.

5 Conclusions

Hydrogen is one of the energy carriers that is applied to replace fossil fuels in internal combustion engines and fuel cell vehicles. This research work revealed a reactor simulation using fully linked mass, energy, and flow equations. The reconditioning of hydrocarbons, such as natural gas, delivers an efficient, low-cost, and extensively applied technique for producing hydrogen, ensuring energy security, and promoting ecologically sustainable fuels in the short and medium term. The computed model found to provide a computer-aided method for building and refining the steam reformer at a considerably higher efficiency and lower cost. It runs a parametric analysis of a steam reformer device using a three-dimensional CFD model. The impacts of the reformer are thoroughly investigated and tested under a variety of operating conditions. This research assisted in the identification of critical criteria and provided insight into the physical processes that influence the performance of a steam reformer under various operating conditions.

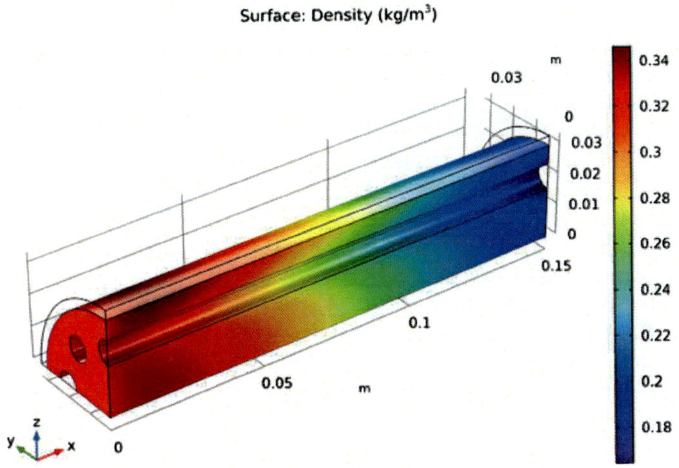

Fig. 11 Overall gas density in the reformer bed

References

1. International Energy Agency (IEA) (2019) [https://www.iea.org/]
2. Al-Baghdadi MA (2013) Alternative fuels research progress. International Energy and Environment Foundation (IEEF). ISBN- 13: 9781484057711
3. Al-Baghdadi MA (2018) PEM fuel cell engines: principles, design, modelling, and analysis. International Energy and Environment Foundation (IEEF). ISBN-13: 9781983474996
4. Gateau P (2007) Design of reactors and heat exchange systems to optimize a fuel cell reformer. In: Proceedings of the COMSOL user's conference Grenoble
5. Yun J, Van Trinh N, Yu S (2021) Performance improvement of methanol steam reforming system with auxiliary heat recovery units. Int J Hydrogen Energy
6. Minutillo M, Perna A, Sorce A (2020) Green hydrogen production plants via biogas steam and autothermal reforming processes: energy and exergy analyses. Appl Energy 277:115452
7. Wang Y, Hong Z, Mei D (2021) A thermally autonomous methanol steam reforming microreactor with porous copper foam as catalyst support for hydrogen production. Int J Hydrogen Energy 46(9):6734–6744
8. Yun J, Kim Y, Yu S (2020) Interactive heat transfer characteristics of 5 kW class shell-and-tube methane steam reformer with intermediate temperature heat source. Int J Hydrogen Energy 45(41):21767–21778
9. Cormos A-M, Dumbrava I, Cormos C-C (2020) Evaluation of techno-economic performance for decarbonized hydrogen and power generation based on glycerol thermo-chemical looping cycles. Appl Therm Eng 179:115728
10. Qi Y, Andersson M, Wang L, Wang J (2021) System behavior prediction by artificial neural network algorithm of a methanol steam reformer for polymer electrolyte fuel cell stack use. Fuel Cells
11. Moon K, Kale GR (2013) Energy analysis in combined reforming of propane. J Eng 2013(Article ID 301265):10 pages. https://doi.org/10.1155/2013/301265
12. Schwiedernoch R, Tischer S, Correa C, Deutschmann O (2003) Experimental and numerical study on the transient behavior of partial oxidation of methane in a catalytic monolith. Chem Eng Sci 58:633

13. Shabanian SR, Rahimi M, Amiri A, Sharifnia S, Alsairafi AA (2012) Computational fluid dynamics modeling of hydrogen production in an autothermal reactor: effect of different thermal conditions. Korean J Chem Eng 29(11):1531–1540
14. Schadel BT, Duisberg M, Deutschmann O (2009) Steam reforming of methane, ethane, propane, butane, and natural gas over a rhodium-based catalyst. Catal Today 142(1–2):42–51
15. Silva PP, Ferreira RA, Nunes JF, Sousa JA, Romanielo LL, Noronha FB, Hori CE (2015) Production of hydrogen from the steam and oxidative reforming of LPG: thermodynamic and experimental study. Brazilian J Chem Eng 32(3):647–662
16. Tran A, Aguirre A, Durand H, Crose M, Christofides PD (2017) CFD modeling of a industrial-scale steam methane reforming furnace. Chem Eng Sci 171:576–598
17. Pashchenko D (2019) Numerical study of steam methane reforming over a pre-heated Ni-based catalyst with detailed fluid dynamics. Fuel 236(15):686–694
18. Ngo SI, Lim YI, Kim W, Seo DJ, Yoon WL (2019) Computational fluid dynamics and experimental validation of a compact steam methane reformer for hydrogen production from natural gas. Appl Energy 236:340–353
19. Dhinakaran V, Siva Shanmugam N, Sankaranarayanasamy K (2017) Experimental investigation and numerical simulation of weld bead geometry and temperature distribution during plasma arc welding of thin Ti-6Al-4V sheets. J Strain Anal Eng Des 52(1):30–44
20. Dhinakaran V, Siva Shanmugam N, Sankaranarayanasamy K (2017) Some studies on temperature field during plasma arc welding of thin titanium alloy sheets using parabolic Gaussian heat source model. Proc Inst Mech Eng Part C J Mech Eng Sci 231(4):695–711
21. Lao L, Aguirre A, Tran A, Wu Z, Durand H, Christofides PD (2016) CFD modeling and control of a steam methane reforming reactor. Chem Eng Sci 148:78–92
22. Tran A, Aguirre A, Durand H, Crose M, Christofides PD (2017) CFD modeling of a industrial-scale steam methane reforming furnace. Chem Eng Sci 171:576–598
23. Schäfer M (2006) Computational engineering: introduction to numerical methods, vol 321. Springer, Berlin
24. Khan A, Faheem M, Raza A (2021) Solution of third-order Emden–Fowler-type equations using wavelet methods. In: Engineering computations
25. Bhattacharya S, Demirci HE, Nikitas G, Prakhya GK, Lombardi D, Alexander NA, Mylonakis G (2021)Physical modeling of interaction problems in geotechnical engineering. In: Modeling in geotechnical engineering. Academic Press, pp 205–256
26. Introduction to COMSOL Multiphysics® (2019) COMSOL INC. [https://www.comsol.com/documentation].

Fuel Cell Power Pack with Integrated Metal Hydride Hydrogen Storage for Powering Electric Forklift

Ivan Tolj[1](✉), Mykhaylo Lototskyy[2], Adrian Parsons[2], and Sivakumar Pasupathi[2]

[1] Department of Thermodynamics, Faculty of Mechanical Engineering and Naval Architecture, University of Split, 21000 Split, Croatia
itolj@fesb.hr

[2] HySA Systems Competence Center, South African Institute for Advanced Materials Chemistry (SAIAMC), University of the Western Cape, Bellville 7535, South Africa

1 Introduction

Fuel cells are gaining more attention in material handling industry as they provide numerous benefits for warehouses, airports, seaports, etc. European Commission (EC) recognized the importance of the fuel cell and hydrogen technology and their inevitable role in reducing greenhouse gas emission by further encouraging hydrogen fuel cells in heavy-duty vehicles in their hydrogen strategy for a climate-neutral Europe [1]. Fuel cell-powered forklifts possess several advantages over the battery-powered ones: operation time up to 8 h on a single tank of hydrogen, constant power throughout an entire shift, lower refueling time, stable operation, etc. Also, fuel cell-powered forklifts allow maximization of warehouse floor as no additional power modules are needed per forklift. For battery-powered forklifts, at least, two to three battery packs are needed—one deployed in the forklift, one on charging, and one on cooling. As each battery has a volume of ~1 m^3—switching to fuel cell-powered modules allows maximizing warehouse space utilization and reducing operational costs. Due to these factors, fuel cell hydrogen-fueled forklifts became a promising early market niche application. At the moment, more than 20,000 fuel cell-powered forklift are been deployed in warehouses, stores, and other manufacturing facilities in the United States, as compared to only 500 in Europe and 100 in Japan [2, 3].

As a rule, fuel cell-powered forklifts utilize a hybrid power train when fuel cell delivers average power, and the peak power is provided by batteries or supercapacitors. The development of a hybrid power train with a 16 kW polymer electrolyte membrane fuel cell system, ultracapacitor modules, and lead-acid battery was reported in [4]. Reference [5] reports about the implementation of the hybrid power train which supplied the required 50 kW peak power in a typical forklift work cycle. Five prototypes fuel cell-powered forklifts were built, tested in laboratory, and delivered to Taiwanese supermarket warehouse. Each hybrid-drive system included 2.7 kW proton exchange membrane fuel cell stack, lithium battery pack, and supercapacitors. The components of the drive system could supply 7 kW of electric power. In Ref. [6], a game theory model was developed to

M. Kolhe et al. (eds.), *Recent Advances in Renewable Energy Systems*,
Lecture Notes in Electrical Engineering 876,
https://doi.org/10.1007/978-981-19-1581-9_2

assess the potential of fuel cell and battery-powered forklifts for reducing greenhouse gas emissions in the Ontario province, Canada. Their analysis showed that battery-powered forklifts are more cost-effective compared to fuel cell-powered ones but only when lower levels of discounted power are available. On other hand with increased social cost of carbon (SCC) and discounted power available—fuel cell-powered forklifts become more cost-effective. Also, it was found that fuel cell-powered forklifts have lower operational and maintenance costs over the battery-powered, and that both technologies are highly effective in reducing greenhouse gas emissions. Economic comparison study between fuel cell and battery-powered forklift was done in Ref. [7] for different forklift power plants. It was found out that fuel cell forklifts are more expensive to purchase and to operate for the facility with fleet of 50 forklifts. As this study is based on relatively old data from 2010 and taking into account that fuel cell power module and refueling station purchase prices are significantly lower than at the time when this study was accomplished—new study preferably with the higher number of deployed forklifts (>100) and considering different refiling station technologies (i.e., metal hydride) urgently must be done. Performance simulation of hybrid PEM fuel cell/battery-powered forklifts was done in Ref. [8]. Study was done for different combinations of lead-acid battery capacity and fuel cell stack sizes with conjunction of two different control strategies to investigate hydrogen consumption and battery state-of-charge. Comparison was done for two different load cycles, and it was found that combination of 110 cells stacked with two 55 Ah batteries provides most economical combination in terms of hydrogen consumption. A similar study has been recently performed [9] for different load cycles (VDI60) and working conditions. It was found out that optimal battery capacity for a 3-ton electric forklift is 10 Ah with fuel cell stack power of 11 kW. So far, almost, all demonstrated FC-powered utility vehicles used compressed H_2 stored in composite gas cylinders at pressures up to 350 bar. This solution is associated with too big volume of the onboard hydrogen storage component (storage density 0.02 kg H_2/L) and its insufficient weight to provide vehicle stability during operation. To mitigate the latter problem, an additional ballast is required for proper vehicle counterbalancing. In addition, the high H_2 storage pressure poses safety concerns and requires expensive refueling infrastructure. The problems can be solved using metal hydrides (MH) for the onboard hydrogen storage [3–5, 10]. HySA Systems in 2012 developed a metal hydride hydrogen storage extension tank which was successfully integrated in a commercial GenDrive 1600-80CEA fuel cell power module (Plug Power Inc.) and tested onboard 3-ton electric forklift from STILL [11]. Since 2015 developed prototype operates at Imapala Platinum in Springs, and so far, only, problems were related to Li-ion battery pack. Test results of this prototype were presented in Ref. [12]. Based on the tests at the Impala Refinery, it was found that forklift can operate ~3 h for heavy-duty operation (VDI60 cycles) and ~7 h for light duty operation. Here, we present a developed power fuel cell pack with MH hydrogen storage which was developed by Hydrogen South Africa Systems [13] and integrated by Hot Platinum (Pty) Ltd, South Africa. Forklift was successfully tested according to "Verein Deutscher Ingenieure" (VDI60) drive cycle protocol to identify optimization ways for improvement of operational stability at high loads, decreasing hydrogen consumption, and balance-of-plant (BoP) power demands.

2 Experimental

2.1 Fuel Cell Power Module

Liquid-cooled PEM fuel cell stack from Ballard with 75 cells and 15 kW power output was integrated into power pack module. Liquid cooled/heated MH tank developed by HySA Systems was used for efficient hydrogen storage and as a ballast to achieve fully rated forklift lifting capacity. Developed novel MH tank allowed decreasing of hydrogen pressure on the high-pressure side of hydrogen subsystem from 13.5 to 3.4 bar and reducing refueling pressure to 100 bar [11].

Power pack specifications are depicted in Table 1.

Table 1 Power pack specifications

Vehicle	STILL RX60-30L
Bus voltage	80 VDC
Output power	~15 kW average, 30 kW peak
Dimensions	840 mm (W) × 1010 mm (D) × 777 mm (H)
Weight	1800, ..., 1900 kg
Stack	14.5 kW closed cathode PEMFC stack (Ballard)
H_2 storage	Integrated MH storage unit, 20 Nm^3
Battery bank	Deep cycle lead-acid, 8, ..., 10 kWh

Power module concept design is presented in Fig. 1.

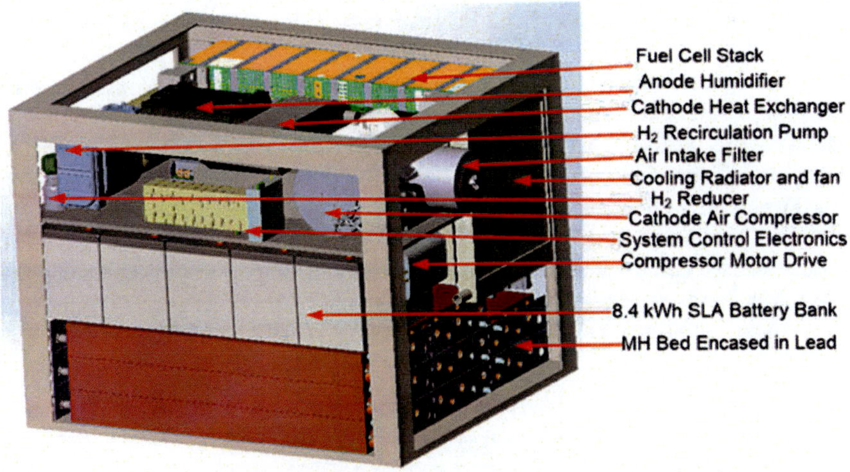

Fig. 1 Power module concept design

2.2 "Zero" Prototype—Bench-Top Version

To optimize balance-of-plant special fuel cell testbed was developed and tested, Fig. 2. In order to reduce system costs, we used standard components mainly from automotive industry such as valves, coolant pumps, heat exchangers, different flow meters, and filters. Individual cell voltage was monitored by specially developed CVM system, and to simulate different operating conditions, load bank resistors have been used.

Fig. 2 "Zero" prototype—bench-top version

2.3 Optimal BoP Design and Electrical Layout

Balance-of-plant design is presented in Fig. 3. The BoP is designed around liquid-cooled closed cathode Ballard 9SSL/75 cell FC stack (14.4 kW rated power; 48.2 V DC at 300 A).

Optimal electrical layout (which was reached after several iterations) is presented in Fig. 4. In order to compensate for the peak power draw of the load, an ultracapacitor (Maxwell, 97.2 V; 83 F) was introduced in parallel with the lead-acid battery pack. Main malfunctions identified during off-boards test of the commercial power module (from Plug Power Inc.) under real environmental conditions were mainly related to Li-ion batteries. To mitigate problems that occurred with the commercial fuel cell-powered module, we decided to integrate a lead-acid battery bank (6 × 12 V/100 Ah deep cycle SLA). Auxiliaries were connected directly to 24 and 48 V branches of the battery bank thus avoiding additional DC-to-DC converters.

3 Results

3.1 Off-Board Tests

Off-board tests were performed by connecting bench-top version with the resistive load bank to emulate a drive cycle based on the data provided from the VDI60 test with a

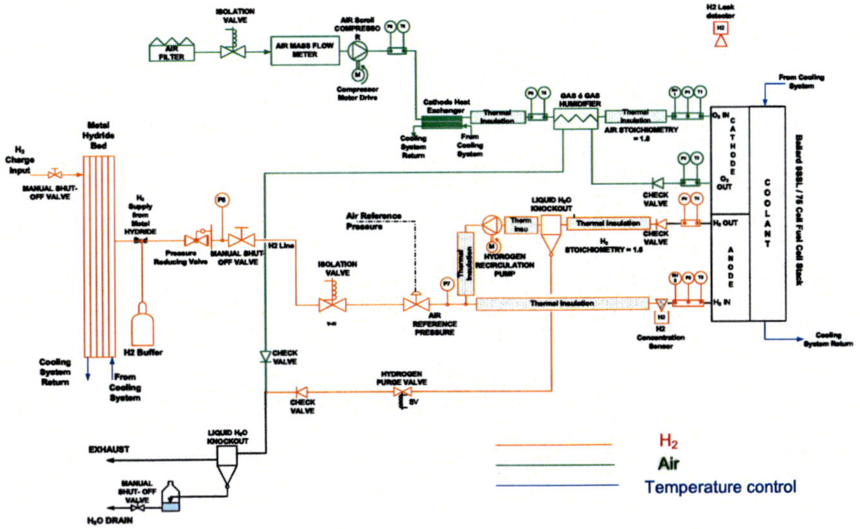

Fig. 3 Balance-of-plant layout

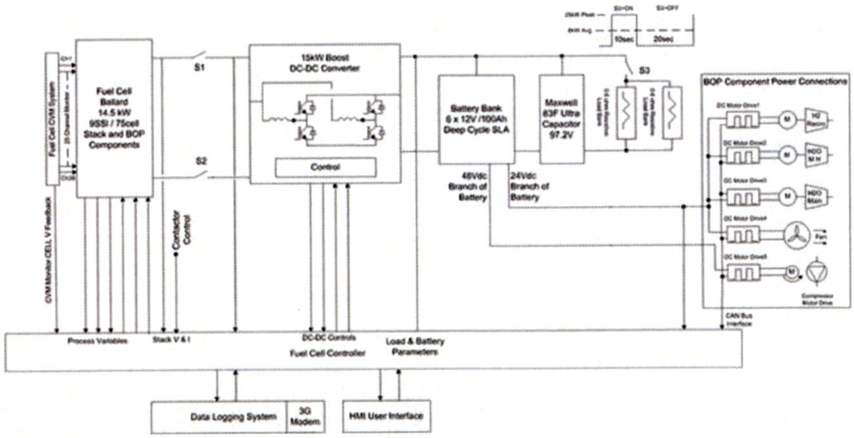

Fig. 4 Optimal electrical layout

commercial power module. Integration of ultracapacitors allowed to decrease battery contribution to the load power during peaks. Also, it was noticed that the battery did not undergo thermal runaway by heating up. During test, power pack was able to deliver stable power at the peaks up to 30 kW, while average fuel cell stack power was 8 kW. Average energy consumption during test was found to be around 9 kWh per hour where average load power was 8 kW. As compared to pure battery system (lead-acid), fuel cell-based ones have higher energy consumption due to power required for powering BoP components. This power was found to be 0.8 kW. Average hydrogen consumption during tests was ~ 900 NL/kWh, Fig. 5.

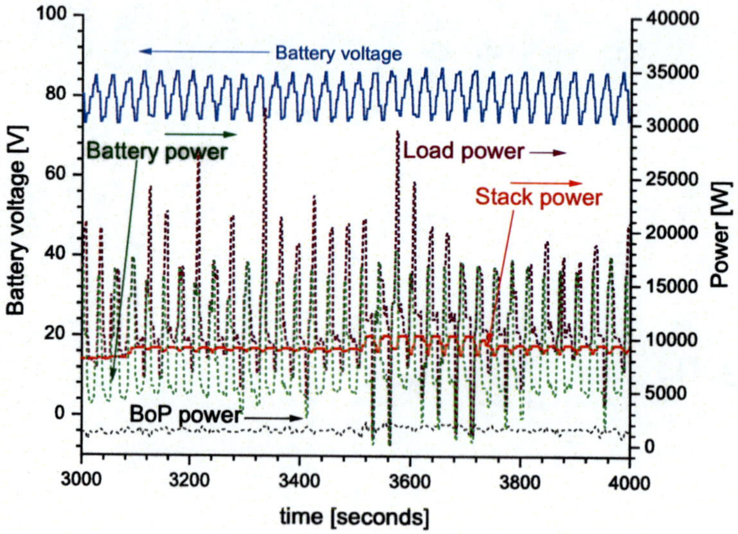

Fig. 5 Off-board test results of power module

3.2 Onboard Tests

Forklift was tested according to VDI60 standard protocol [14]. VDI60 testing protocol consists of 45 cycles in 60 min. Each cycle consists of forklift at starting position "1" holding 70% of rated load capacity (our test were performed with 100% rated load capacity of 3 ton), forward driving to position "2" (1.5 m distance) and lifting load up to 2 m, lowering the load and reverse driving between position "2" and "3" (distance between 30 m), forward driving from position "3" to "4" (1.5 m distance) followed with lifting load up to 2 m, lowering the load and reverse driving from position "4" to "1." Tests were carried out at Hot Platinum (Pty) Ltd premises in Athlone, Cape Town, South Africa, Fig. 6.

Developed power pack was able to provide stable operation during VDI60 tests, Fig. 7.

Power pack was tested at ambient temperature of 26.5 °C. energy consumption during VDI60 tests was 9.564 kWh/h compared to 7.5 kWh/h claimed by forklift manufacturer that indicates more aggressive driving (60 cycles in 60 min with 100% load capacity). Peak load power during test was ~35 kW, and fuel cell power was ~14 kW. Even though balance-of-plant components were optimized, BoP power consumption during VDI60 cycle test was still relatively high ~5.1 kWh/h. The main reason for this high BoP consumption is air compressor which requires further air mass flow and pressure optimization. During the tests, MH tank temperature was monitored. It was found that MH tank temperature reaches 50 °C indicating efficient heat transfer between fuel cell coolant and MH tank, Fig. 8.

Fig. 6 Left—shows developed 15 kW—PEMFC power pack before installation in STILL electric forklift. Top right—3-ton concrete blocks used as a load, bottom right—forklift captured during VDI60 driving tests

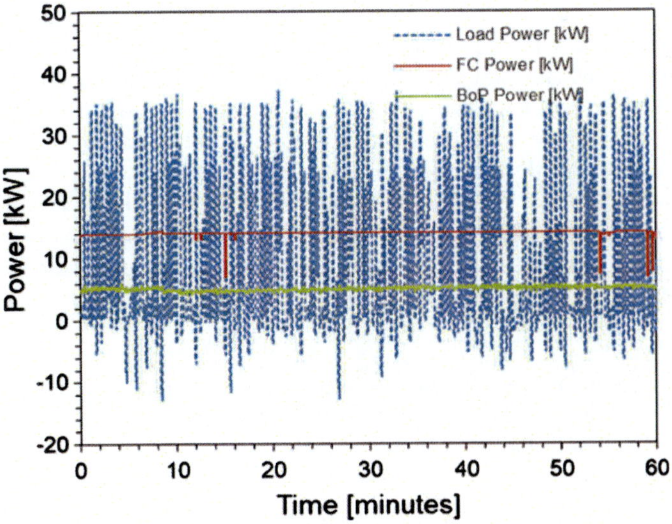

Fig. 7 Load power, FC power, and BoP power during VDI60 test

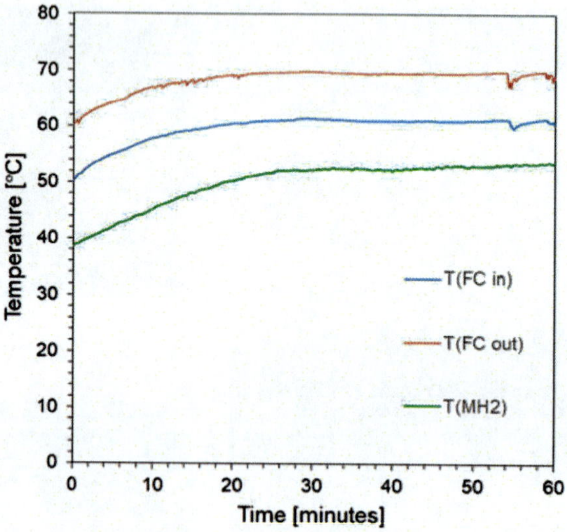

Fig. 8 Metal hydride tank and fuel cell stack coolant temperatures during VDI60 tests

4 Conclusion

In this work, development of power pack with original metal hydride hydrogen storage was presented. Power pack was integrated onboard commercial electric forklift from STILL. Before onboard integration, developed system was tested in order to optimize system components. This allows us to use OEM components resulting in system costs reduction. Integration of MH tank allowed to decrease refueling and operating pressures by allowing similar useable hydrogen storage capacity and refueling time compared to the commercial power pack. Developed power pack was tested onboard electric forklift according to VDI60 standard protocol, and it was found that power pack can maintain continuous power during complete 60 VDI test cycles. Moreover, temperatures of fuel cell stack coolant and MH tank were monitored, and it was found that MH tank temperature reaches 50 °C indicating efficient heat transfer from the stack coolant to the MH. Energy consumption was found to be higher than claimed by the forklift manufacturer (9.564 kWh/h vs. 7.5 kWh/h) indicating more aggressive driving. Balance-of-plant consumption during tests was as high as ~5.1 kWh/h mainly due to air compressor. In order to decrease BoP consumption to below 1.5 kWh/h, future compressor optimization is needed.

Acknowledgements. This work received financial support from EU Horizon 2020/RISE project "Hydrogen fueled utility vehicles and their support systems utilizing metal hydrides—HYDRIDE4MOBILITY" (project number: 778307).

References

1. Communication from the Commission to the European Parliament, the Council, the European Economic and Social Committee and the Committee of the Regions: A hydrogen strategy for a climate-neutral Europe, Brussels, 8 July 2020. https://eur-lex.europa.eu/legal-content/EN/TXT/PDF/?uri=CELEX:52020DC0301&from=EN
2. DOE hydrogen and fuel cells program record #18002, 30 July 2018. https://www.hydrogen.energy.gov/pdfs/18002_industry_deployed_fc_powered_lift_trucks.pdf
3. Yartys VA, Lototskyy MV, Linkov V, Pasupathi S, Davids MW, Tolj I, Radica G, Denys RV, Eriksen J, Taube K, Bellosta von Colbe J, Capurso G, Dornheim M, Smith F, Mathebula D, Swanepoel D, Suwarno S (2021) Hydride4mobility: an EU HORIZON 2020 project on hydrogen powered fuel cell utility vehicles using metal hydrides in hydrogen storage and refuelling systems. Int J Hydrogen Energy (in press). https://doi.org/10.1016/j.ijhydene.2021.01.190
4. Keränen TM, Karimäki H, Viitakangas J, Vallet J, Ihonen J, Hyötylä P, Uusalo H, Tingelöf T (2011) Development of integrated fuel cell hybrid power source for electric forklift. J Power Sources 196:9058–9068
5. Hsieh C-Y, Pei P, Bai Q, Su A, Weng F-B, Lee C-Y (2021) Results of a 200 hours lifetime test of a 7 kW Hybride-Power fuel cell system on electric forklifts. Energy 214:118941. https://doi.org/10.1016/j.energy.2020.118941
6. Haghi E, Shamsi H, Dimitrov S, Fowler M, Raahemifar K (2020) Assessing the potential of fuel cell-powered and battery-powered forklifts for reducing GHG emissions using clean surplus power; a game theory approach. Int J Hydrogen Energy 45:34532–34544. https://doi.org/10.1016/j.ijhydene.2019.05.063
7. Renquist JV, Dickman B, Bradley TH (2012) Economic comparison of fuel cell powered forklifts to battery powered forklifts. Int J Hydrogen Energy 37:12054–12059. https://doi.org/10.1016/j.ijhydene.2012.06.070
8. Hosseinzadeh E, Rokni M, Advani SG, Prasad AK (2013) Performance simulation and analysis of a fuel cell/battery hybrid forklift truck. Int J Hydrogen Energy 38:4241–4249. https://doi.org/10.1016/j.ijhydene.2013.01.168
9. Radica G, Tolj I, Lototskyy M, Pasupathi S (2020) Control strategy of Fuel cell-Battery hybrid system for optimizing Lift truck load cycle. In: 2020 5th International conference on smart and sustain technology, SpliTech 2020, 23 Sept 2020, Article number 9243836. https://doi.org/10.23919/SpliTech49282.2020.9243836
10. Lototskyy MV, Tolj I, Pickering L, Sita C, Barbir F, Yartys V (2017) The use of metal hydrides in fuel cell applications. Progr Natur Sci 27:3–20
11. Lototskyy MV, Tolj I, Davids MW, Klochko YV, Parsons A, Swanepoel D, Ehlers R, Louw G, van der Westhuizen B, Smith F, Pollet BG, Sita C, Linkov V (2016) Metal hydride hydrogen storage and supply systems for electric forklift with low-temperature proton exchange membrane fuel cell power module. Int J Hydrogen Energy 41:13831–13842
12. Lototskyy MV, Tolj I, Parsons A, Smith F, Sita C, Linkov V (2016) Performance of electric forklift with low-temperature polymer exchange membrane fuel cell power module and metal hydride hydrogen storage extension tank. J Power Sources 316:239–250
13. Lototskyy M, Tolj I, Klochko Y, Davids MW, Swanepoel D, Linkov V (2020) Metal hydride hydrogen storage tank for fuel cell utility vehicles. Int J Hydrogen Energy 45:7958–7967. https://doi.org/10.1016/j.ijhydene.2019.04.124
14. Type Sheets for Industrial Trucks (2012) VDI 2198, Verein Deutscher Ingenieure e.V., Düsseldorf

Testing an In-House CFD Code for Solving Gas–Solid Flow with Different Simulation Parameters

Is Bunyamin Suryo[1(✉)], Tri Yogi Yuwono[1], and Uwe Schnell[2]

[1] Department of Mechanical Engineering, Institut Tekhnologi Sepuluh Nopember, Sukolilo, Surabaya 60111, Indonesia
isbunyamin@gmail.com

[2] Institut Für Feuerungs- Und Kraftwerkstechnik (IFK), Universitaet Stuttgart, Stuttgart, Germany

1 Introduction

The fluidization process is widely used in the commercial operation or industrial reactors. One of the reactor types which implement the fluidization process in their operation is fluidized beds. The history of Bubbling Fluidized Bed development began at the first time as the BFB reactor was introduced by Fritz Winkler of Germany, December 16, 1921 [1]. In 1939 at the Massachusetts Institute of Technology, Warren Lewis and Edwin Gilliland developed a high-velocity fluidization process for fluid catalytic cracking (FCC) [2]. The Circulating Fluidized Bed concept for solid fuel combustion was initiated by Metallgesellschaft AG, a German company in the 1970s. Researchers have identified that numerical simulation methods have an important role in studying gas–solid hydrodynamics in fluidized beds [3]. By the development of advanced numerical techniques and high-performance computing, CFD served as a significant tool to analyse the gas–solid flow. The objective of this research is to implement the Eulerian–Eulerian (EE) approach to an in-house CFD code, called AIOLOS, for solving gas–solid two-phase flow in a riser. The code is then examined to predict the behaviour of gas–solid flow in a 3D-cylindrical riser. In this paper, the simulation case is based on the experiments of K.M. Luo which were reported in his doctoral thesis [4] and its simulation by Cabezas-Gomez et al. [5].

2 Mathematical Model of Two-Fluid Method

Anderson and Jackson [6] and Ishii [7] are attributed by most researchers as the pioneers for introducing the governing equations of the Two-Fluid Model of gas–solid two-phase flow. The mathematical model derivation for the Two-Fluid Model by Anderson and Jackson [6], Jackson [8], and Ishii [7] was based on the averaging of a local instantaneous of the equations of motion of a single solid particle and the Navier–Stokes equations for fluid motion over a region. A review work of the Two-Fluid Model for fluidization application by Enwald et al. [9] is also a good recommendation for the reader. The

M. Kolhe et al. (eds.), *Recent Advances in Renewable Energy Systems*,
Lecture Notes in Electrical Engineering 876,
https://doi.org/10.1007/978-981-19-1581-9_3

comparison of Jackson's model [8] and Ishii's model [7] was performed by van Wachem et al. [10]. There are two differences between Ishii's and Jackson's Two-Fluid Model. Van Wachem et al. concluded that Jackson's model is more suitable for the gas–solid two-phase problem, whereas the developed model by Ishii is more appropriate for gas–liquid two-phase flow [10]. By assuming that the simulated system is isothermal, no mass transfer between gas and solid phases, mono-sized and perfectly spherical particle, the Navier–Stokes equations consist of the conservation equation of mass and momentum for gas and solid phases. The conservation equations of mass for gas and solid phases are as follows:

Mass conservation of gas phase:

$$\frac{\partial}{\partial t}(\varepsilon_g \rho_g) + \frac{\partial}{\partial x_i}(\varepsilon_g \rho_g U_{gi}) = 0 \tag{1}$$

Mass conservation of solid phase:

$$\frac{\partial}{\partial t}(\varepsilon_m \rho_m) + \frac{\partial}{\partial x_i}(\varepsilon_m \rho_m U_{mi}) = 0 \tag{2}$$

Volume fractions comply with:

$$\varepsilon_g + \varepsilon_m = 1 \tag{3}$$

Conservation of momentum for gas phase:

$$\frac{\partial}{\partial t}(\varepsilon_g \rho_g U_{gi}) + \frac{\partial}{\partial x_j}(\varepsilon_g \rho_g U_{gj} U_{gi}) = -\varepsilon_g \frac{\partial P_g}{\partial x_i} + \frac{\partial \tau_{gij}}{\partial x_j} - I_{gmi} + \varepsilon_g \rho_g g_i \tag{4}$$

Conservation of momentum for solid phase:

$$\frac{\partial}{\partial t}(\varepsilon_m \rho_m U_{mi}) + \frac{\partial}{\partial x_j}(\varepsilon_m \rho_m U_{mj} U_{mi}) = -\frac{\partial P_m}{\partial x_i} - \varepsilon_m \frac{\partial P_g}{\partial x_i} + \frac{\partial \tau_{mij}}{\partial x_j} + I_{gmi} + \varepsilon_m \rho_m g_i \tag{5}$$

In the above equations, ρ, ε, τ, U, P_g, g, P_m, and I_{gm} are density, volume fraction, stress tensor, velocity, gas pressure, gravity, solid pressure, and the momentum exchange between solid and gas phases, respectively. The indices g stand for the gas phase, while m stands for the solid phase. The indices i and j are both from 1 to 3 which represent the direction of coordinates x, y, and z.

Two different ways of modelling the important flow parameters of the solid viscosity and solid pressure could be employed. The first approach is by assuming the flow parameter as a constant or by using an empirical model based on the solid properties and local solid volume fraction [9, 11–13]. The second approach is by using the kinetic theory of granular flow (KTGF). The theory is based on the kinetic theory of gases by Chapman and Cowling [14]. The theory presents a property called granular temperature to represent the random motion of solid particles [15]. In the Eulerian–Eulerian approach, the interplay between gas and solid phases is accounted by the gas–solid momentum transfer which is determined by a momentum exchange coefficient multiplied by the gas–solid relative velocity. Some models could be used to calculate the coefficient, such as Wen-Yu model [16], Gidaspow model [15], and Syamlal et al. model [17].

3 Numerical Simulation

In this study, the simulation case is based on the experiments of Luo [4] and the simulation by Cabezas-Gomez et al. [5]. The simulation is a 3D-cylindrical riser, and the geometry of the simulation test case is shown in Fig. 1. The inlet boundary conditions are listed below:

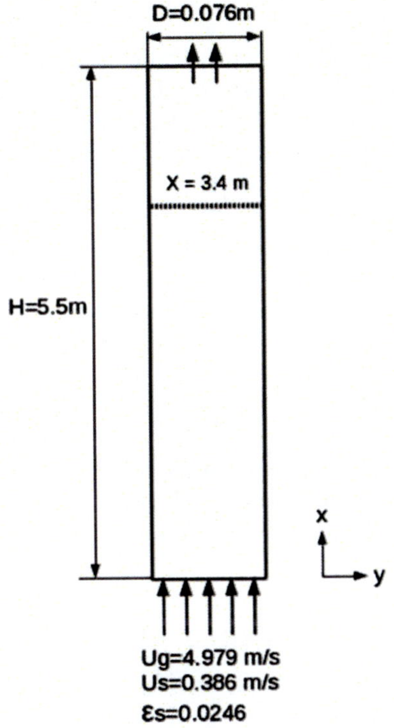

Fig. 1 Simulation boundary conditions (presented in 2D plane)

Gas phase	Solid phase
$\mu_g = 1.82 \times 10^{-5}\,\text{kg}\,\text{m}^{-1}\,\text{s}^{-1}$	$d_m = 520 \times 10^{-6}\,\text{m}$
$\rho_g = 1.225\ \text{kg}\,\text{m}^{-3}$	$\rho_m = 2620\ \text{kg}\,\text{m}^{-3}$
$\varepsilon_g = 0.9745$	$\varepsilon_m = 0.0246$
$U_{g,x} = 4.979\,\text{m}\,\text{s}^{-1}$	$U_{m,x} = 0.386\,\text{m}\,\text{s}^{-1}$

The outlet boundary condition is a zero gradient for any property. A no-slip wall boundary condition is used for both phases.

4 Simulation Results

4.1 The Influence of Meshing

Determining a proper mesh for the computation is really important. The trade-off between the meshing scheme which closely represents the actual condition of the computation domain, and the required computing resource is always arising in any simulation. In this study, using Gidaspow drag model, Carnahan-Starling radial distribution function model, restitution coefficient of 0.7, and time step of 0.0001 s, two different meshes are examined in order to find an optimum point. These two meshes are the standard meshing consisting of 168,600 cells and the fine meshing consisting of 948,480 cells. In order to compare the effect of different meshes, a simulation up to 10 s is carried out on a single computer and its results are then compared with respect to solid velocity and solid volume fraction properties. The comparison of the solid velocity profile is shown in Fig. 2a. After simulation of 10 s, either solid velocity profile or solid volume fraction profile for both meshes shows a qualitatively good agreement. In Fig. 2a, the deviation of solid velocity between the two meshes is significant enough in the riser centre but in the area close to the riser wall, the deviation can be ignored. Furthermore, the comparison of computation resources in terms of calculation time and the fine meshing is much more time demanding than the standard meshing. While the difference of cell numbers is about 779,880 cells between the two different meshes, the fine meshing needs 38 days of computation time while the standard meshing only demands 10 days for the simulation up to 10 s. Taking into consideration the simulation results and the computation time, the standard meshing looks more appropriate than the fine meshing for the discretization of the simulation domain. Thus, the standard meshing is employed for a full 40 s simulation.

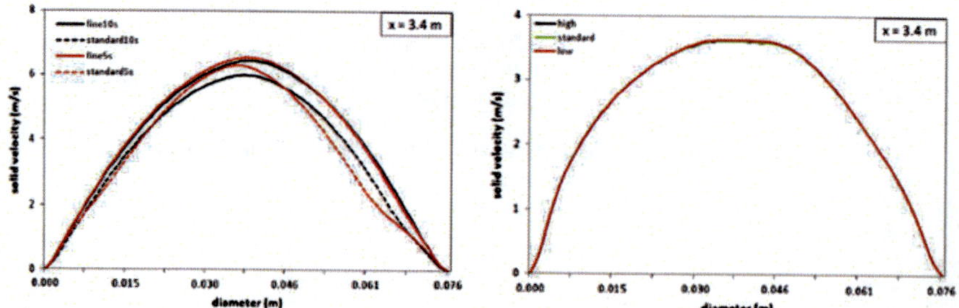

(a) (b)

Fig. 2 Solid velocity profile obtained by: **a** two meshes: standard and fine, **b** different initial conditions

4.2 The Influence of Initial Condition

In this section, different initial conditions are examined in order to understand their effect on the simulation of the gas–solid flow in the riser. Three different initial conditions are employed to simulate the gas–solid flow in the riser up to 6 s using Syamlal et al. drag model, Syamlal et al. radial distribution function model, coefficient of restitution of 0.84, and time step of 0.0001 s. The first is "high" initial value which means the axial gas velocity, the axial solid velocity, and the solid volume fraction are determined as 3 m/s, 2.5 m/s, and 0.0002, respectively. The second initial condition is "low" which means the axial gas velocity, the axial solid velocity, and the solid volume fraction are fixed as 0.002 m/s, 0.001 m/s, and 0.0002, respectively. The "standard" initial condition assigns the axial gas velocity, the axial solid velocity, and the solid volume fraction as 3 m/s, 0.001 m/s, and 0.0002, respectively. After 6 s of calculation, it can be concluded that the initial conditions have no significant effect on the hydrodynamics of the gas–solid flow in the riser. Figure 2b shows quantitatively good agreement of solid velocity for three different initial conditions. Therefore, it could be deduced that the initial condition effect in the current simulations tends to disappear after a certain calculation time. This conclusion shows consistency with the result reported by Benyania et al. [14]. The "standard" initial condition is used to simulate a full 40 s simulation.

4.3 The Variation of Simulation Parameters

The simulation parameters which are examined in this study are 3 different drag models, 2 different radial distribution function models, and 2 different coefficient of restitution values. The combination of these 3 parameters yields 12 variations of the simulation parameters as listed in Fig. 3. To understand the effects of these parameters on the gas–solid flow, these 12 variations are simulated with different time steps, 0.0001 and 0.00015 s; thus in total, there are 24 variations of the simulations. Due to the limitation of the computational resource, these 24 variations are simulated up to 6 s and its results are compared with each other. The objective of this comparison is in order to categorize those 24 variations into several groups which will afterwards be fully simulated up to 40 s.

				Drag model		
				Gidaspow [G]	Syamlal et al. [S]	Wen-Yu [W]
Radial Distribution Function [D]	Carnahan-Starling [D1]	Coefficient of Restitution [E]	0.84 [E2]	GD1E2	SD1E2	WD1E2
			0.70 [E3]	GD1E3	SD1E3	WD1E3
	Syamlal et al. [D2]	Coefficient of Restitution [E]	0.84 [E2]	GD2E2	SD2E2	WD2E2
			0.70 [E3]	GD2E3	SD2E3	WD2E3

Fig. 3 The combination of simulation parameters

For the time step of 0.0001 s, the 12 variations of the simulation can be categorized into 5 typical groups as shown by Fig. 4a. The solid velocity and solid volume fraction contours are at a height of 3.4 m from the riser inlet. It can be seen that group II and group III have a similar solid volume fraction contour in which the solid phase occupies one of the sidewalls and almost half of the riser cross-section area. Whereas group IV and group V results tend to have a similar solid volume fraction. The solid phase in these groups scatters almost all of the cross-section area of the riser. A contrasting result is shown by group I which has a less distributed solid phase. However, all of the group results show the same trend where the solid phase only exists in one of the sidewalls and its value is the maximum value of the solid volume fraction. For the solid velocity contour, all variations show that in the riser centre the solid velocity is maximum. Comparing the maximum solid velocity in the riser centre obtained by all variations, group I has the highest maximum solid velocity, while group V has the lowest maximum solid velocity. Comparing the contours of the solid volume fraction and the contour of the solid velocity, it shows that the contours of the solid volume fraction are very poor in terms of symmetricity. The parameter combination of the radial distribution function of Syamlal et al. (D2) and the coefficient of restitution of 0.7 (E3) tends to achieve a more uniform solid-phase distribution in the riser cross-section area. While the parameter combination of the radial distribution function of Carnahan-Starling (D1) and the restitution coefficient of 0.84 (E2) tends to obtain a poor solid-phase distribution in the riser cross-section area. Thus, the parameter combination of D2E3 may accelerate the spread of the solid phase inside the riser.

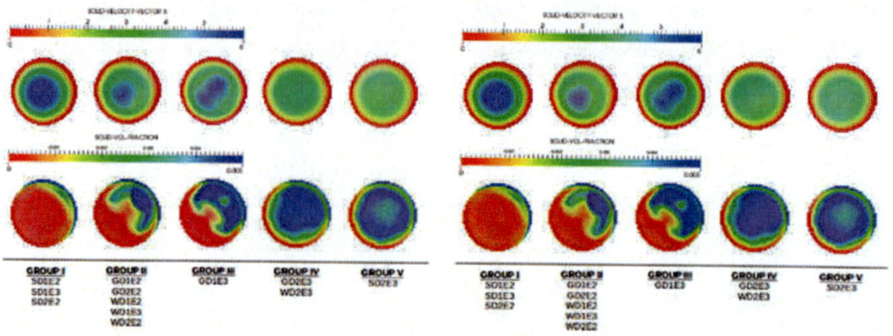

(a) (b)

Fig. 4 The solid velocity and solid volume fraction contour for different model combinations at time step: **a** 0.0001 s, **b** 0.00015 s

Figure 4b shows the solid volume fraction and solid velocity contour for 12 variations of the simulation with time step of 0.00015 s. The results could be clustered into 4 groups in which group III and group IV look very similar to each other. In both groups, the solid phase occupies almost all of the cross-section area of the riser and its solid velocity in the riser centre is lower than the solid velocity in the riser centre of the other groups, group I and group II. The group II shows a more distributed solid phase than group I in which the solid phase only exists on one side of the riser cross-section.

For a full 40 s simulation, the time step of 0.00015 s, the radial distribution function of Syamlal model and coefficient of restitution of 0.70 are used. Using these parameters, the simulation variations are GD2E3, SD2E3, and WD2E3. Figure 5 shows the radial profiles of the axial solid velocity. Comparing the simulation results to the experimental data of Luo, it can be recognized that the result obtained by Syamlal et al. model is quantitatively close to the experimental data. However, the radial profiles of the axial solid velocity resulted by the numerical simulations show an asymmetrical profile.

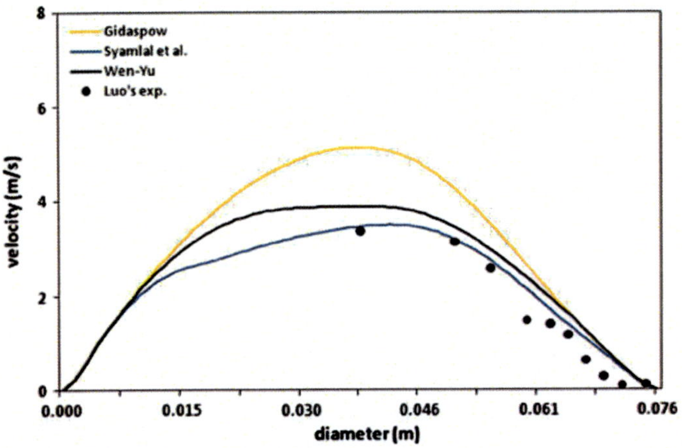

Fig. 5 Solid-phase velocity profiles for simulations with different drag models

5 Conclusions

The simulation parameters which are employed in this study are three different drag models, two different radial distribution function models, two different coefficient of restitution values, and two different time steps of 0.0001 s and of 0.00015 s. In total, 24 variations of the simulations are calculated up to 6 s and its results are compared with each other. From the simulations, the numerical results can be categorized into several groups. For a full 40 s simulation, the simulations using time step of 0.00015 s, the radial distribution function of Syamlal model, coefficient of restitution of 0.70, with three different drag model, Gidaspow, Syamlal, and Wen-Yu, show that the result obtained by Syamlal et al. drag model has a quantitatively close to the experimental data. However, either the radial profiles of the axial solid velocity or the solid volume fraction resulted by the numerical simulations shows an asymmetrical profile.

References

1. Basu P (2006) Combustion and gasification in fluidized beds. CRC Press, Florida
2. Lewis WK, Gilliland ER (1940) Conversion of hydrocarbons with suspended catalyst. US Patent 2,498,088, Original Application 1940, Patented 1950

3. Gidaspow D, Jung J, Singh RK (2004) Hydrodynamics of fluidization using kinetic theory: an emerging paradigm: 2002 flour-daniel lecture. Powder Technol 148(2):123–141
4. Luo KM (1987) Dilute, Dense-phase and maximum solids-gas transport. Ph.D. thesis, Illinois Institute of Technology, Chicago
5. Cabezas-Gomez L, da Silva RC, Navarro HA, Milioli FE (2008) Cluster identication and characterization in the riser of a circulating fluidized bed from numerical simulation results. Appl Math Model 32(3):327–340
6. Anderson TB, Jackson R (1967) Fluid mechanical description of fluidized beds: equations of motion. Ind Eng Chem Fundam 6(4):527–539
7. Ishii M (1975) Thermo-fluid dynamic theory of two-phase flow, volume 22 of direction des ´etudes et recherches d´electricit´e de france. Eyrolles, Paris
8. Jackson R (1997) Locally averaged equations of motion for a mixture of identical spherical particles and a Newtonian fluid. Chem Eng Sci 52(15):2457–2469
9. Hans Enwald, Eric Peirano, and A-E Almstedt. Eulerian two-phase flow theory applied to fluidization. *International Journal of Multiphase Flow*, 22:21–66, 1996.
10. Van Wachem BGM, Schouten JC, Van den Bleek CM, Krishna R, Sinclair JL (2001) Comparative analysis of cfd models of dense gas–solid systems. AIChE J 47(5):1035–1051
11. Yuan P Tsuo and Dimitri Gidaspow. Computation of flow patterns in circulating fluidized beds. *AIChE Journal*, 36(6):885–896, 1990.
12. Charles S Campbell and David G Wang. Particle pressures in gas-fluidized beds. *Journal of Fluid Mechanics*, 227:495–508, 1991.
13. Kuipers JA, Van Duin KJ, Van Beckum FP, Van Swaaij WP (1992) A numerical model of gas-fluidized beds. Chem Eng Sci 47(8):1913–1924
14. Benyahia S, Arastoopour H, Knowlton TM, Massah H (2000) Simulation of particles and gas flow behavior in the riser section of a circulating fluidized bed using the kinetic theory approach for the particulate phase. Powder Technol 112(1):24–33
15. Gidaspow D (1994) Multiphase flow and fluidization: continuum and kinetic theory descriptions. Academic Press, California
16. Wen CY (1966) Mechanics of fluidization. In: Chemical engineering progress symposium series, vol 6, pp 100–101
17. Syamlal M, O'Brien TJ (1989) Computer simulation of bubbles in a fluidized bed. In: AIChE symposium series, vol 85, pp 22–31

A Numerical Study for the Prediction of Unmanned Aerial Vehicle Aerodynamic Performance Based on Dihedral and Tip-Twist Angles of the Wing

Adi Susanto(✉) and Arif Wahjudi

Department of Mechanical Engineering, Institut Teknologi Sepuluh Nopember, Surabaya 60111, Indonesia
adisst@gmail.com

1 Introduction

UAV technology has become a necessity in various elements of society, ranging from hobbies to meeting the needs of military reconnaissance. Advances in electronics allow UAVs to carry devices such as cameras, sensors, or parts of sensor systems so UAVs can assist in detecting physical quantities such as pressure, humidity, heat, light, pollution levels, and so on. The information obtained can be used for agricultural management [1] or disaster detection [2].

UAVs have various types based on their gross weight, one of which is micro UAVs which weigh between 250 g and 2 kg [3]. The wing is the main part that allows fixed-wing UAVs to lift the load carried. Wing geometry is closely related to the aerodynamic performance of the UAV. There are two kinds of wing geometry, the first is global geometry which includes airfoil and planform, while the second is local geometry such as dihedral angle and tip-twist [4]. The dihedral is the angle formed between the wing axis and the horizontal axis; in addition, the tip-twist is the angle formed between the airfoil chord lines on the tip relative to the root chord.

One of the aerodynamic performances is $\max(C_L/C_D)$ which is the ability of a UAV to lift loads while resisting the resistance that occurs due to fluid motion around it. The UAV design will be better if $\max(C_L/C_D)$ is higher. Another aerodynamic performance is $C_D|_{\alpha=0}$. This performance shows the drag received by the UAVs when exploring the air at an angle of attack of 0°, so it has a direct effect on the amount of energy needed to cruise.

The artificial neural network is a reliable method to find the relationship between aerodynamic variables. An artificial neural network has been used to find the relationship between geometric variables and aerodynamic coefficients. Network training involves collecting data from numerical simulations or experiments. The prediction results form an equation that can be used to interpolate new datasets. The results obtained indicate that a simple network structure can solve non-linear behavior caused by changes in the

M. Kolhe et al. (eds.), *Recent Advances in Renewable Energy Systems*,
Lecture Notes in Electrical Engineering 876,
https://doi.org/10.1007/978-981-19-1581-9_4

Reynolds Number, Mach number, and wing geometry [5], both in the steady or non-steady analysis [6]. Comparison of neural network predictions with numerical simulation results and experiments shows good accuracy [7, 8].

This study seeks to find a relationship between changes in the geometry of the dihedral wing and tip-twist on the aerodynamic performance of the UAV, respectively, $\max(C_L/C_D)$ and $C_D|_{\alpha=0}$, using a numerical method, artificial neural network. ANN programming is done by using the MATLAB Toolbox, with the Levenberg–Marquardt method chosen as the backpropagation algorithm. Symbols summarized in Table 1 used to describe UAV wing geometries and parameters.

Table 1 Symbols description

Symbol	Description	Symbol	Description
C_L	Lift coefficient	$C_D\|_{\alpha=0}$	Drag coefficient at the angle of attack 0o
C_D	Drag coefficient	c_r	Root chord
α	Angle of attack	c_t	Tip chord
$\max(C_L/C_D)$	The maximum value of the ratio of lift to drag coefficient	ϕ	Tip-twist angle
		Γ	Dihedral angle

2 Methods

The training process with ANN begins with data collection through a series of simulations based on experimental design. The constant variables include the global geometry of the wing represented by the airfoil and planform. The airfoil for the wing is CAL2263M, and this airfoil has the highest score based on the weighted sum method (WSM) in terms of aerodynamic criteria such as lift coefficient at 0° attack angle, maximum lift coefficient, minimum drag coefficient, maximum range parameter, maximum flight resistance parameter, and coefficient pitching moments [9]. Baseline wing planform includes root chord information $c_r = 189$ mm and tip chord $c_t = 160$ mm, span 400 mm, and tip-offset 93 mm according to Fig. 1. Other information needed in the simulation includes wing mass and the location of the relative center of gravity of the wing concerning the coordinates (0,0,0) are summarized in Table 2.

The independent variable that is regulated or also known as factor consists of Γ varied in the range 0°–8° with an increment of every 1° and ϕ varied from −2° to 2° with an increment of every 0.5°. This setting indicates that the simulation will be performed on two factors and nine levels on each factor, so there are 81 level pairs to be simulated.

Numerical simulation using XFLR5, an analysis tool for airfoils, wings, and planes operating at low Reynolds Numbers, which can produce fast and relatively accurate aerodynamic performance analysis at low Reynolds number [10]. Simulation with XFLR5 includes two steps of each airfoil analysis carried out by setting the top and bottom

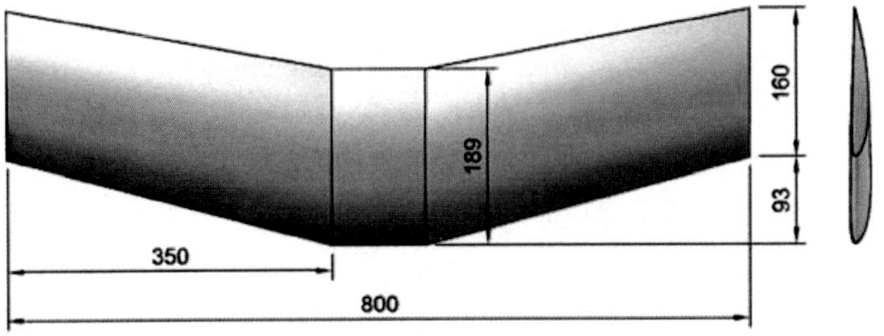

Fig. 1 The baseline for wing planform

Table 2 UAV components dimension for XFLR5 simulation

Components (section)	Position (mm)			Mass (g)
	X	*Y*	*Z*	
Additional masses				
Servo1	130	90	0	12
Servo2	130	−90	0	12
Fuselage	110	0	−60	200
Rotor	−70	0	−56	35
Battery	28		−100	200

transition location ($x/c = 1$). The next step is the wing and plane design with ring vortex analysis type, polar type fixed speed 22.2 m/s, on 3467 VLM panels.

Response data was obtained by extracting the polar simulation results containing the C_L and C_D values for the UAV in the range of angles of attack specified to the spreadsheet software, where the $\max(C_L/C_D)$ value can be calculated and $C_D|_{\alpha=0}$ observed. Both factors proved to have a significant effect on the response based on ANOVA at the 95% confidence level. The response data is then used as a target in the neural network training process. The preliminary step taken is to normalize the target data so that it has values that are spread within the range of -1 to 1 according to the data range required by the transfer function used [11]. The transfer function, Tansig, was chosen to connect the input-hidden1 and hidden1-hidden2 layers because this transfer function provides a wide and generic range of output values so that the output is easy to use in the future post-processing.

The target data was randomly divided into three subsets, each training set, validation set, and testing set with a division of 70%, 15%, and 15% representing 57, 12, and 12 data for each subset, respectively. The training process starts on a network structure with two neurons in a single hidden layer. The MSe value in this structure is stored. Followed by training on the structure of three neurons in a single hidden layer, the MSe in this structure is compared with the previous MSe, the smallest value is stored, this step is repeated until it reaches the smallest MSe value from the structure consisting of a maximum of two hidden layers, and the number of neurons in each hidden layer reaches ten. A simple network structure and a large number of training points are expected to avoid overfitting. The training process will stop at the network structure with several neurons in each hidden layer that gives the smallest MSe.

3 Discussion

3.1 Training $\max(C_L/C_D)$

The $\max(C_L/C_D)$ training resulted in the smallest MSe of 8.4757×10^{-7} on the best network structure at 2–5–7–1. Prediction and target values are plotted according to Fig. 2a. The prediction coincided with the target shows values that are not much different, indicating a qualitatively good training process.

$$a_1 = \frac{2}{1 + \exp[-2(lW1\_1 * Xp1 + b1)]} - 1 \tag{1}$$

$$b1 = [-1.51 \; -1.29 \; -1.38 \; -1.82 \; 2.05]^{\mathrm{T}} \tag{2}$$

$$\mathrm{IW1\_1} = \begin{bmatrix} 0.12 & 0.460 & 0.37 & 0.21 & 0.77 \\ 1.39 & -1.9 & 0.002 & -3.45 & -0.67 \end{bmatrix}^{\mathrm{T}} \tag{3}$$

$$a_2 = \frac{2}{1 + \exp[-2(LW2\_1 * a1 + b2)]} - 1 \tag{4}$$

$$b2 = [-2.08 \; -0.78 \; -0.029 \; 0.11 \; -0.24 \; -2.13 \; 2.04]^{\mathrm{T}} \tag{5}$$

$$\mathrm{LW2\_1} = \begin{bmatrix} 0.48 & 0.42 & -1.66 & -0.97 & 1.22 \\ 0.98 & -0.19 & 0.39 & -1.59 & -0.89 \\ -0.27 & -1.51 & 2.6 & 0.48 & 2.43 \\ 0.26 & 0.82 & -1.56 & -0.58 & -0.38 \\ 0.83 & -1.14 & -1.28 & 0.82 & -0.20 \\ -2.64 & 0.45 & 0.61 & -0.23 & 0.34 \\ 0.22 & -0.10 & 0.7 & -1.83 & -1.18 \end{bmatrix} \tag{6}$$

$$a_3 = LW3\_2 * a2 + b3 \tag{7}$$

$$b3 = [-0.31]; \tag{8}$$

$$\mathrm{LW3\_2} = [1.10 \; -1.76 \; -0.73 \; -0.91 \; 1.01 \; 1.04 \; -0.27] \tag{9}$$

The output value of the first layer a_1 is the result of the sigmoid hyperbolic tangent function to the corrected input value with weights (b1) and bias (lW1_1), shown in Eq. 1. The second hidden layer also uses the sigmoid hyperbolic tangent transfer function to calculate the output a_2. The input is a_1 corrected by bias (b2) and weight (LW2_1) shown in Eq. 4. The last layer output is the result of the purelin function to the second layer output which has been corrected by bias (b3) and weight (LW3_2), according to Eq. 7.

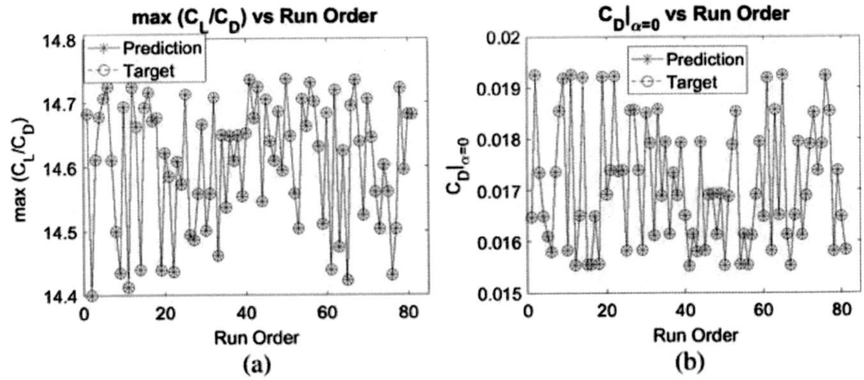

Fig. 2 Comparison for target and neural network prediction for **a** $\max(C_L/C_D)$, and **b** $C_D|_{\alpha=0}$

3.2 Training $C_D|_{\alpha=0}$

Similar to the previous training process, the training for input Γ and ϕ to the response $C_D|_{\alpha=0}$ resulted in the smallest MSe of 1.952×10^{-8} obtained in the network structure 2-4-9-1. Fig. 2b shows the location of the predictions and targets that coincide with each other, indicating good training quality. The relationship between input, hidden layer, and output is shown by Eqs. 1, 4, and 7. Bias and weights for each layer are written in matrix form in Eqs. 10–15.

$$b1 = \begin{bmatrix} 2.59 & 0.97 & -0.47 & 0.98 \end{bmatrix}^{\mathrm{T}} \tag{10}$$

$$\mathrm{IW1\_1} = \begin{bmatrix} -0.95 & -0.001 & -0.08 & -0.02 \\ -2 & 1.39 & -0.18 & -0.85 \end{bmatrix}^{\mathrm{T}} \tag{11}$$

$$b2 = \begin{bmatrix} -2.37 & -2.4 & 1.53 & -1.27 & -0.52 & 0.2 & -0.6 & 2.18 & 2.65 \end{bmatrix}^{\mathrm{T}} \tag{12}$$

$$\mathrm{LW2\_1} = \begin{bmatrix} -0.08 & -0.56 & -0.41 & 1.25 & 0.3 & 0.004 & -2.14 & -0.082 & 1.42 \\ 1.87 & -1.74 & 1.15 & 1.33 & 1.02 & -0.51 & -0.66 & -0.18 & -1.61 \\ 0.17 & 0.61 & -0.22 & 0.75 & 0.37 & -2.282 & -0.9 & 0.865 & 0.4 \\ -1.76 & -1.29 & -0.42 & -0.91 & -0.68 & -0.25 & -0.26 & -2.33 & 1.03 \end{bmatrix}^{\mathrm{T}} \tag{13}$$

$$b3 = [-0.56]; \tag{14}$$

$$LW3\_2 = \begin{bmatrix} 1.26 & -0.99 & 0.36 & -0.13 & 0.8 & 0.37 & 0.24 & 0.50 & 0.44 \end{bmatrix} \tag{15}$$

3.3 Training Results

Figure 3 shows the relationship between factor Γ and the ϕ response $\max(C_L/C_D)$. Responses range from 14.4 to 14.7. Factor ϕ plays a bigger role in changing the response value than factor Γ. The Γ causes the lift force to split into vertical and horizontal components. The lift force is the component of the vertical force which decreases in value as the dihedral angle increases. Increasing the ϕ will increase the lift at the wingtip as the angle of attack increases, thereby increasing the overall lift coefficient, but at the same time increasing the induced drag value due to the increased reference area. Changes in $C_D|_{\alpha=0}$ response to changes in dihedral and tip-twist factors are presented in Fig. 4. The smallest response value is 0.0155, while the largest is 0.0192. Changes in the value of ϕ provide the largest contribution to the change in responses. Induced drag caused by an increase in ϕ plays a major role in this response.

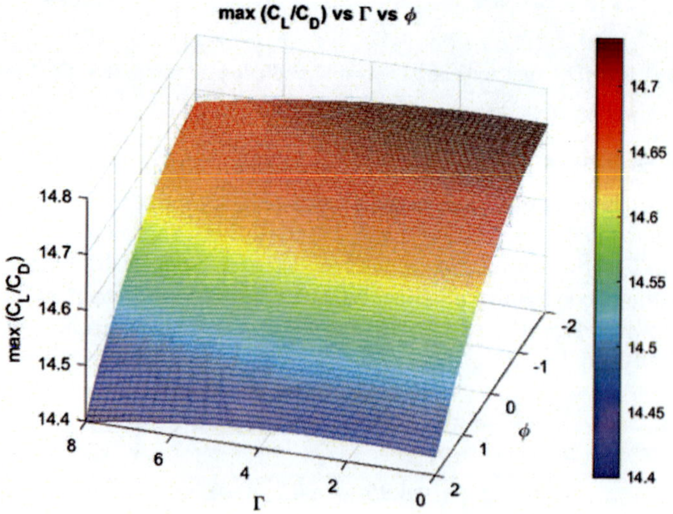

Fig. 3 Relation of Γ and ϕ to response $\max(C_L/C_D)$

3.4 Network Usage

The use of the network is done by selecting random arbitrary Γ and ϕ values within the trained range, acting as input to the MATLAB equation according to Eqs. 1–15, and the prediction results are compared with the simulation results, according to Table 3. The largest relative error was calculated equal to 0.20609%, as a result of $\Gamma = 7.2$o and $\phi = 1.4$. These relative error values were found to be less than 5% indicating that the MATLAB equation is generalized well to interpolate data.

4 Conclusion

The relationship between factor and response $\max(C_L/C_D)$ is represented by ANN with a structure of 2-5-7-1. The lowest MSe obtained is 8.4757e-07. The addition of the dihedral value will reduce the lift coefficient, while the increase in tip-twist will increase

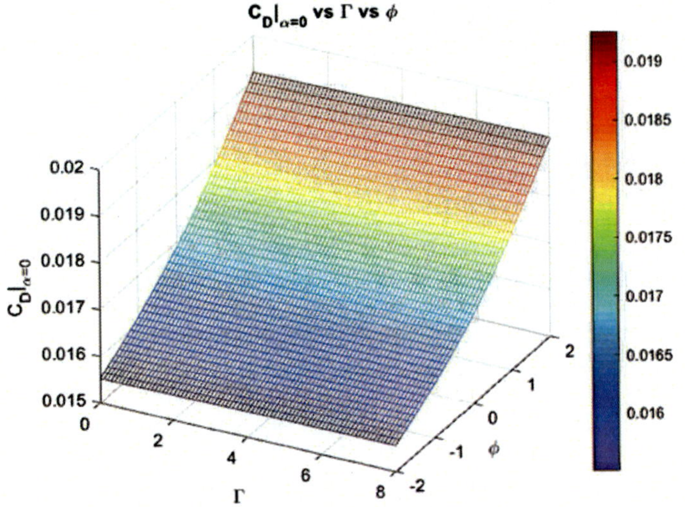

Fig. 4 Relation of Γ and ϕ to response $C_D|_{\alpha=0}$

Table 3 Comparison prediction to validation simulation

Γ	ϕ	$\max(C_L/C_D)$			$C_D\|_{\alpha=0}$		
		predict	Val. simulation	Error (%)	predict	Val. simulation	Error (%)
1.9	−1.2	14.71327	14.7161	0.019265	0.015992	0.016	0.050025
4.6	0.8	14.57068	14.5703	0.00261	0.017701	0177	0.005649
7.2	1.4	14.48302	14.4823	0.004949	0.018438	0.0184	0.206096

the lift coefficient but also induces the drag coefficient to increase as well. Between factors and response, $C_D|_{\alpha=0}$ is represented by ANN with structure 2-4-9-1. The lowest MSe obtained is 1.9517e−08. The increase in induced drag due to the increase in tip-twist contributes more to the value of this response than the effect of the dihedral changes. The comparison of the prediction results to the confirmation simulation results in an error value is smaller than 5%. Both ANN structures indicate that the relationship between wing geometries to aerodynamic performances can be represented by a simple network. These networks are generalizing well to interpolating data.

References

1. Boursianis AD, Papadopoulou MS, Diamantoulakis P, Liopa-Tsakalidi A, Barouchas G, Salahas P, Karagiannidis G, Wan S, Goudos SK (2020) Internet of Things (IoT) and agricultural unmanned aerial vehicles (UAVs) in smart farming: a comprehensive review. Internet of Things 12:1–17
2. Akhloufi MA, Couturier A, Castro NA (2021) Unmanned aerial vehicles for Wildland fires: sensing, perception, cooperation and assistance. Drones 5(15):1–25

3. Chamola V, Kotesh P, Agarwal A, Naren N, Gupta N, Guizani M (2020) A comprehensive review of unmanned aerial vehicle attacks and neutralization techniques. Ad Hoc Netw 1–26:11
4. Güzelbey İH, Eraslan Y, Doğru MH (2018) Effects of taper ratio on aircraft wing aerodynamic parameters: a comparative study. In: International Mediterranean science and engineering congress, vol 3. IMSEC, Adana, pp 1–6
5. Rajkumar T, Bardina J (2002) Prediction of aerodynamic coefficients using neural network for sparse data, 2002. In: FLAIRS 2002, AAAI, vol 2, pp 242–246, Florida
6. Wallach R, de Mattos BS, Giradi RM (2006) Aerodynamic coefficient prediction of a general transport aircraft using neural network. In: ICAS 2006, vol 25, pp 1–15, ICAS, Hamburg
7. Adique M, Amiralaei MR, Alighanbari M (2010) Application of artificial neural networks in aerodynamics prediction of low-Reynolds-number figure-eight motion of an airfoil. In: AIAA 2010, vol 48, pp 1–8, Toronto
8. Ignatyev D, Khrabrov A (2018) Experimental study and neural network modeling of aerodynamic characteristics of canard aircraft at high angles of attack. Aerospace 5(26):1–27
9. Hieu NK, Thien LH (2016) Airfoil selection for fixed wing of small unmanned aerial vehicle. Lecture Notes Electr Eng 371:881–890
10. Prisacariu V (2018) Analysis of UAV's flight characteristics. Rev Air Force Acad 3(38):29–36
11. Secco NR, de Mattos B (2017) Artificial neural networks to predict aerodynamic coefficients of transport airplanes. Aircraft Eng Aerosp Technol Int J 89(2):211–230

A Numerical Study for Prediction of Unmanned Aerial Vehicle Aerodynamic Performance Based on Chord Tip and Offset of the Wing

Firiana Firdaus(✉), Arif Wahjudi, and Wawan Aries Widodo

Institut Teknologi Sepuluh Nopember, Surabaya, JawaTimur 60119, Indonesia
ittofirdaus@gmail.com

1 Introduction

In recent years, the use of unmanned aerial vehicle (UAV) in several sectors has been widespread, ranging from the military, agriculture and plantations, mining, civil, rescue missions, border guarding, and even a hobby among certain circles [1, 2].

UAVs like airplanes have the same flying principle to fly and hover in the air. The part that plays the most role in this is the wings. The wings have a cross-section called an airfoil. The shape of the airfoil profile of a wing will affect the drag and lift, which will affect the aerodynamic performance of the UAV. The aerodynamic performances include the C_L/C_D and C_D-0. C_L/C_D is a value that indicates the efficiency of the airfoil. Then, C_D-0 is the coefficient of drag acting on the wing at an angle of attack of 0 degrees. The geometry of a wing, such as a taper ratio and swept angle, affects its aerodynamic performance [3, 4]. However, changing the sweep angle does not have a significant influence on the aerodynamics of the UAV but rather on its stability [5].

XFLR5 software is open-source to help design a UAV. XFLR5 uses Xfoil as the database base. XFLR is effectively used to analyze aerodynamics and UAVs at low Re. By comparing various variations of the wing geometry model and displaying the results in graphs or numerical data as well as 2D animation and 3D results obtained [6].

Artificial neural network (ANN) is an intelligence algorithm designed to classify or find patterns of data that can later be used to build a model of the relationship between input and output data [7]. The input data is aerodynamic performance obtained from XFLR5, while the output data is an adjective function [8].

So the purpose of this paper is to find the relationship between input in the form of aerodynamic data and changes in the wing's geometry. That is by adjusting the taper ratio or adjusting the width of the wing's chord tip (C_t) and adjusting the swept angle's location so that the offset value is obtained. Where every change in C_t and offset will indicate a different aerodynamic performance value, so with the help of ANN, a relationship can be found between the changes in the dimensions of the wing and its aerodynamic performance.

M. Kolhe et al. (eds.), *Recent Advances in Renewable Energy Systems*,
Lecture Notes in Electrical Engineering 876,
https://doi.org/10.1007/978-981-19-1581-9_5

2 Method

2.1 Preliminary Design

A small UAV Cessna 186 is used as a baseline model to obtain each component's overall dimension and weight. However, the wing adopted an airfoil shape of CAL2263M with a maximum wingspan of 800 mm. This study selected the taper ratio by arranging the chord tip (C_t) and swept angle will set the offset distance (see Fig. 1) parameters as the primary geometry variation to optimize the wing performance. The taper ratio varies from 55 to 85% from the cord root [8], while the sweep angle varies from 0° to 15° [5]. There is nine level with two factors that eventually give 81 sample data using a full factorial model shown in Table 1. Factors and levels of simulations.

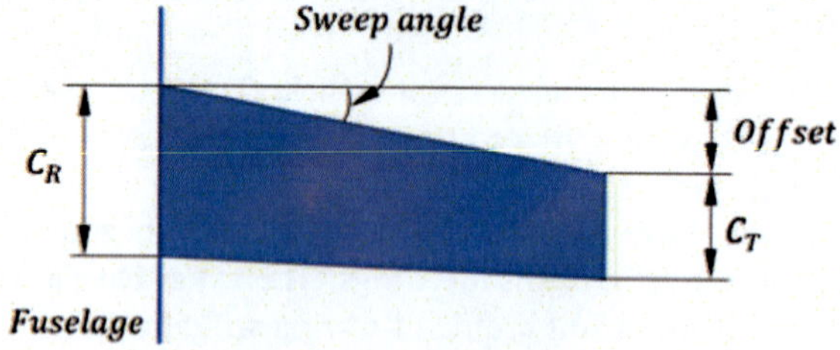

Fig. 1 Wing geometry

Table 1 Factors and levels of simulations

No.		First factor		Second factor	
		C_t		Sweep angle	Offset
		% Of C_R mm		o (deg)	mm
Level	1	55	103.95	0	0.000
	2	58.75	111.0375	1.875	11.458
	3	62.5	118.125	3.75	22.940
	4	66.25	125.2125	5.625	34.472
	5	70	132.3	7.5	46.078
	6	73.75	139.3875	9.375	57.785
	7	77.5	146.475	11.25	69.619
	8	81.25	153.5625	13.125	81.609
	9	85	160.65	15	93.782

2.2 XFLR5 and ANOVA

After defining the input design experiment, we simulated the entire design combination using XLFR5 based on CFD. The output performance that we are interested in is the lift coefficient to drag coefficient maximum ratio (C_L/C_D max) and drag coefficient at zero angle of attack. The constant parameters are defined, such as the overall dimension, weight, and speed of the UAV at 22 m/s. After we obtained the complete sample data, we conducted a statistical test of ANOVA to ensure both input geometries have a significant effect on the defined output of C_L/C_D max and C_D-0.

2.3 ANN

A preliminary set up of the training process is constructed as follows. The initial ANN structure consisted of input and output layers with two hidden layers with a maximum of ten neurons. We used the Lavenberg-Marquardt as a training algorithm. A typical input and output data would have a different order; thus, we used a scaling process called the mapminmax function to normalize it. The entire training data was divided into 70% for the training process, 15% for validation, and 15 for testing. The activation function used in the hidden layer is tansig, and in the output layer is purelin. We evaluate the training process based on the mean square error. Figure 2 above shows the illustration of the ANN structure.

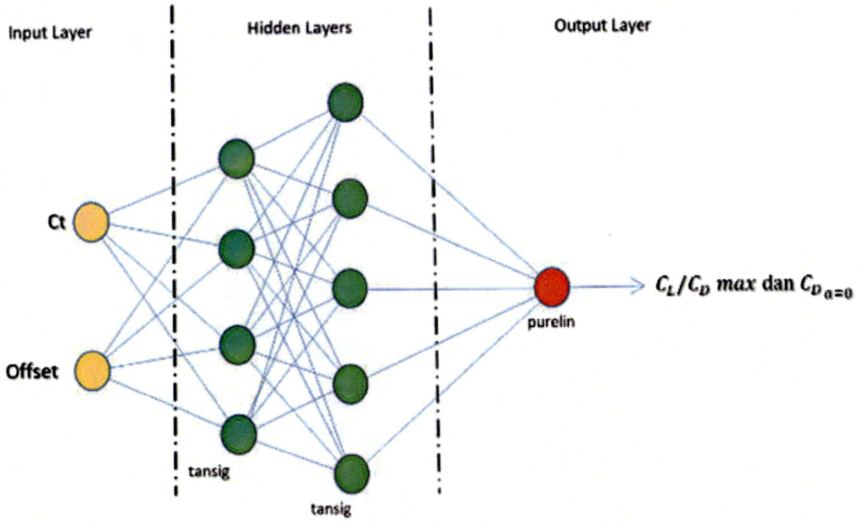

Fig. 2 The architecture of neural network

3 Discussion

This training process aims to predict the hidden relationship between the input and output. The quality of the training process can be examined from the MSE value. A small MSE value means the training process is successful. Moreover, the relationship predicted by ANN is presented in the form of its structure based on the calculation of each parameter as followed in Eq. (1):

$$a1 = \text{tansig}(IW1\_1 * Xp1 + b1) = \frac{2}{1 + \exp(-2(IW1\_1 * Xp1 + b1))} - 1 \quad (1)$$

The result of the output layer is obtained from the input value, which has been transformed by the activation function of the tangent sigmoid (tansig), then, the result from the first layer is passed into the second layer, and the same process is calculated according to this Eq. (2)

$$a2 = \text{tansig}(LW2\_1 * a1 + b2) = \frac{2}{1 + \exp(-2(LW2\_1 * a1 + b2))} - 1 \quad (2)$$

Finally, the result from the second output layer is activated using purelin function based on this Eq. (3)

$$a3 = \text{purelin}(LW3\_2 * a2 + b3) = LW3\_2 * a2 + b3 \quad (3)$$

Based on the training process, we obtained the smallest MSE for C_L/C_D max response was 3.6566×10^{-7} when the ANN structure has seven neurons at the first hidden layer and two neurons at the second hidden layer. The training process was stopped when the gradient reached 0.000182777 after 14 epochs. The C_D-0 respond reported having a maximum MSE value of 1.8591×10^{-7} while having the same ANN structure as the C_L/C_D. The training process was stopped when the gradient achieved 1.2586×10^{-9} after 15 epochs.

The quality of the prediction model is presented by injecting the initial data training back to the optimum ANN model, as depicted in Figs. 3 and 4. The ANN model has been fully tracked down the complex variation of the initial data compare to the prediction values.

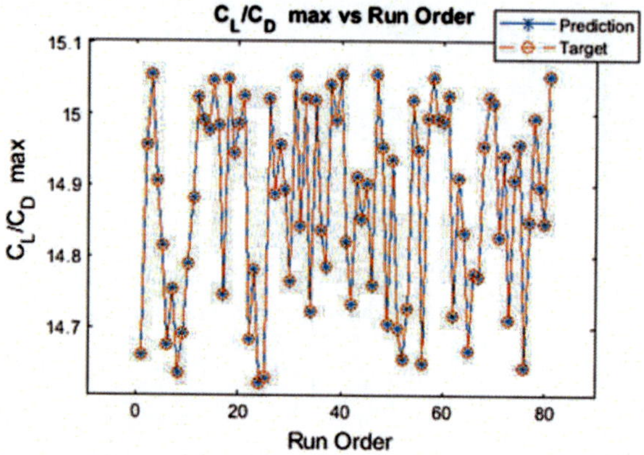

Fig. 3 Prediction graph of C_L/C_D max

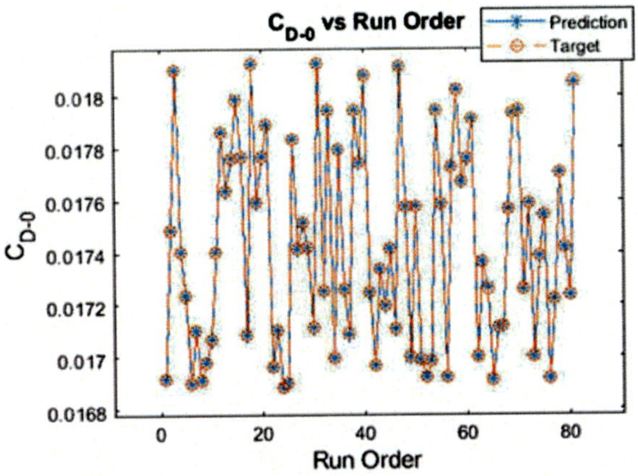

Fig. 4 Prediction graph of C_D-0

Figure 5 shows a graph of the relationship between C_t and offsets as the input layer to the maximum C_L/C_D max as the output layer. If the C_t and the offset value are increased, it will be decreasing the C_L/C_D max value. Where the value of C_L and C_D are both very influential on the cross-sectional area. So that increasing CT and offset will increase the cross-sectional area and result in a decrease in the value of C_L/C_D max.

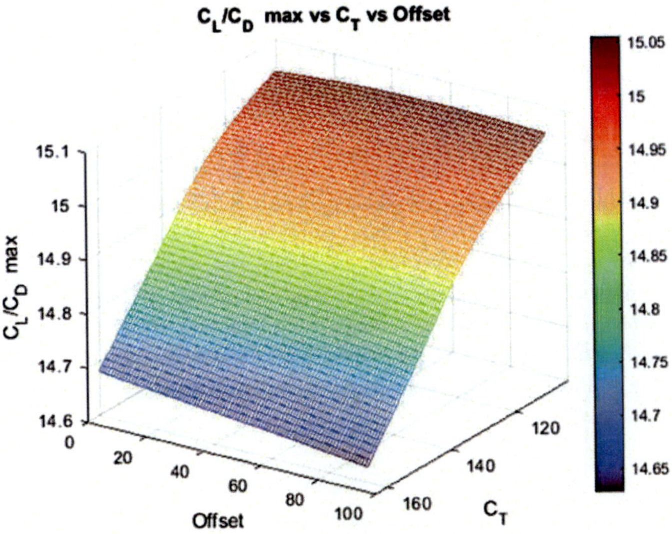

Fig. 5 Response surface graph of C_L/C_D max

From Fig. 6 obtained, the C_D-0 would have a minimum value by increasing the C_t, while the effect of wing offset was less visible. The drag coefficient at zero angles of attack depends on the air pressure variation along the UAV body, while the aerodynamic force changed according to the air pressure at the wing surface; thus, as the C_t value increase, the surface area of the wing also increases.

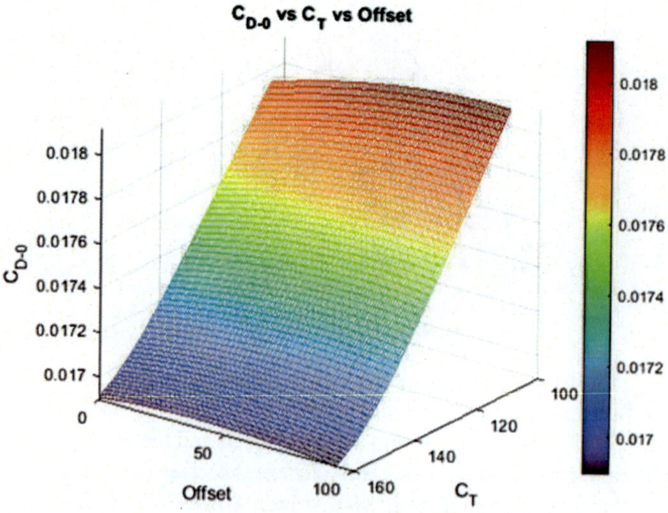

Fig. 6 Response surface graph of C_D-0

4 Conclusion

From the ANN training of C_L/C_D maximum, the smallest MSE value is 3.958×10^{-7} with seven neurons in the first hidden layer and three neurons in the second hidden layer. And from the ANN training of C_D-0, the smallest MSE is 1.8591×10^{-7} with also seven neurons in the first hidden layer and three neurons in the second hidden layer.

Dimensional change of C_t affects the aerodynamic performance. If the C_t increases, it will decrease the value of its aerodynamic performance, either C_L/C_D or C_D-0. Changes in the C_t and offset dimensions will increase the surface area, affecting the aerodynamic performance. C_t changes have a significant effect on aerodynamic performances. However, the change in offset does not provide a significant change in the aerodynamic performance of the UAV.

References

1. Aleksander KC (2018) Military use of unmanned aerial vehicles—a historical study. Saf Def 4:17–21
2. Hildebrand JM (2018) Situating hobby drone practices. Digit Cult Soc 3(2):207–218
3. Sadraey MH (2013) Aircraft design : a systems engineering approach., 1(1). India
4. Nita M, Scholz D (2012) Estimating the oswald factor from basic aircraft geometrical parameters. Dtsch Luft- und Raumfahrtkongress 281424:1–19

5. Güzelbey İH, Eraslan Y, Doğru MH (2019) Effects of taper ratio on aircraft wing aerodynamic parameters: a comparative study. Eur Mech Sci 3(1):18–23
6. Prisacariu VL (2018) Analysis of Uavs flight characteristics. Rev Air Force Acad 16(3):29–36
7. Zupan J (1994) Introduction to artificial neural network (ANN) methods: what they are and how to use them. Acta Chim Slov 41(September):327–327
8. Boutemedjet A, Samardžić M, Rebhi L, Rajić Z, Mouada T (2019) UAV aerodynamic design involving genetic algorithm and artificial neural network for wing preliminary computation. Aerosp Sci Technol 84:464–483

Potential of a Grid-Tied PV System: A Field Study in Hot and Sunny Climate Region

I Nyoman Suamir(✉), I Wayan Temaja, and I Nengah Ardita

Mechanical Engineering Department, Politeknik Negeri Bali, Badung, Bali 80364, Indonesia
nyomansuamir@pnb.ac.id

1 Introduction

Along with policies in most countries in the world including Indonesia, to encourage the use of renewable energy, various modifications of solar power technology were applied to buildings with building-integrated photovoltaic systems [1, 2]. Such system can integrate energy source from renewable energy with energy from the national electricity grid. Braun and Ruther [3] have also reported their research on applying building integrated photovoltaic system for commercial buildings. Their research is very beneficial, especially in hot climate regions.

Photovoltaic (PV) systems are prospective for a rural area with low electrification. The applications of PV systems are economically more attractive due to their price tends to decline. The systems can have a substantial impact on countryside development with various applications such as for lighting and for PV home system. The PV systems were also widely applicable for productive events with the power range from 130 to 250 Wp (watt-peak) [4].

This paper presents a field study of a lab-scale PV system installed at the educational building located in Bali Island Indonesia. The investigation was based on Bali Island weather conditions which have intensive value of solar irradiation ranging from 4.6 to 7.2 kWh m^{-2}. The region had high potential solar energy sources [5, 6]. The utilization of solar energy in the country was only 2.78 MW (0.03% of the total energy use and about 0.41% of the total renewable energy utilization) [7]. The share of solar PV energy sources in Indonesia and also in Bali Island was nonetheless insignificant. Therefore, this paper that presents the advantages of PV systems in terms of energy efficiency and reduction of environmental impacts due to the replacement of the use of fossil fuel-sourced electricity can contribute to the commitment of energy policy-makers in improving the application of solar energy source in the country.

2 Materials and Methods

2.1 Description of the Grid-Tied PV System

The grid-tied PV system is located at Politeknik Negeri Bali south region of Bali Island. The schematic of the PV system is presented in Fig. 1. The PV system includes: power

M. Kolhe et al. (eds.), *Recent Advances in Renewable Energy Systems*,
Lecture Notes in Electrical Engineering 876,
https://doi.org/10.1007/978-981-19-1581-9_6

generation, power distribution, integration to the grid, and a protection system. The power generation system comprises an array of 10 solar PV panels of 310 Wp each and a grid inverter of 5 kW. The PV panels used are monocrystalline which has higher efficiency compared with the polycrystalline type. The power distribution and grid integration system is completed with some breakers and analog energy meters to record renewable energy production, and also, one digital power meter for monitoring energy exported to the grid. PV system also incorporates a protection system to prevent damages from both direct and indirect lightning. The protection system consists of DC and AC Surge Protective Devices (SPDs).

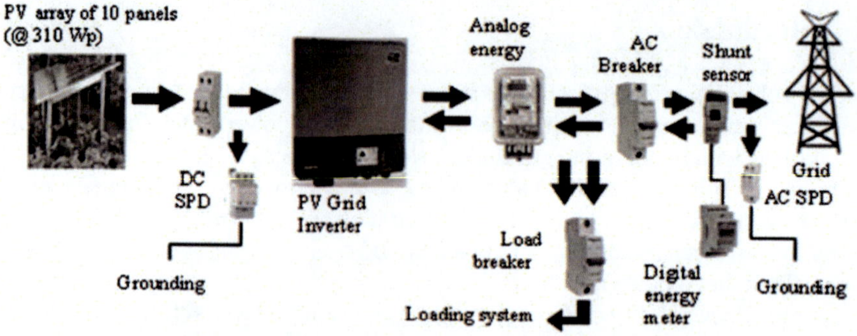

Fig. 1 Schematic of the PV system and its integration to the grid

The PV system employs a fixed array-type installation with the tilt angle of 17° and azimuth direction facing north. The tilt angle applied is in the range of optimum tilt angle based on Bali solar radiation conditions. The optimum tilt angle in the region was reported ranging from 10° up to 18° with azimuth direction facing north [8]. This is also in agreement with the PV system array installed in East, Central, and West Java regions. The regions are located in the west direction of Bali Island. The optimum tilt angle of PV array for azimuth direction facing north of those three regions is, respectively, from 0° to 40°, from 1° to 34° at optimum 18° and 10° [9].

The PV panel has a maximum power of 310 Wp at maximum voltage and a current of 33 V and 9.4 A, respectively. Each PV panel contains 60 PV cells. The PV cell type is monocrystalline silicon of size 0.150 m × 0.150 m each; thus, the PV cell area of each panel becomes 1.35 m^2. Based on the standard test conditions (irradiance 1000 W m^{-2} at 25 °C), the PV system is specified for maximum efficiency of 22.96%.

2.2 Data Gathering Method

The PV system incorporates an Internet of Things-based (IoT) monitoring system which is embedded with sensors, software, and other technologies for connecting and exchanging data with other devices and systems over the Internet. Comprehensive data (hourly, daily, monthly, and yearly) related to system performance can be accessed through the Internet. Recorded data in 2020 were used for the analysis.

Some data were also recorded on site which included solar radiation, energy generated, and exported to the grid. Solar irradiance was measured by using a Lutron solar power meter SPM-1116SD of 10 W m^{-2} accuracy. For irradiance lower and higher than 1000 W m^{-2}, the resolutions can reach 0.1 W m^{-2} and 1 W m^{-2}, respectively. The energy meters used have measurement accuracy in the range of ±1%. Recorded data from the measurement system were processed using spreadsheet software. Performance parameters of the PV system were identified and calculated. The parameters included energy generated by the solar PV system, electricity intensity, and solar PV cell efficiency. The environmental impact reduction due to the replacement of fossil fuel-based electricity was also established.

2.3 Energy Performance Analysis

The energy performance parameters considered in this paper involve renewable energy generation, capacity factor, and efficiencies. Capacity factor (*CF*) is the ratio of AC energy (E_{AC} in kWh) produced to the installed capacity (P_{max} in kW) during the analysis period in (h). The *CF* in (%) can be calculated from Eq. (1) [10].

$$CF(\%) = \frac{E_{AC} \text{ in Analysis Period (kWh)}}{P_{max}\text{(kW)}x \text{ Analysis Period (h)}} x100 \quad (1)$$

Efficiencies of the PV system comprise efficiency of the PV cells, the efficiency of the inverter, and overall efficiency. The efficiency of the PV cells (η_{PV}) in (%) can be calculated from Eq. (2).

$$\eta_{PV}(\%) = \frac{P_{DC}}{GHI} x100 \quad (2)$$

where P_{DC} is DC electrical power intensity of the PV cell (kW m^{-2}), GHI is global horizontal irradiance (kW m^{-2}) measured at the same period as reported in [11, 12].

The efficiency of grid inverter (η_{Inv}) in (%) can be calculated from Eq. (3).

$$\eta_{Inv}(\%) = \frac{P_{AC}}{P_{DC}} x100 \quad (3)$$

where P_{AC} is AC electrical power intensity (kW m^{-2}). The overall PV system efficiency (η_{Sys}) can be determined from Eq. (4).

$$\eta_{Sys}(\%) = \frac{P_{AC}}{GHI} x100 \quad (4)$$

Data of site solar irradiation applied for energy efficiency calculations refer to irradiance data of the region as reported in [11, 13].

2.4 Environmental Analysis

Reduction of the environmental impact through replacement of the fossil fuel-based electricity sources was estimated according to BS EN 378-1 standard [14], while the greenhouse gases (GHG) emissions factor for electricity production in Indonesia used an emissions factor of 0.84 tCO_{2e} MW h^{-1} as reported in [15].

3 Results and Discussion

3.1 Renewable Energy Generation

Results of the renewable energy generation and a capacity factor of the PV system are presented: (i) in Fig. 2 for a one-year period, (ii) in Fig. 3 for daily generation in August (the highest generation month). From Fig. 2, it can be seen monthly electrical energy generation of the PV system for one year in 2020. The figure also shows a one-year variation of the PV system capacity factor. Monthly, electrical energy generation varies from 300 to 499 kWh with an average of 412 kWh. Annual energy generation of the PV system can reach 4.95 MWh. The utilization potential of the PV system stated as a capacity factor can reach an annual average value of 38.7%. The maximum monthly utilization potential with a capacity factor value of 46.1% occurs in August, while the minimum value of 26.1% happens in December.

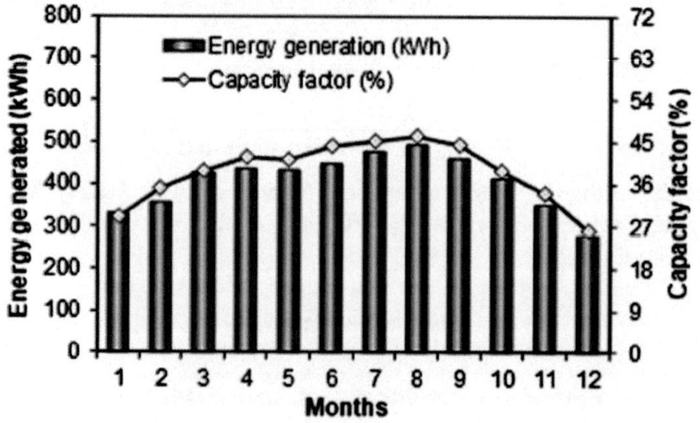

Fig. 2 Monthly solar energy generation and capacity factor for the investigated year

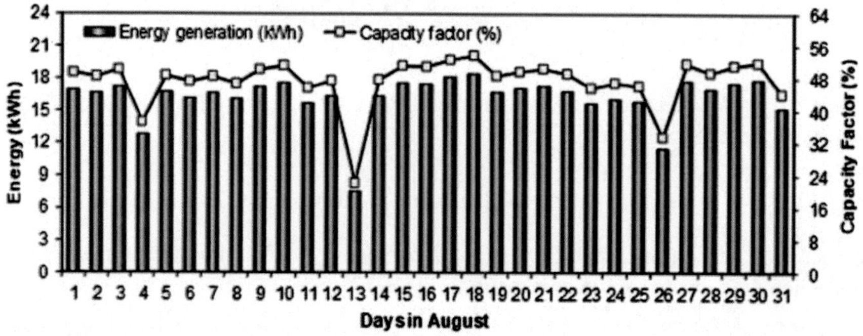

Fig. 3 Daily solar energy generation and capacity factor in August (month with the highest energy generation)

The utilization potential of the PV system where the electrical energy generation and capacity factor are above the annual average is found to occur from March to October.

This shows eight months of the year the PV system can show good energy performance and utilization potential. For the remaining four months, the potential utilization is relatively decreased due to rainy weather conditions.

More thorough investigation can identify energy production and potential utilization of the PV system in the months with the highest energy generation in a year as shown in Fig. 3. The maximum daily energy generation and PV system utilization of the highest potential month August (Fig. 3) are, respectively, 18.5 kWh and 53.7% while the lowest potential month occurs in December with maximum daily energy as high as 11.8 kWh with a capacity factor of 34.3%. The average daily energy generation in August is 16.6 kWh and December 9.0 kWh with a difference of around 45.8%.

3.2 Power Generation and Energy Efficiency Analysis

Figure 4 shows the variation of the AC and DC power generation together with electrical energy generation during the production period of the selected day in August (the highest production month). The time of production is 11 h starting from 6.40 to 17.40. The average instant AC and DC power generations during the production period are 1662 W and 1858 W, respectively, while the maximum AC and DC power generations, which occur at around 12 o'clock, can reach, respectively, as high as 2552 W and 2811 W. AC and DC cumulative energy generation of the PV system in the selected highest day can correspondingly reach 18.5 kWh and 20.8 kWh, while the AC electrical power generation during the lowest production day in December occurs in about 10 h starting from 6.30 in the morning to 4.30 in the afternoon. It is one hour shorter than the time of production in August. Average instant power generation during the production period is 274 W, and maximum power is 741 W, while cumulative AC electrical energy generation in the selected lowest day of production is only 2.82 kWh.

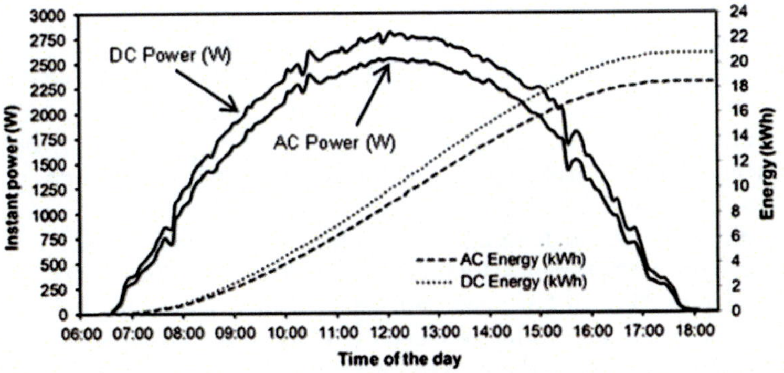

Fig. 4 AC and DC electrical power as well as energy generation during the highest production day in August

Figure 4 also clearly shows that the AC power or AC energy generation is lower than the DC power or DC energy generated from the PV system. This happens due to losses in the grid inverter which also means the AC power, or AC energy generation is directly influenced by the efficiency of the PV grid inverter.

Variation of the PV system efficiencies can be seen in Fig. 5. The efficiencies comprise PV cell efficiency and grid-inverter efficiency. In general, the efficiencies tend to increase from the start of the production time at 6.40 until 8.30, afterward relatively flatten up to 15.00 and then decrease until the end of the production time. The PV cell efficiency can reach a maximum value of 21.8% and of about 18.6% on average. The installed PV cells can perform with cell efficiency that is very close to the specified maximum efficiency of 22.96%. Whereas, the grid-inverter efficiency of the PV system is 89.4% on average with a maximum value is as high as 93.4%. Overall efficiency of the PV system spans from 11.2 to 19.7% with an average value of 16.7%.

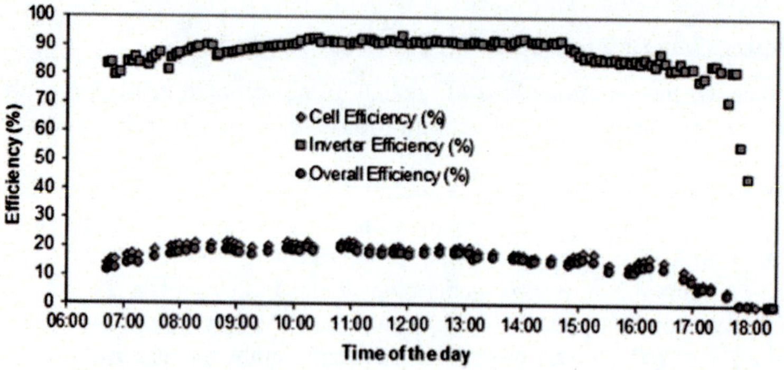

Fig. 5 Energy efficiencies of solar PV system investigated at a clear and sunny day in August

3.3 Environmental Impact Reduction

Figure 6 illustrates the variation of environmental impact reduction (emissions reduction) of the investigated PV system. Emissions reduction due to the replacement of fossil fuel-based electrical energy varies following the quantity of solar PV energy generation. The maximum impact reduction occurs in August and the minimum one in December. The PV system is found to have a potential annual contribution to environmental impact reduction with equivalent to CO_2 emissions reduction of 4.12 tons.

As previously discussed, the utilization potential of the investigated PV system of a maximum of 46.1% is significantly higher than overland and floating PV systems that have been reported in [10]. Such systems have a maximum utilization potential of 17.6% and 20.8%, respectively. This means that the utilization potential of the investigated PV system (46.1%) is higher by about 162% than the overland PV system and 122% higher than the floating PV system. The exceptional utilization potential together with the potential reduction in annual environmental impact make the grid-tied PV system as an incredibly prospective alternative for implementing renewable energy sources to power commercial buildings in Indonesia, especially in Bali Island.

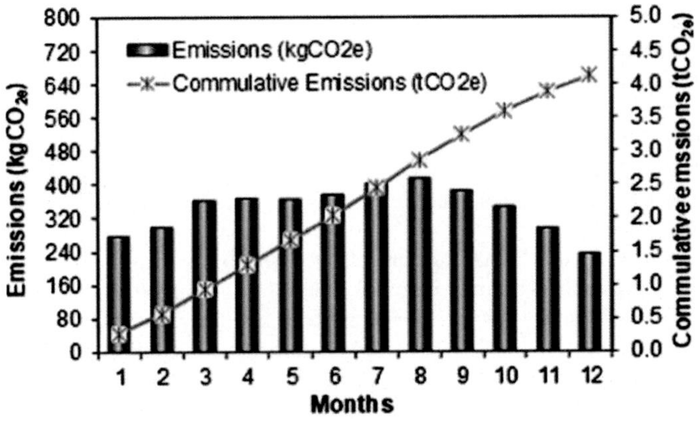

Fig. 6 Reduction of CO_2 emissions due to the use of renewable energy from the PV system

4 Conclusions

A grid-tied PV system has been developed and investigated for renewable energy generation potential in the southern region of Bali Island, Indonesia. The investigation results have shown that monthly solar electrical energy generation can reach 499 kWh and about 4.95 MWh annually. The PV system also has an average utilization potential of 38.7%. Efficiencies of the PV system were found to vary from 11.2% to 19.7%. It has also been proven that the PV system has a potential contribution to environmental impact reduction of about 4.12 tCO_2e per year due to the replacement of fossil fuel-based electrical energy. Substantial reduction in the environmental impact and excellent utilization potentials can make the grid-tied PV system as a prospective alternative of renewable energy source for commercial buildings in Indonesia.

Acknowledgements. Authors acknowledge the financial support from the Ministry of Research, Technology, and Higher Education of the Republic of Indonesia and Politeknik Negeri Bali through applied research scheme.

References

1. Yoon JH, Song J, Lee SJ (2011) Practical application of building integrated photovoltaic (BIPV) system using transparent amorphous silicon thin-film PV module. Sol Energy 85:723–733
2. Good C, Andresen I, Hestnes AG (2015) Solar energy for net zero energy buildings—a comparison between solar thermal, PV and photovoltaic–thermal (PV/T) systems. Sol Energy 122:986–996
3. Braun P, Rüther R (2010) The role of grid-connected, building-integrated photovoltaic generation in commercial building energy and power loads in a warm and sunny climate. Energy Convers Manage 51:2457–2466
4. Feron S (2016) Sustainability of off-grid photovoltaic systems for rural electrification in developing countries. A Rev Sustain 8:1326

5. Rumbayan M, Abudureyimu A, Nagasaka K (2012) Mapping of solar energy potential in Indonesia using artificial neural network and geographical information system. Renew Sustain Energy Rev 16:1437–1449
6. Veldhuis AJ, Reinders AHME (2015) Reviewing the potential and cost-effectiveness of off-grid PV systems in Indonesia on a provincial level. Renew Sustain Energy Rev 52:757–769
7. Yudiartono A, Sugiyono A, Wahid LMA (2018) Adiarso: Indonesia energy outlook 2018. Center for energy resources development technology. agency for the assessment and application of technology, p 117
8. Sugirianta IBK, Sunaya IGAM, Saputra IGNAD (2020) Optimization of tilt angle on-grid 300 Wp PV plant model at Bukit Jimbaran Bali. J Phys: Conf Ser 1450:012135
9. Handoyo EA, Ichsani D (2013) Prabowo: the optimal tilt angle of a solar collector. In: International conference on sustainable energy engineering and application (ICSEEA). Energy procedia 32, 166–175
10. Choi Y (2014) A study on power generation analysis of floating PV system considering environmental impact. Int. J Softw Eng Appl 8:75–84
11. Suamir IN, Wirajati IGAB, Santosa IDMC, Susila IDM, Tri Putra IDGA (2020) Experimental study on the prospective use of PV panels for chest freezer in hot climate regions. J Phys: Conf Ser 1569:032042
12. Suamir IN (2014) Solar driven absorption chiller for medium temperature food refrigeration, a study for application in Indonesia. Appl Mech Mater 493:167–172
13. Kumar S (2016) Assessment of renewables for energy security and carbon mitigation in Southeast Asia: the case of Indonesia and Thailand. Appl Energy 163:63–70
14. BS EN 378-1 (2016) Refrigerating systems and heat pumps—safety and environmental requirements. BSI Standards Limited, p 70
15. Ministry of Energy and Mineral Resources: Emission Factors (2016) http://jcm.ekon.go.id/en/index.php/content/Mzg%253D/emission_factor. Assessed 7 Oct 2021

Simulation and Dynamic System Modeling in an Elastically Supported Rigid Cylinder for Vibration Energy Harvesting

Subekti[1,4](✉), Harus Laksana Guntur[2], Vivien S. Djanali[3], and Achmad Syaifudin[2]

[1] Graduate Program of Engineering Department, Institut Teknologi Sepuluh Nopember, Surabaya, Indonesia
subekti@mercubuana.ac.id

[2] Mechanical Engineering Department, Institut Teknologi Sepuluh Nopember, Surabaya, Indonesia

[3] Department of Mechanical Engineering, Institut Teknologi Sepuluh Nopember, Kampus Keputih-Sukolilo, Jl. Arif Rahman Hakim, Surabaya 60111, Indonesia

[4] Mechanical Engineering Department, Universitas Mercu Buana, West Jakarta, Indonesia

1 Introduction

Southeast Asia is heavily affected by global energy use. Energy demand has increased by 60% in the last 15 years. Abundant renewable energy potential (RE) is spread throughout the region. The cost of renewable energy, especially solar panels and windmills, has decreased significantly in recent years. This promises cost competitiveness with conventional generation technologies. The ASEAN Plan of Action for Energy Cooperation (APAEC) 2016–2025 sets a RE target of 23% by 2025 [1]. Energy demand is very high today; it is necessary to increase renewable energy sources to avoid the use of commercial energy sources, poverty, hunger, epidemics, and the dangers of radiation [2].

The effort of creating alternative energy is currently being developed by utilizing flow-induced vibration. The fluid used in the utilization of vibration due to flow can be in the form of fluid flow in the form of gas, air, or water. Fluid flow and structure are interactive systems, and their interactions are dynamic. This system is a coupling system of the forces acting on the structure caused by the fluid around it. The force of the fluid causes the structure to deform. When the structure is deformed, it can change its orientation to the fluid flow so that the next time, it is possible to change the fluid force. This is called flow-induced vibration (FIV). Research on the use of vibration due to the flow of water into electrical energy has been carried out a lot [3–7]. Things that must be considered in designing FIV into alternative energy are influenced by the shape of the cylinder which will greatly affect the magnitude of the resulting amplitude ratio [8, 9], the amount of attenuation produced to be used as fluid kinetic energy and convert it into electrical energy [10, 11], the Reynolds number needed to increase the lift force and the resulting amplitude and synchronization range, and the resulting variable electrical load resistance to optimize the harnessed energy generated.

M. Kolhe et al. (eds.), *Recent Advances in Renewable Energy Systems*,
Lecture Notes in Electrical Engineering 876,
https://doi.org/10.1007/978-981-19-1581-9_7

In this paper, the fluid flow used is very different from the research that has been done; the fluid that will be used is airflow. A lot of airflows is generated in tall buildings, especially the flow in the ducting area, shown in Fig. 1. By utilizing airflow, as a new renewable energy source, home energy can be developed.

Fig. 1. Air flow in the ducting area in the apartment

The harnessed energy used in the dynamic system modeling is electrodynamic vibration energy harvesting as a result of the tension in the wire and beam vibrating due to the turbulence that is formed resulting in an external force that causes the cylinder to insulate transversely to the direction of the fluid. The technique of harvesting energy from structural vibrations using piezoelectric transducers is increasingly popular nowadays [12–15].

2 Dynamic System Modeling for Vibration Energy Harvesting

The utilization of vibration due to flow as alternative energy actually uses the force generated from the collision. The force that occurs in the bluff body is influenced by drag and lift. The drag force acts in the direction of the fluid flow. The lift force, on the other hand, acts perpendicular to the fluid flow.

Figure 2a is an elastically supported rigid cylinder system model installed by electrodynamic vibration energy harvesting on the beam (point A) and wire (point B). Energy harvesting in the beam is produced as a result of vibrations in the beam caused by the resulting oscillations, while in the wire, it occurs due to oscillations in the cylinder which causes the wire to move following the cylinder movement which will produce tension in the wire. The free-body diagram of the elastically supported rigid cylinder system model is shown in Fig. 2b. Here, we show that M and m are the masses of the cylinder and wire; C is the damping constants of an elastically supported rigid; K and k are the linear constants of the springs elastically supported rigid and wire; l is the length of the wire; T is the tension in the wire produced from an excitation force of $F\,(t)$, respectively. The system can be derived as follows:

$$\left(Ml^2 + \frac{2}{3}m\right)\ddot{\theta} + Cl^2\dot{\theta} + \left(Kl^2 + 2k\right) + 2T = F_{\mathrm{L}}(t) \tag{1}$$

The flow velocity and the shape of the bluff body greatly affect the lift force $F_{\mathrm{L}}\,(t)$ and the drag force $F_{\mathrm{D}}\,(t)$. In this paper, we will model the effect of the excitation force originating from the lift force $F_{\mathrm{L}}\,(t)$ as follows [10]

$$F_{\mathrm{L}}(t) = 0.5\rho U_o^2 DLC_{\mathrm{L}}(t) \tag{2}$$

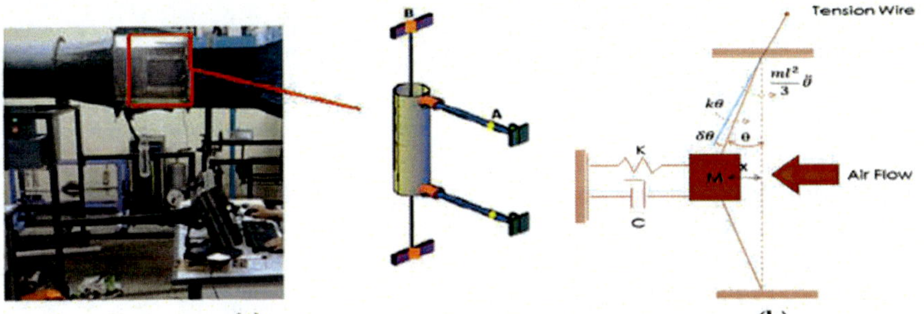

(a) (b)

Fig. 2. **a** Elastically supported rigid cylinder system model with electrodynamic vibration energy harvesting at beam (point A) and wire (point B), **b** force body diagram an elastically supported rigid cylinder

where ρ is the density of water, U_0^2 is the wind velocity, D is the diameter of the cylinder, L is the length of the cylinder, and C_L (t) is the coefficient of lift. The coefficient of lift on the system can be seen as follows [10]

$$C_L(t) = C_L^o \sin(2\pi f_s t) \tag{3}$$

where C_L^o is amplitude of elevator, usually uses a value of 0.3, $f_s = S_t U_o / D$, where S_t is a Strouhal number, usually uses a value of 0.2.

At point A (beam), the vibration energy conversion system uses a commonly used model, namely a standard system consisting of a spring and mass [16], as shown in Fig. 3a. In this system, the mass transfer z (t) is caused by the input displacement y (t). Damping includes mechanical damping and electric induction damping. Here, the power from the mechanical system generated by the electric induction damping can be converted into electrical power, while mechanical damping always causes a loss of system energy. The excitation force y (t) in the vibration energy harvesting system comes from the longitudinal vibration of the beam caused by the lift force that occurs on the cylinder. The lift force on the cylinder occurs due to the excitation force originating from the wind velocity applied to the system.

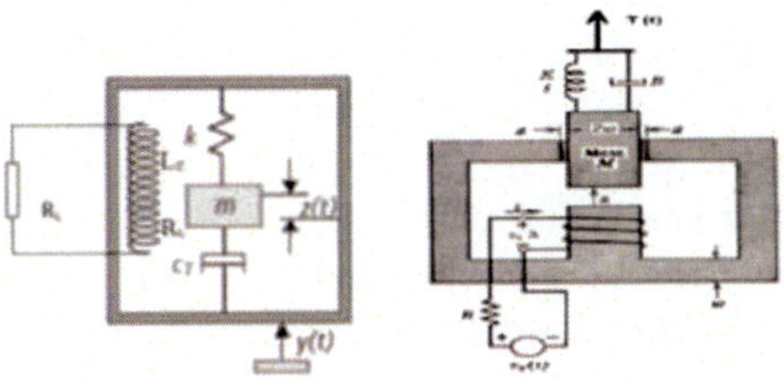

(a) point A (b) point B

Fig. 3. Schematic of the vibration energy harvester

For a mechanical system, the excitation force $y\ (t)$ on the vibration energy harvester is given as $y\ (t) = Y\ \sin\ \omega t$; the system can be derived as [17],

$$m\ddot{z}(t) + (b_e + b_m)\dot{z}(t) + kz(t) = m\omega^2 Y\ \sin(\omega t) \tag{4}$$

where k is the spring constant, b_e dan b_m are the electrically induced damping coefficients and the mechanical damping coefficients, respectively.

For an electrical system, on permanent electromagnetic energy, following Faraday's law of induced voltage as [18],

$$E(t) = -N\frac{d\phi}{dz} = -NA\frac{dB}{dz}\frac{dz}{dt} \tag{5}$$

where N and A are the turns of the coil and the area of the coil, respectively. In the schematic of the vibration energy harvester at point A (Fig. 3a), it is shown that there is the magnetic damping force originating from the electrically induced damping coefficient ($F_M = b_e\dot{z}(t)$), so it can be expressed as [18],

$$F_M = b_e\dot{z}(t) = NA\frac{dB}{dz}\frac{E(t)}{R_{Coil} + R_{Load}} \tag{6}$$

where R_{Coil} and R_{Load} are the impedance of the coil and external load, respectively.

At point B (wire), the vibration energy conversion system uses a commonly used model, namely a standard system consisting of a spring and mass [16], as shown in Fig. 3b. In this system, the mass transfer $x\ (t)$ is caused by the input displacement. $T\ (t)$, so the system can be derived as

$$T(t) - \frac{1}{2}\frac{L_o i^2}{g\left(1 + \frac{x}{g}\right)^2} = M\ddot{x} + B\dot{x} + K(x - l) \tag{7}$$

To determine the performance of a configuration system of an elastically supported rigid cylinder system model that is installed with electrodynamic vibration energy harvesting in a model, modeling that includes two aspects is needed, as follows:

- System dynamics
- Generator system.

Dynamic modeling. An elastically supported rigid cylinder system model fitted with electrodynamic vibration energy harvesting was carried out, based on Eqs. (1) and (4)–(7). Dynamic modeling uses the Simulink model which is built based on the frequency converted into electrical power (watts). The simulation model can be directly set up in Simulink Matlab.

The excitation force that occurs in the system comes from the air velocity that occurs in the elastically supported rigid cylinder system, as the resulting lift force. In the model simulation, variable parameters are made, such as wind velocity in the ducting and the diameter of the cylinder used, which can be seen in Table 1.

Table 1 Simulation parameter

Wind velocity (m/s)	Cylinder of diameter (mm)	Cylinder of massa (mm)
10	100	40
8	80	30
3	50	20

3 Result and Discussion

The entire system has been modeled using the MATLAB Simulink R2017b software as shown in Fig. 4, where this system consists of a dynamics system and a generator system that occurs due to the excitation force originating from the lift force, $F_L(t)$. The dynamics modeling in this paper is carried out to see the effect of the given wind velocity and cylinder diameter so that it will affect the given lift force. All simulations were run with the ode15s solver.

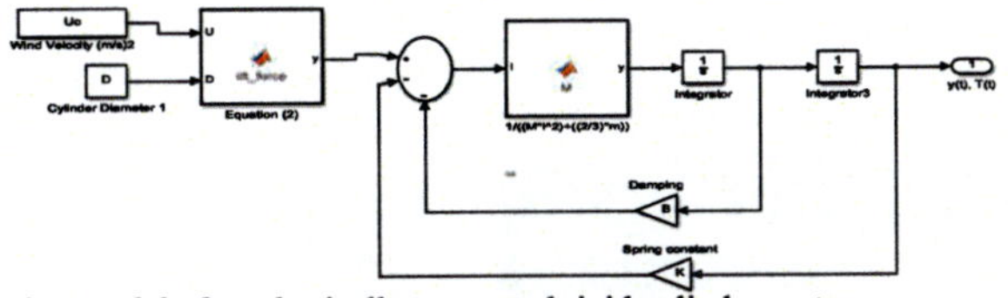

(a) Simulation model of an elastically supported rigid cylinder system

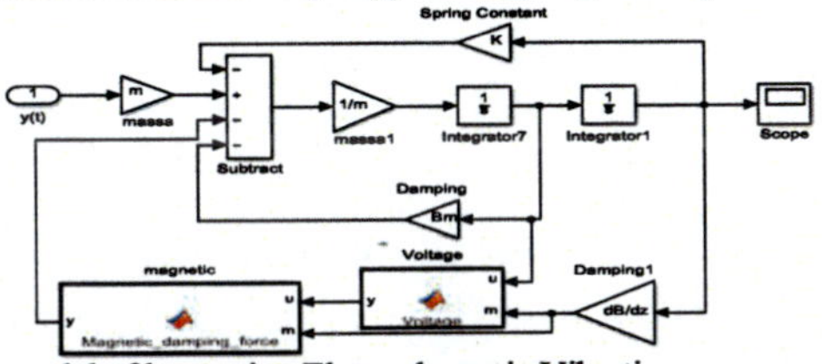

(b) Simulation model of harvesting Electrodynamic Vibration energy system at point A

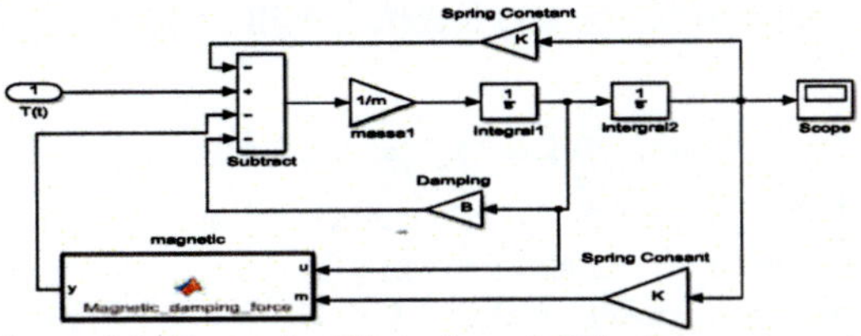

(c) Simulation model of harvesting Electrodynamic Vibration energy system at point B

Fig. 4. Simulation modeling dynamic system

One can be directly set up a simulation model in Simulink of Matlab, shown in Fig. 4a. by using these modules according to the movement of an elastically supported rigid cylinder system model system differential Eq. (1). While Fig. 4b and c are matlab simulinks for the movement of harvesting electrodynamic vibration energy system at point A and B in the differential Eqs. (4)–(7).

3.1 The Effects of Wind Velocity

Analysis of the effect of wind velocity on an elastically supported rigid cylinder system model system installed with electrodynamic vibration energy harvesting on a cylinder diameter of 100 mm. The simulation results are carried out at point A and point B.

3.1.1 Electrodynamic Vibration Energy Harvesting at Point A

The response of the harvesting force given in Fig. 5a shows that the greater the speed of the given airflow is, the smaller the power gets. At a speed of 3 m/s, the power produced is unstable.

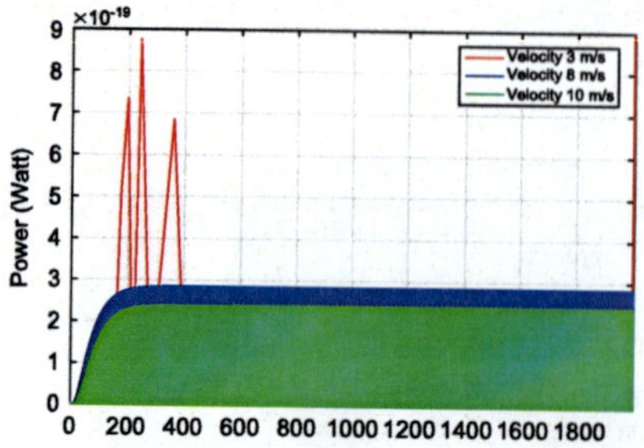

(a) Point A

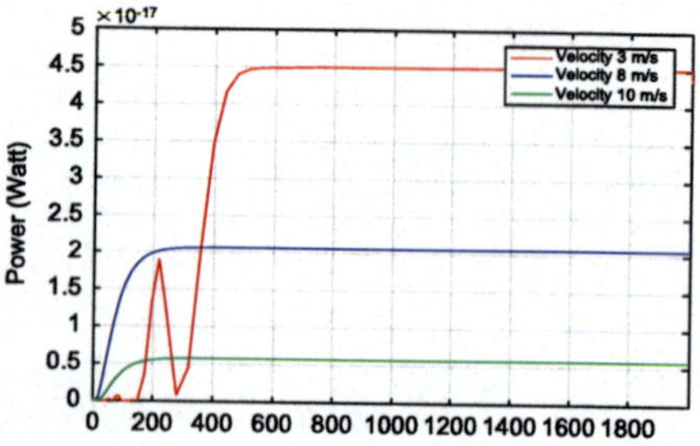

(b) Point B

Fig. 5. Effects of flow velocity on the force generated

For a higher excitation force, it is shown that it produces a linear system and shows a stable force behavior at a certain time. Thus, the time response to the force is expanding. In addition, the fluid flow velocity as an excitation force in a certain system shows that the greater the excitation force, the less force that will be generated in the system.

3.1.2 Electrodynamic Vibration Energy Harvesting at Point B

The phenomenon that occurs at point B is observed when power is compared to time, as shown in Fig. 5b. At point B, it is shown that at a fluid flow velocity of 3 m/s; after 180 s, the power slowly increases until it reaches 1.9×10^{-17} watts. Then, there is a decrease and increase again to 4×10^{-17} watts. After 4×10^{-17} watts, the power gradually stabilizes.

While at an airflow rate of 8 m/s, it is shown that the power increases rapidly over time until it reaches 2×10^{-17} watts. At a flow rate of 10 m/s, the power increases slower and stabilizes faster at 0.5×10^{-17} watts.

3.2 The Effects of Cylinder Diameter

Analysis of the influence of changes in cylinder diameter on an elastically supported rigid cylinder system model system installed with electrodynamic vibration energy harvesting. The simulation is carried out at point A and point B.

3.2.1 Electrodynamic Vibration Energy Harvesting at Point A

The response of the harvesting power given in Fig. 6a shows that the larger the diameter, the greater the power produced at the same airflow velocity. At a diameter of 100 and 50 mm, the power is shown to be very stable. While at a diameter of 80 mm, the power produced fluctuates and is less stable. The less stable power requires additional equipment to stabilize the power produced.

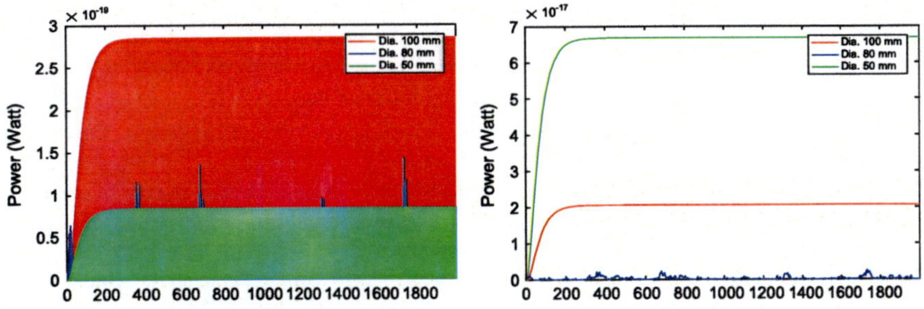

(a) Point A (b) Point B

Fig. 6. Effects of cylinder diameter on the power generated

3.2.2 Electrodynamic Vibration Energy Harvesting at Point B

Due to the effect of changing the diameter on the cylinder diameter on the Electrodynamic Vibration Energy Harvesting at Point B, it is shown that the larger the given diameter is, the greater the power that is generated at the same airflow velocity. At a diameter of 100 and 50 mm, the power is shown to be very stable. While at a diameter of 80 mm, the power produced fluctuates and is less stable and has a very small power close to 0. For more details, it can be seen in Fig. 6b.

4 Conclusions

A simulation using a simulator on an elastically supported rigid cylinder system model system that is installed with electrodynamic vibration energy harvesting shows that the airflow velocity and given diameter greatly affect the amount of power generated.

At an airflow rate of 8 m/s, it is shown that the power increases rapidly over time until it reaches 2×10^{-17} watts. At a flow rate of 10 m/s, the power increases slower and stabilizes faster at 0.5×10^{-17} watts. The larger the given diameter is, the greater the power that is generated at the same airflow velocity. At a diameter of 100 and 50 mm, the power is shown to be very stable, while at a diameter of 80 mm, the power produced fluctuates and is less stable. The phenomenon that occurs due to the placement of electrodynamic vibration energy harvesting shows that placement at point B has a faster power than the placement at point A.

References

1. Asean Renewable Energy Integration, IEA (2019), Yoshihisa S (2007) Energy consumption: an environmental problem. In: Transactions on electrical and electronic engineering, IEEJ, pp 12–16
2. Bernitsas MM, Ben-Simon Y, Raghavan K, Garcia EMH (2008) VIVACE vortex induced vibration aquatic clean energy: a new concept in generation of clean and renewable energy form fluid. ASME J Offshore Mech Arct Eng 130(4)
3. .Bernitsas MM, Raghavan K (2004) Converter of current/tide/wave energy. Provisional Patent Application, U. S. Patent and Trandemark Office Serial No. 60/628
4. Bernitsas MM, Raghavan K (2005) Fluid motion energy converter, internasional. Provisional Patent Application, U. S. Patent and Trandemark Office Serial
5. Subekti HA, Surjosatyoe A (2018) The use of flow-induced vibration as an alternative resouce of new power plant in Indonesia. E3S Web Conf 67
6. Biantoro A, Iskendar I, Subekti S, bin Muhd Noor NH (2021) The effects of water debit and number of blades on the power generated of prototype turbines propeller as renewable electricity. Jurnal Rekayasa Mesin 12(1):203–215
7. Jijian L, Xiang Y, Fang L, Juan Z (2017) Analysis on flow induced motion of cyinder with different cross sections and the potential capacity of energy transference from the flow. Hindawi, Shock Vibration
8. Subekti YK, Yohei J, Takanori E (2009) Identification of nonlinearity of flow-induced vibration for structure having nonlinear property by using wavelet transform. In: ICROSS-SICE international joint conference, August 18–21, Fukuoka, Japan
9. Khalak A, Williamson CHK (1997) Investigation of relative effects of mass and damping in vortex-induced vibration of a circular cylinder. J Wind Eng Indus Aerodyn 69–71:341–350
10. Feng CC (1968) The measurement of vortex-induced effects in a flow past a stationary and ocilalating and D-section Cylinders. University BC, Vancouver-Canada
11. Covaci C, Gontean A (2020) Piezoelectric energy harvesting solutions: a review. Sensors 20
12. Rosman MN, Azhan NH (2019) Piezoelectric transducer applications for sound energy harvesting: a case study of passing road vehicle. In: AIP conference procedding 2129
13. Jiang J, Liu S, Feng L, Zhao D (2021) A review of piezoelectric vibration energy harvesting with magnetic coupling based on different structural characteristics. Micromachines 12(4):436
14. Odetoyan AO, Ede AN (2021) Energy harvesting from vibration of structures-a brief review. IOP Conf Ser Mater Sci Eng 1107
15. Williams CB, Yates RB (1996) Analysis of a micro-electric generator for microsystems. Sens Actuators A, Phys 52(1–3):8–11
16. Han M, Yuan Q, Sun X, Zhang HX (2014) Design and fabrication of integrated magnetic MEMS energy harvester for low frequency applications. J Microelectromech Syst 23(1):204–212
17. El-hami M, Glynne-Jones P, White NM, Hill M, Beeby S, James E, Brown AD, Ross JN (2001) Design and fabrication of a new vibration-based electromechanical power generator. Sens Actuators A, Phys 92(1–3):335–342

Parameters Analysis of Vortex Formation on Conical Basin of Gravitational Water Vortex Power Plant (GWVPP)

Erna Septyaningrum, Ridho Hantoro, Sarwono, and Ester Carolina(✉)

Sepuluh Nopember Institute of Technology, Surabaya, Indonesia
estercarolina.line2@gmail.com

1 Background

The gravitational water vortex power plant (GWVPP) is a type of micro-hydro power plant having an output that does not exceed 100 kW. GWVPP can be installed at low river speeds, starting from 0.5 m/s [1]. Since it operates in low-head resource, approximately 0.8–2 m, GWVPP seems to be a good alternative for rural area energy provision; furthermore, it is also an easy to operate and maintenance technology. The additional advantage of GWVPP is its ability to enhance the oxygen concentration in the water; hence, it provides a good impact on the river/canal ecosystem [2]. Due to the low head, It does not work on the pressure differences principle, however utilizes vortex flow called eddies. Vortex flow takes an important role in GWVPP performance. The vortex is described as a rotating motion of a multitude of material (in this case is water) around a common center, which is the flow strength and the rotation is governed by the Coriolis effect [3–5]. The development of GWVPP in the research field also encourages the development of GWVPP prototypes, aiming to commercialize this technology. This technology is developed in several countries, including Indonesia. 13 kW GWVPP had been installed in Bali to supply a school nearby the Ayung River [6].

The performance of GWVPP is determined by the basin used in the system since the basin has a major role in vortex formation. The previous work developed the basin geometry. The common type of basins for GWVPP is cylindrical dan conical basins. The utilization of non-curved geometry (such as rectangular) is avoided since it has more pressure loss compared with curved geometry (conical and cylindrical basin). The pressure loss leads to vortex strength reduction [7, 8]. Research conducted by [2] shows that the conical basin performs a stronger vortex than the cylindrical basin.

The tangential velocity that affects vortex strength in vortex flow is an important parameter for designing GWVPP [9]. In addition, increasing vortex height will increase vortex strength [10]. The basin geometry has an important role in defining the performance of the vortex flow. The parameters of basin geometry that affect the vortex formation are the size of the basin opening, basin diameter, notch angle, canal height, and cone angle [2]. The basin geometry will affect the tangential flow that enters the canal or basin. This water flow forms a strong vortex and exits the outlet at the bottom of the shallow center of the basin. The maximum tangential velocity of the vortex flow affects

M. Kolhe et al. (eds.), *Recent Advances in Renewable Energy Systems*,
Lecture Notes in Electrical Engineering 876,
https://doi.org/10.1007/978-981-19-1581-9_8

the performance of the turbine to produce power output or mechanical power [11]. The research of [12] found that the optimum vortex strength in a conical basin occurs for a basin with an outlet to inlet diameter ratio of 30%. Moreover, the size enlargement of the conical basin is expected to have a great effect on vortex formation.

Depth analysis of basin geometry effect of vortex formation is enquired for determining the optimum GWVPP basin design and size. In this current work, computational fluid dynamics (CFD) simulation was carried out to get the correlation between the and dimension enlargement to the vortex formation in the conical basin. This study analyzes the performance of vortex flow which will affect the efficiency of the runner. Since it is a robust and economical method, a numerical simulation based on CFD is often used by several studies for the preliminary study. This method utilizes mathematical physics (mass conservation, energy conservation, other fundamental, and auxiliary equation) together with programming to analyze fluid flow [2]. This current work is expected to find the geometry of the basin which leads the better vortex performance, based on variations in basin type, outlet diameter, and basin enlargement. This work will give some recommendations which could be considered for GWVPP basin design.

2 Research Method

2.1 Conical Basin Design

As the conical basin performs better than others, this current work develops this type of basin by carrying out depth analysis on the geometry effect. Some literatures are recognized for determining the variation of geometry also the inlet flowrate to analyze vortex height and vortex strength.

A 3D basin geometry design was made with a scale of 1:1, shown in Fig. 1. It is also used for conducting CFD simulation. The variations in this work involve the outlet to inlet diameter ratio, dimensional enlargement, and inlet flowrate. This work varies the outlet diameter (d_0) of the basin, while the inlet diameter (D) is kept constant. The height to inlet diameter ratio of the conical basin is 2, with 20° of cone angle. The d_0/D is varied to be 0.2 and 0.3. Moreover, the dimensional enlargement analysis is also conducted using the variation of 1X; 1.1 X, and 1.2X.

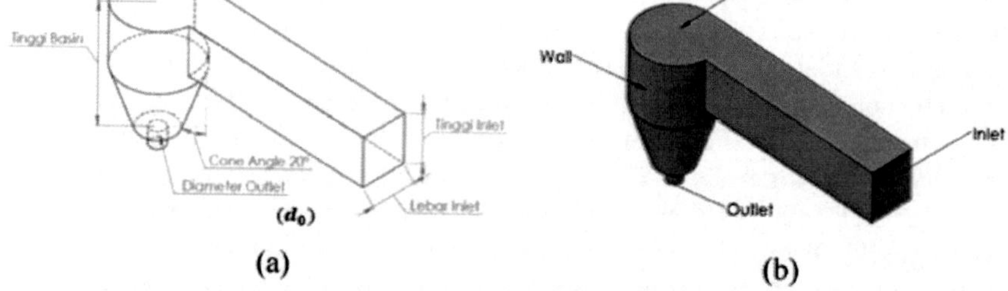

Fig. 1. **a** Conical basin geometry **b** boundary condition

2.2 Simulation Setup

In the CFD simulation procedure, the meshing process takes an important role in affecting the simulation result. The curvature and proximate meshing methods together with the suitable sizing methods are handed to the domain, aiming to get the high-quality grid. Due to the dimensional enlargement, the meshing performed in this simulation has different parameter elements. To overcome this issue, the grid-independent study for all dimensions (i.e., 1X; 1.1X and 1.2X) was conducted to get the appropriate mesh size, as given in Fig. 2.

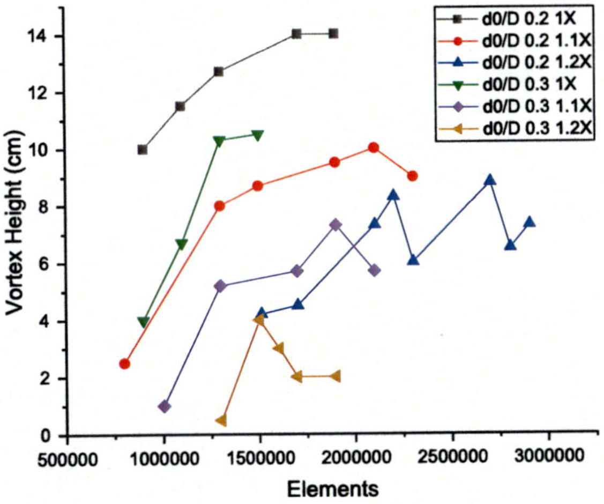

Fig. 2. Grid independence of enlargement of basins

The simulation of the CFD was carried out by setting up the required parameter values for the vortex flow simulation. The boundary condition and turbulence model were chosen to determine the simulation result. Both setting should meet the requirement to represent the real condition. The opening is used to represent the free surface in the upper area of the basin and the outlet, while the mass flow inlet is set in the inlet part. The shear stress transport (SST) $k - \omega$ turbulence model is employed in this simulation with the steady condition, homogeneous, free surface, and two phase of water and air. References [13]–[15] suggested the shear stress transport (SST) $k - \omega$ for simulating the vortex flow. Meanwhile, the simulation validation is conducted by comparing the experimental dan simulation result for several number of elements, shown in Table 1. It is confirmed that the boundary condition and turbulence model, which are used, can represent the real condition.

3 Result and Discussion

Vortex height experienced by the conical basin at the ratio d_0/D 0.2 and the ratio d_0/D 0.3 tends to increase with the increase in the inlet flow rate. As the mass flow received by the

Table 1 Simulation validation

Number of element	Vortex height (cm)		Error (%)
	Simulation	Experiment	
1,566,230	123.5	125	1
1,730,950	115		8
1,891,031	110.5		12

basin is getting bigger, the vortex height will increase until it reaches a steady condition at each given inlet flowrate (the flowrate variation used are 2.41, 2.65, 2.89, 3.13, 3.37 kg/s). This phenomenon occurs in all d_0/D variation and dimensional enlargement, as depicted in Fig. 3. Figure 3 confirms that higher d_0/D tends to have lower vortex height. The lowest vortex height value is 2 cm at a diameter ratio of d_0/D 0.3 with an enlargement of 1.2X, while the highest vortex height in both types of the conical basin is 23.8 cm in a cylindrical basin with a ratio of d_0/D 0.2 enlargement 1X using an inlet flow rate of 3.37 kg/s. From the comparison of the six conical basins, it was found that the conical basin with a ratio of d_0/D 0.2 with enlargement 1X experienced the highest vortex height compared to other conical basins.

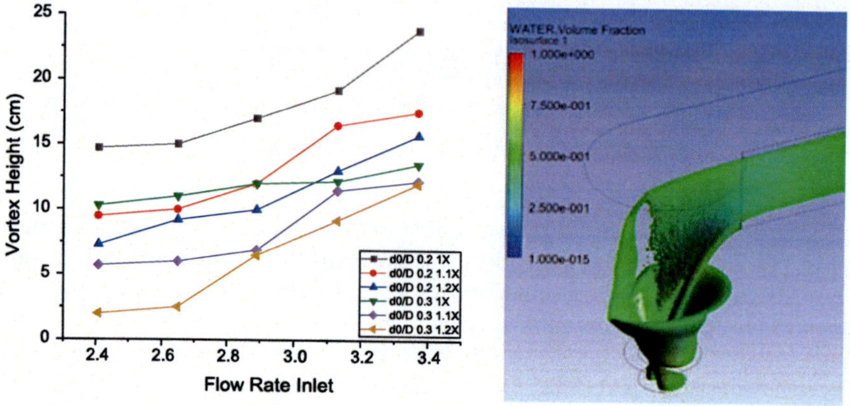

(a) (b)

Fig. 3. Simulation result for **a** vortex height **b** iso-surface volume fraction

The iso-surface volume fraction of the conical basin shown in Fig. 3b is an iso-surface with a volume fraction of water 0.5 which shows the interface between water and air. The iso-surface shows the formation of an air core or air vortex. As the air pressure gradually decreases to below atmospheric pressure and sucks the air out of the outlet, the air core appears in the middle of the basin. The pressure on the upper surface basin is set according to atmospheric pressure, which is 1 atm or 101,325 Pa. This decreased air pressure in the water core causes an increase in the velocity of the water flow. This is used to install a turbine because of the increasing velocity of the water flow. Basin shape has different air pressure reduction performance. Based on Fig. 4, the water core in the conical basin with the ratio of d_0/D 0.2 has a higher pressure drop gradient than a ratio of d_0/D 0.3. At a ratio of d_0/D 0.2, it decreases from atmospheric pressure of 101,325 Pa to about 99,180 Pa, while the ratio of d_0/D 0.3 is enough to decrease from atmospheric pressure of 101,325–100,900 Pa.

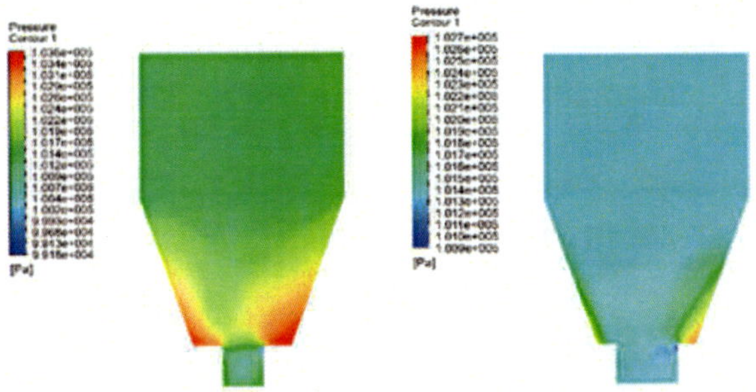

(a) (b)

Fig. 4. Pressure contour of ratio d_0/D **a** 0.2 **b** 0.3

Moreover, the tangential velocity of water is one of the important parameters that must be considered for determining the basin design. Tangential velocity represents the velocity of water in which is always along the tangential line at any given point. It indicates the rotational motion of the vortex flow in the basin. Figures 5 and 6 show the tangential velocity distribution in the conical basin with the ratio of d_0/D 0.2 and 0.3, respectively. For both d_0/D and enlargement (1X, 1.1X, and 1.2X), the tendency of the tangential velocity values is almost similar. In the first few radii to the basin diameter (r/d_0), the tangential velocity increases to a maximum. Due to the wall friction impact (viscous flow), the tangential velocity decreases when approaching the basin wall.

The vortex strength of the vortex flow in the conical basin is calculated at a steady state. For all variations used in this work, there is a tendency of increasing the vortex height when the radius is increasing, until it reaches a maximum value. The vortex strength near the air core has minimal value. Based on Figs. 7 and 8, the best vortex strength performance is shown in the basin conical ratio d_0/D 0.3 enlargement 1.2X for the inlet flow rate of 2.65 kg/s. At this condition, the vortex strength is approximately 0.77 m^2/s.

In the conical basin, water falls from the inlet of the canal and hits the basin wall and forms a vortex. The flow vector shown in Fig. 9a is a tangential projection of the water velocity of the vortex surface. The velocity vector indicates that the water flows through the vortex. It also can be seen that the center of the vortex is not in the center of the basin. Figure 9b is a velocity streamline in the conical basin. It can be seen that the flow forms a vortex when it falls into the basin. The velocity becomes faster when the flow is approaching the outlet as it has a smaller opening area and progressively lower pressure, following the Bernoulli law. The presence of an inlet along an open rectangular channel and the curved shape of the tank or basin creates a strong vortex flow around the center

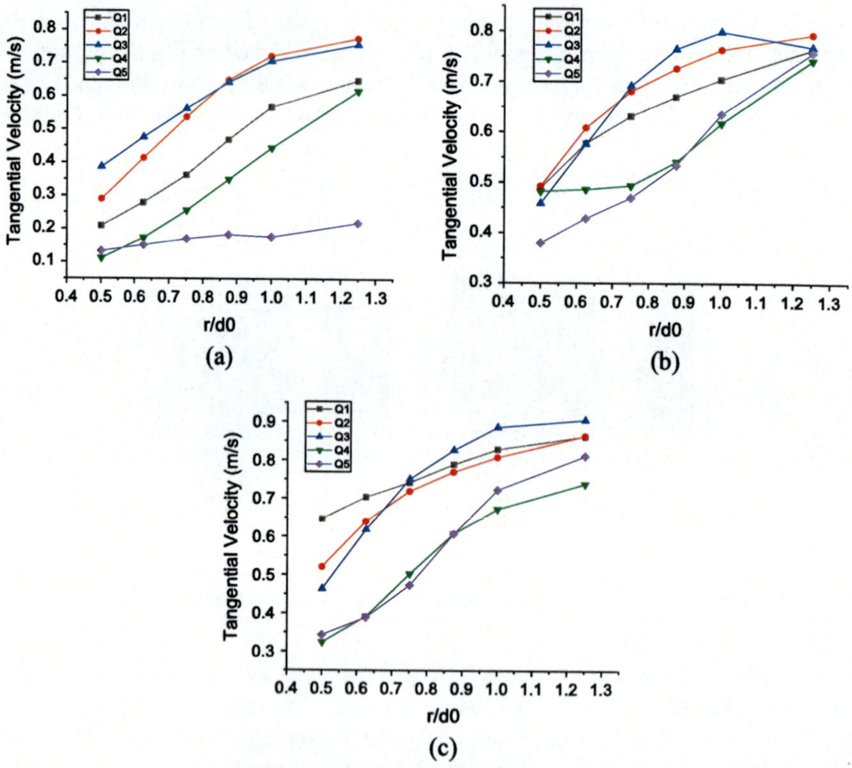

Fig. 5. Distribution of tangential velocity in basin conical d_0/D 0.2 **a** enlargement 1X; **b** enlargement 1.1X; **c** enlargement 1.2X

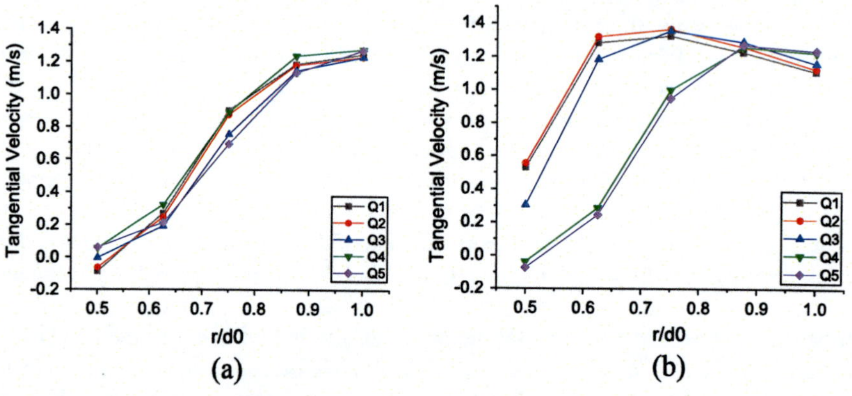

Fig. 6. Distribution of *tangential velocity* in basin *conical* d_0/D 0.3 **a** enlargement 1X; **b** enlargement 1.1X; **c** enlargement 1.2X

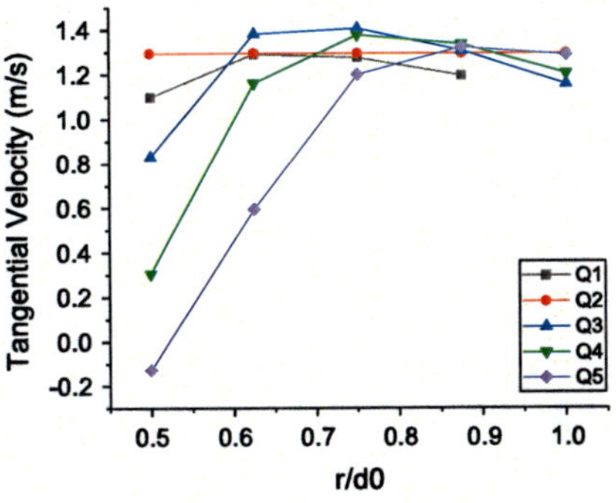

(c)

Fig. 6. continued

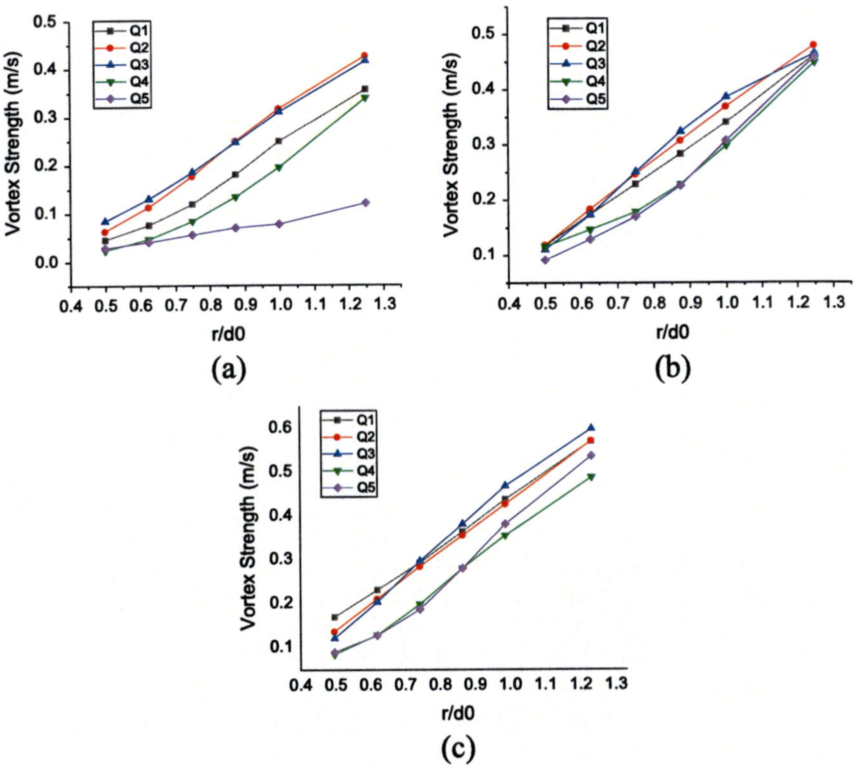

(c)

Fig. 7. Distribution of *vortex strength* in basin *conical* d_0/D 0.2 **a** enlargement 1X; **b** enlargement 1.1X; **c** enlargement 1.2X basin when water exits through the outlet. The vortex flow is also said to be strong if the air core is formed as shown in Fig. 9a.

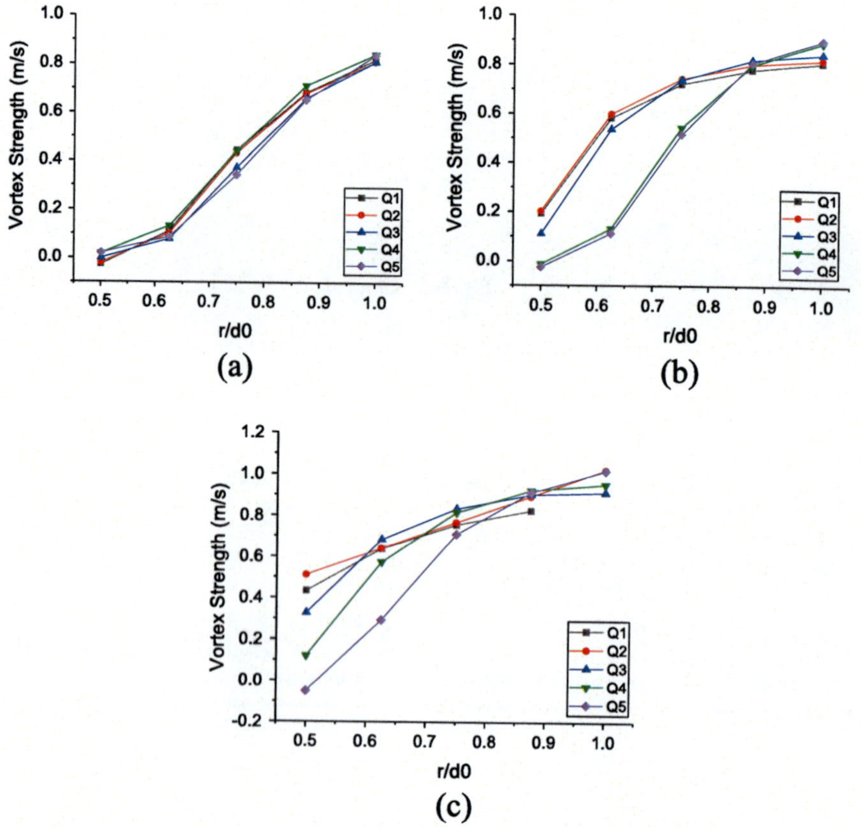

Fig. 8. Distribution of *vortex strength* in basin *conical* d_0/D 0.3 **a** enlargement 1X; **b** enlargement 1.1X; **c** enlargement 1.2X

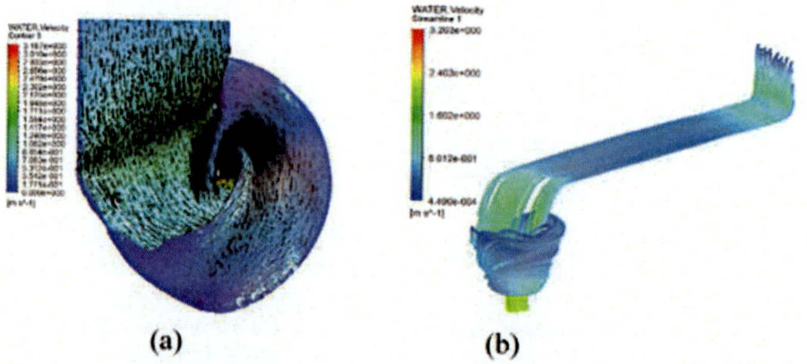

Fig. 9. **a** Vortex flow vector of water conical **b** streamline speed of conical

Overall, the conical basin with a ratio of d_0/D 0.3 enlargement 1.2X at an inlet flow rate of 2.65 kg/s has the maximum tangential velocity with a value of 1.295 m/s. It is the same with the tangential velocity that in the conical basin, it is found that the maximum vortex strength value is in the conical basin with a ratio of d_0/D 0.3 enlargement 1.2X at the inlet flow rate of 2.65 kg/s. The vortex flow performance affects the turbine efficiency. Based on experiments in the [12], turbines installed in conical basins with d_0/D 0.3 ratios have higher efficiency than of d_0/D ratios 0.1, 0.2, 0.4, and 0.5. In this study, the basin that has the best vortex flow performance based on numerical analysis is the conical type with a ratio of 0.3. This is indicated by the value of the tangential velocity and vortex strength in the conical basin with a ratio of d_0/D 0.3 which is the largest among others. The conical basin used in this work has a tapper in the bottom, causing some disturbance that inhibits the vortex flow in the outlet side [2].

Enlargement of the basin size also affects the performance of vortex flow, by seeing the tangential velocity and vortex strength. The basin enlargement of 1.2X has a good tangential velocity and vortex strength compared to 1X and 1.1X. The enlargement of the basin wall circle certainly affects the tangential velocity for water entering from the inlet and flowing toward the outlet. As a result, the vortex strength increased together with the tangential velocity. The vortex height that occurs in the basin with an enlargement of 1.2X is indeed not higher than the other enlargements, and it is enough to use an inlet flow rate that is not too large, which is 2.65 kg/s. This shows that the gravitational water vortex power plant (GWVPP) can operate in a fairly large basin and with a fairly small inlet flow rate.

4 Conclusion

A simulation study had been carried out to analyze the influence of geometry on the vortex formation of the conical basin. It shows that both inlet to outlet diameter ratios (d_0/D) and the size enlargement have a significant impact on the value of vortex height, tangential velocity, and vortex strength. The vortex height tends to be higher for small d_0/D dan size. Moreover, it also verifies that conical basin with d_0/D 0.3 has better tangential velocity and vortex strength compared to that of d_0/D 0.2. The enlargement also significant impact as the biggest basin size used in this study (1.2X) has better vortex strength compared to others. The enlargement of the basin wall affects the tangential velocity for water entering from the inlet and flowing toward the outlet and increasing the value of vortex strength. But, the bigger the basin size, the lower vortex height. This issue should be considered for designing runners. In conclusion, this work suggests using a conical basin with a ratio of d_0/D 0.3 for GWVPP.

Acknowledgements. The authors would like to express their sincere gratitude to the Directorate of Research and Community Service (DRPM) Institut Teknologi Sepuluh Nopember, which funded the research under a scheme called Department Research with contract number 1542/PKS/ITS/2021. The authors also thank Energy Engineering and Environmental Conditioning Laboratory for supporting this research.

References

1. Alzamora Guzmán VJ, Glasscock JA, Whitehouse F (2019) Design and construction of an off-grid gravitational vortex hydropower plant: a case study in rural Peru. Sustain Energy Technol Assessments 35:131–138. doi: https://doi.org/10.1016/j.seta.2019.06.004
2. Dhakal S et al (2015) Comparison of cylindrical and conical basins with optimum position of runner: Gravitational water vortex power plant. Renew Sustain Energy Rev 48:662–669. https://doi.org/10.1016/j.rser.2015.04.030
3. Timilsina AB, Mulligan S, Bajracharya TR (2018) Water vortex hydropower technology: a state-of-the-art review of developmental trends. Clean Technol Environ Policy 20(8):1737–1760. https://doi.org/10.1007/s10098-018-1589-0
4. Mulligan S, Casserly J, Sherlock R (2016) Effects of geometry on strong free-surface vortices in subcritical approach flows. J Hydraul Eng 142(11):04016051. https://doi.org/10.1061/(asce)hy.1943-7900.0001194
5. Lugt HJ, Gollub JP (1985) Vortex flow in nature and technology. Am J Phys 53(4):381–381. https://doi.org/10.1119/1.14177
6. Turbulence, "Turbulent|TECHNOLOGY"
7. Zardin B, Cillo G, Borghi M, Adamo AD', Fontanesi S (2017) Pressure losses in multiple-elbow paths and in V-bends of hydraulic manifolds. Energies 10(6). doi: https://doi.org/10.3390/en10060788
8. Kalpakli A (2012) Experimental study of turbulent flows through pipe bends. Eng Sci Royal Inst Technol (KTH)
9. "Analytical Solution for a Strong Free-Surface Vortex Describing flow in a Full-Scale Gravitationla Vortex Hydropower System.pdf"
10. Kueh TC, Beh SL, Rilling D, Ooi Y (2014) Numerical analysis of water vortex formation for the water vortex power plant. Int J Innov Manag Technol 5(2):111–115. https://doi.org/10.7763/IJIMT.2014.V5.496
11. Saleem AS et al (2020) Parametric study of single-stage gravitational water vortex turbine with cylindrical basin. Energy 200:117464. https://doi.org/10.1016/j.energy.2020.117464
12. Sreerag S, Raveendran CK, Jinshah BS (2016) Effect of outlet diameter on the performance of gravitational vortex turbine with conical basin. Int J Sci Eng Res 7(4):457–463. Available: http://www.ijser.org
13. Chen Y, Wu C, Wang B, Du M (2012) Three-dimensional numerical simulation of vertical vortex at hydraulic intake. Proc Eng 28:55–60. https://doi.org/10.1016/j.proeng.2012.01.682
14. Shah SR, Jain SV, Patel RN, Lakhera VJ (2013) CFD for centrifugal pumps: a review of the state-of-the-art. Proc Eng 51:715–720. https://doi.org/10.1016/j.proeng.2013.01.102
15. Khan HH (2016) Blade optimization of gravitational water vortex turbine. Ghulam Ishaq Khan Inst Eng Sci Technol 6(2):1689–1699

Solar Canopy with IoT-Based Single-Axis Solar Tracking System as a Solution for Utilizing Urban Open Parking Area

Radix Kautsar Ramadhan(✉), Hafiz Rayhan Gunawan, and Galang Adi Saputro

Institut Teknologi Sepuluh Nopember, Surabaya, Indonesia
radixkautsar9f24@gmail.com

1 Introduction

Indonesia has a very abundant solar energy potential reaching 4.8 kWh/m^2/day [1] or about ten times solar energy potential of Germany [2]. According to the Chair of the Indonesian Solar Energy Advocacy and Education Association, this potential can almost be used to meet all national electricity needs [3]. However, the utilization of this potential has only reached 0.05% [4]. One of the reasons is limited land.

As an alternative solution, Indonesia can imitate Rutgers University that has built many solar canopies to utilize its campus open parking area [5]. Unfortunately, Chase Weir from TruSolar in his research stated that the output power of fixed solar canopy is not worth the investment cost [6]. Even though the investment cost is now much lower than before, further innovation is still needed to increase its output power. Actually, there are various methods to increase the output power of the solar canopy, one of which is the use of a solar tracking system.

Although now a dual-axis solar tracking system has been developed, which is capable to produce greater output power than a single-axis solar tracking system [7], the complexity and additional investment costs are still not commensurate with the increase in output power. On the contrary, the use of a dual-axis solar tracking system is less effective when used in Indonesia, considering that the intensity of sunlight in the equatorial region tends to be evenly distributed with the angle of the sun being not too extreme. Then, based on other studies, the scheduled-based solar tracking system is proven to be more efficient than the sensor-based solar tracking system [8]. Based on those background, the purpose of this research is to make a model of solar canopy with IoT-based single-axis solar tracking system, as well as to know the performance and prospects of its implementation in terms of efficiency and investment costs.

2 Methodology

2.1 Literature Study

The literature study was carried out by searching for basic theories in designing model starting from the right programming algorithm to process the system database, how the

M. Kolhe et al. (eds.), *Recent Advances in Renewable Energy Systems*,
Lecture Notes in Electrical Engineering 876,
https://doi.org/10.1007/978-981-19-1581-9_9

microcontroller works in driving the motor, the method of placing relays in their function to control and supply electricity to the motor, the amount of electrical energy produced by the solar canopy, and how the motor works.

2.2 System Planning

The following is a system planning flowchart. The top flowchart is an astronomical data processing flowchart. The user's GPS, time, and date data will be inputted to the website application and converted with NREL database to obtain zenith, elevation, and azimuth angle. After data validation, the angle data will be inputted to solar tracking system. The bottom flowchart is a solar canopy mechanism flowchart. The angle data from previous process will be projected using microcontroller to obtain single-axis angle. If the canopy is not in the right position, the motor will move the canopy to determined angle (Fig. 1).

2.3 Design Planning

The model design is shown in Fig. 2.

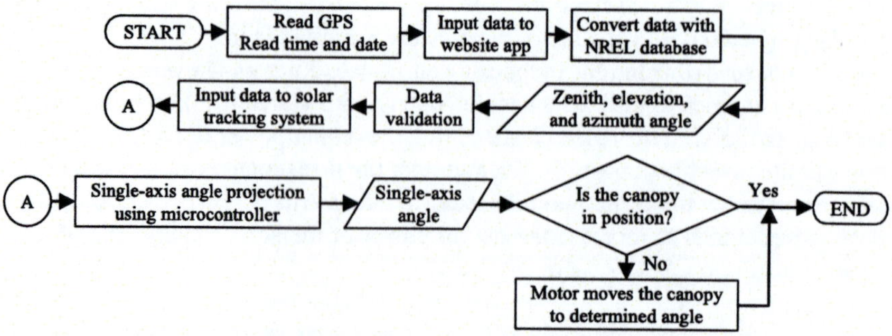

Fig. 1. Astronomical data processing and solar canopy mechanism flowchart

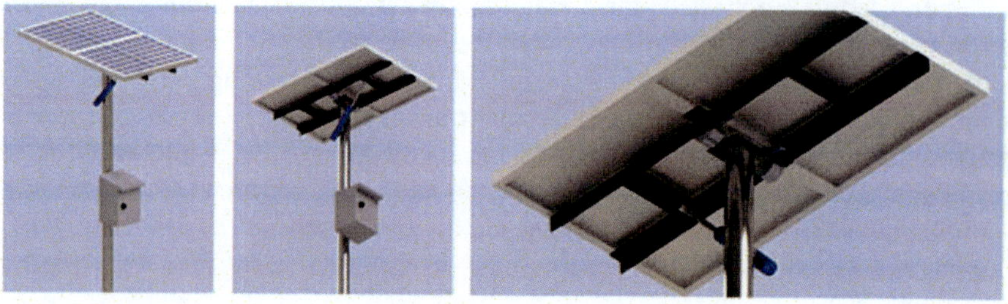

Fig. 2. Model design

2.4 Model Making

The model making process is shown in Fig. 3.

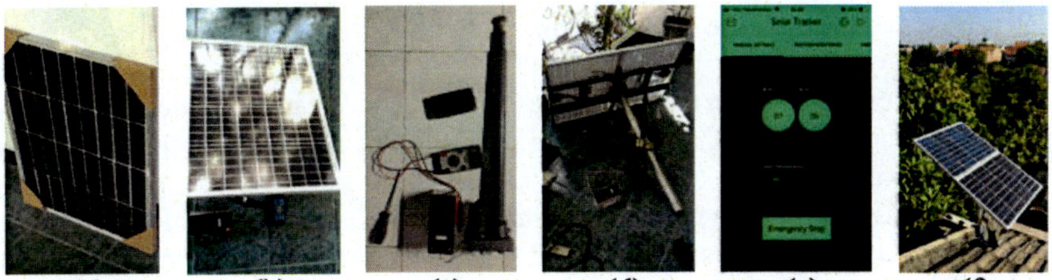

(a) (b) (c) (d) (e) (f)

Fig. 3. **a** Collection of components. **b** Assembly of solar charger. **c** Assembly of motor control. **d** Frame structure making. **e** Application development. **f** Product model

2.5 Model Testing

Model testing process is shown in Fig. 4.

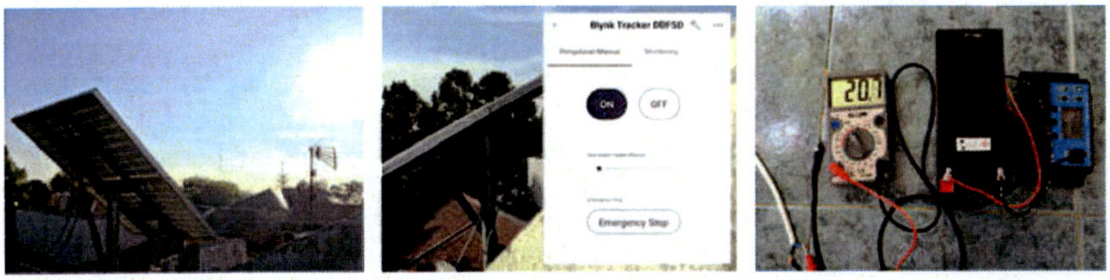

(a) (b) (c)

Fig. 4. **a** Mechanism testing. **b** Control system testing. **c** Output power testing

2.6 Analysis and Evaluation

At this stage, the performance of the model will be analyzed as a material for estimating the prospects for large-scale implementation, both in terms of efficiency and investment cost.

2.7 Report Making

Reporting is carried out after all planned stages have been completed. The report will convey the progress of the implementation of the process as well as describe the results and performance of the model.

3 Result and Discussion

Based on mechanism testing, the result shows that the model can work well according to user's time and location. Mechanism testing is done by measuring the range of angles that can be achieved by the solar tracking system. The angle that can be achieved is 45°–135°.

Based on control system testing, the result also shows that the model can work well according to application commands. When the angle is not reached, then the actuator will move toward the angle to be achieved. Tests are also carried out, whether at night and windy weather conditions, the solar tracker will still reach the corner.

The output power test is done by comparing the fixed model with the single-axis model. The test is carried out by measuring the voltage and amperage produced and connected to the battery, and carried out every hour from morning to evening. So, we get the amount of output power. Based on testing the output power of the model with an area of 0.7 m^2 in fixed model and single-axis model conditions, the test results are obtained as given in Table 1.

Table 1 Fixed and single-axis model output power test results

Time	Fixed model 0.7 m^2			Single-axis Model 0.7 m^2		
	V(V)	I(A)	P(W)	V(V)	I(A)	P(W)
07:00	20.4	1.52	31.008	21.6	3.65	78.84
08:00	20.7	2.18	45.126	20.6	4.35	89.61
09:00	20.7	4.2	86.94	20.4	4.6	93.84
10:00	20.7	4.28	88.596	20.5	4.71	96.555
11:00	20.4	4.79	97.716	20.6	4.91	101.146
12:00	20.7	5.05	104.535	20.7	5.05	104.535
13:00	20.6	4.75	97.85	20.6	4.93	101.558
14:00	20.6	4.27	87.962	20.5	4.77	97.785
15:00	20.5	2.8	57.4	20.3	4.53	91.959
16:00	19.7	0.82	16.154	20.4	4.2	85.68
17:00	19	0.14	2.66	20.2	3.53	71.306

Based on these results, a graph of the comparison of the output power of the fixed and single-axis models is obtained as shown in Figs. 5 and 6.

Analysis, evaluation, and prospect of product implementation are explained through the following implementation example as given in Table 2.

Based on the implementation example and previous estimated calculations, it can be seen that the average output power of the single-axis model is 41% greater than the fixed model. Even though it has a 6-month break-even point difference, but looking at the prospect of profit to be obtained, the single-axis model is still superior. In addition, the investment cost of solar panel tends to decrease every year and there will be many solar panel innovations so that its efficiency will definitely increase. Now, the challenge is how to convince people to start investing in solar canopy.

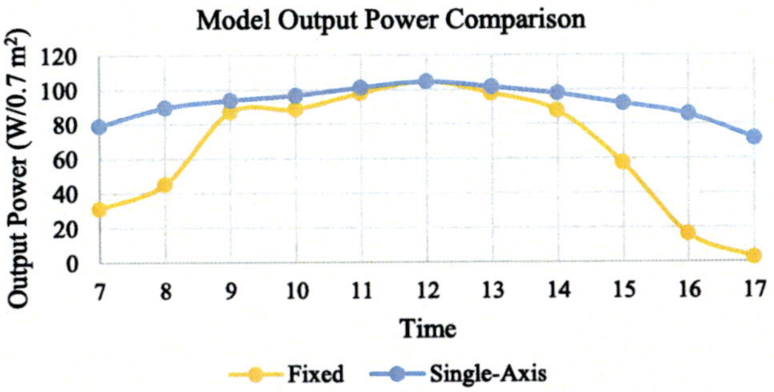

Fig. 5. Fixed and single-axis model output power comparison

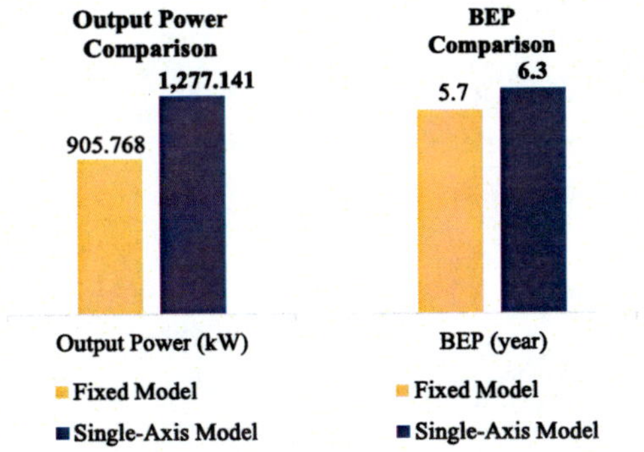

Fig. 6. Model comparison

4 Conclusion

Based on previous results and discussion, it can be concluded that the model can work well according to user's time and location as well as application commands. Then, the output power of single-axis model is 41% greater than fixed model. Even though the single-axis model has a 6-month BEP difference, it is still superior for some reasons. The conclusions here are conclusions drawn from the implementation examples with assumptions as previously mentioned given that we do not make actual products.

Table 2 Analysis of model implementation example

Aspect	Fixed model	Single-axis model
Efficiency		
Average output power (W/m^2)	92.980	131.534
Assumption of the airport's open parking area (m^2)	10,000	10,000
Assumption of model dimension (m^2)	1.303 × 2.384	1.303 × 2.384
Estimated output power/module (W)	288.829	408.591
Assumption of the number of module/parking lot	2 × 28 module	2 × 28 module
Assumption of motor/parking lot	–	75 W/motor 1 motor
Estimated output power/parking lot (kW)	16.174	22.806
Assumption of the number of parking lot	56 parking lots	56 parking lots
Estimated total output power (kW)	905.768	1277.141
Investment cost		
Operation time assumption (h)	6 09:00–15:00	6 09:00–15:00
Estimated energy output/day (kWh)	5,434.608	7,662.846
Estimated energy output/year (kWh)	1,983,631.92	2,796,938.79
Electricity price assumption/kWh	Rp 1444.7 R-1/TR 1300 VA	Rp 1444.7 R-1/TR 1300 VA
Estimated income/year	Rp 2,865,753,034.824	Rp 4,040,737,469.913
Investment cost assumption/kW	Rp 18,000,000	Rp 20,000,000
Estimated investment cost	Rp 16,303,824,000	Rp 25,542,820,000
Estimated break-even point (years)	5.7	6.3

References

1. Secretariat General of National Energy Council: Indonesia Energy Outlook (2019) Secretariat general of national energy council, Jakarta
2. Potensi Energi Tenaga Surya RI 10 Kali Lebih Besar dari Jerman, https://www.liputan6.com/bisnis/read/2493191/potensi-energi-tenaga-surya-ri-10-kali-lebih-besar-dari-jerman. Accessed 13 Feb 2021
3. Pemanfaatan Potensi PLTS Terhambat Masalah Ini, https://ekonomi.bisnis.com/read/20201215/44/1331312/pemanfaatan-potensi-plts-terhambat-masalah-ini. Accessed 13 Feb 2021
4. Peluang Besar Kejar Target EBT Melalui Energi Surya 2019, https://ebtke.esdm.go.id/post/2019/09/26/2348/peluang.besar.kejar.target.ebt.melalui.energi.surya. Accessed 13 Feb 2021
5. Rutgers Board of Governors Approves 32-Acre Solar Canopy Project, https://www.rutgers.edu/news/rutgers-board-governors-approves-32-acre-solar-canopy-project. Accessed 13 Feb 2021
6. The best idea in a long time: Covering parking lots with solar panels, https://www.washingtonpost.com/news/energy-environment/wp/2015/01/28/the-best-idea-in-a-long-time-covering-parking-lots-with-solar-panels/. Accessed 13 Feb 2021
7. Dhanabal R, Bharathi V, Ranjitha R, Ponni A, Deepthi S, Mageshkannan P (2013) Comparison of efficiencies of solar tracker systems with static panel single-axis tracking system and dual-axis tracking system with fixed mount. Int J Eng Technol 5(2):1925–1933
8. Kuttybay N, Saymbetov A, Mekhilef S, Nurgaliyev M, Tukymbekov D, Dosymbetova G, Meiirkhanov A, Syanbayev Y (2020) Optimized single-axis schedule solar tracker in different weather conditions. Energies 13(19):1–18

Investigation of the Four Runner Blade Arrangement Against the Power of Kaplan Turbine

Sirojuddin(✉), Nadia Sari Dewi, and Ragil Sukarno

State University of Jakarta, Jakarta 13220, Indonesia
sirojuddin@unj.ac.id

1 Introduction

The small-sized Kaplan water turbine is suitable for Micro-Hydro Power Plant especially for rural areas. Kaplan turbine is an axial flow reaction turbine that utilizes low head and large water discharge to generate power [1]. This turbine has an adjustable runner blade, and the blade flow angle can be varied to get the best efficiency [2]. The runner blade has a complex design so that it is manufactured using the NACA 2412 Airfoil [3]. The diameter of the hub and the number of blades are depending on specific speed, and the higher head makes specific speed lower and the hub diameter and number of blades greater [4]. Research on the Kaplan turbine runner design to determine power and efficiency based on the number of blades had been carried out. The research used variations of 3, 4, and 5 runner blades, and the results show that four runner blades produce better power and efficiency than other variants [5]. Another research by designing the Kaplan turbine runner blade using the Gottingen (GOE) series and Airfoil.com software, the head is 8 m, and the flow rate is 0.3 m^3/s using computational fluid dynamics (CFD) software to predict the fluid flow that passes through the turbine runners [6]. The results show the power generated is approximately 20 kW and the turbine speed is 900 rpm. A simulation study of a Kaplan turbine with four blades, head of 6 m, and flow rate of 5 m^3/s using ANSYS showed an efficiency of 50.98% [7]. The manufacture of Kaplan turbine runner blades with four blades, head 17 m, and flow rate 33.31 m^3/s by dividing the blades into six sections and calculating the velocity triangle of each section. The design produces power of 5 MW [8].

This study aims to investigate the four runner blade arrangement on Kaplan turbine with variations in the height of the inner blade profile of −5 mm, parallel, + 5 mm, + 10 mm, and +15 mm from the height of standard blade against the power efficiency obtained, when the turbine is in a momentary stop and in rotating condition approach.

M. Kolhe et al. (eds.), *Recent Advances in Renewable Energy Systems*,
Lecture Notes in Electrical Engineering 876,
https://doi.org/10.1007/978-981-19-1581-9_10

2 Materials and Methods

In this study, the runner blade was designed based on net head of 5 m, water discharge of 0.125 m^3/s, and theoretically turbine power of 6.131 kW. The runner blade profile was made using Airfoil NACA 2412 and 2D geometry with AutoCAD software. The 3D models were created using Inventor to get optimum thickness and SolidWorks for flow simulation.

2.1 Governing Equation

The preliminary design calculations were determined by the following formula [10].

$$P_{\mathrm{TH}} = \rho \times g \times Q \times H_{\mathrm{N}} \tag{1}$$

$$P_{\mathrm{T}} = \rho \times g \times Q \times H_{\mathrm{N}} \times \eta_{\mathrm{T}} \tag{2}$$

$$N_{\mathrm{s}} = \frac{N\sqrt{P}}{H_{\mathrm{N}}^{5/4}} \tag{3}$$

$$ku = 0.79 + 1.61 \times 10^{-3}(N_{\mathrm{S}}) \tag{4}$$

$$D_{\mathrm{M}} = \frac{ku \times \left(60\sqrt{2 \times g \times H_{\mathrm{N}}}\right)}{\pi \times N} \tag{5}$$

$$D_{\mathrm{m}} = \left(0.25 + \frac{94.64}{N_{\mathrm{S}}}\right) \times D_{\mathrm{M}} \tag{6}$$

$$D_{\mathrm{cp}} = \sqrt{\frac{D_M^2 + D_m^2}{2}} \tag{7}$$

The description of the symbols is displayed in Table 1. The power of turbine according to drag force and torque approach are calculated as follows.

$$P_{\mathrm{df}} = z \times F_{\mathrm{dh}} \times \frac{\pi \times D_{\mathrm{cp}} \times N}{60} \tag{8}$$

$$P_{\mathrm{tq}} = T_{\mathrm{h}} \times \frac{2 \times \pi \times N}{60} \tag{9}$$

$$\eta_T = \frac{P_{\mathrm{df}}}{P_{\mathrm{TH}}} \times 100\% \tag{10}$$

$$\eta_{\mathrm{T}} = \frac{P_{\mathrm{df}}}{P_{\mathrm{TH}}} \times 100\% \tag{11}$$

where P_{df} = power by drag force approach in static/stop condition (Watt), P_{tq} = power by torque approach in rotating condition (Watt), and η_{T} = turbine efficiency (%).

Table 1 Design parameter and dimensions of turbine [10]

Design parameter	Design dimensions	Description
z	4	Number of runner blades
Q	0.125 m^3/s	Water discharge
H_G	5.25 m	Gross head
H_N	5 m	Net head
η_T	95%	Estimated turbine efficiency
N	1000 rpm	Turbine speed
P_{TH}	6,1312 kW	Theoretical turbine power
P_T	5.8246 kW	Actual turbine power
N_S	322.79 m-kW	Specific speed
D_M	0.24 m	Runner diameter
D_m	0.14 m	Runner hub diameter
D_{cp}	0.19	Center point diameter
P	0.1103625 MPa	Pressure
A	0.0289 m^2	Projected area of blade

Table 2 Boundary conditions

Parameter	Input data
Turbulence model	K-epsilon
Analysis type	Internal
Fluids	Water (H_2O)
Density	1000 kg/m^3
Water discharge	0.125 m^3/s
Gravity	9.81 m/s^2
Roughness	10 μm
Pressure	101,325 Pa
Temperature	293.2 K

2.2 Parameter and Boundary Conditions

The design parameter and dimensions can be seen in Table 1. The simulations were done using boundary conditions set forth in Table 2.

The blade thickness was optimized based on von Mises stress analysis using materials JIS G3221—SFCM60S [9].

$$\sigma_{red} = \sqrt{(\sigma_B + \sigma_T)^2 + \alpha(\tau + \tau_S)^2} \tag{12}$$

$\sigma_{red} \geq \sigma_a \sigma_{red} \geq \sigma_a$ where $\sigma_a = \frac{\sigma_y}{SF} \sigma_a = \frac{\sigma_y}{SF}$.

σ_{red} = reduction stress, σ_B = bending stress, σ_T = tension stress, τ = torsion stress, τ_S = shear stress, $\alpha = 3$ for Heuber Mises-Henckey, and $\alpha = 4$ for Tresca-Guest.

2.3 Blade Profile

Figure 1 shows the blade velocity triangle of outer and inner blade. The dimensions of the velocity triangle are based on design parameter as given in Table 3.

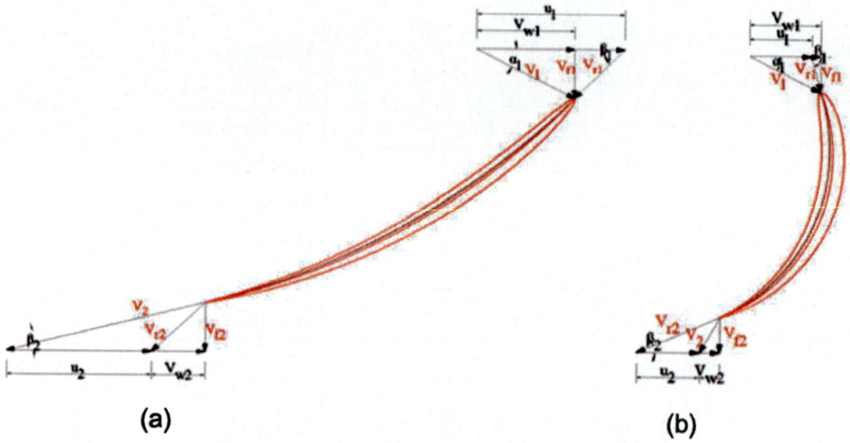

Fig. 1 Blade profile **a** outer blade **b** inner blade

Table 3 Design parameter of velocity triangle [1]

Design parameter	Outer	Inner	Description
$u_1 = u_2$	12.56 m/s	7.33 m/s	Tangential velocity
V_1	9.70 m/s	9.70 m/s	Absolute velocity
$V_{f1} = V_{f2}$	4.19 m/s	4.19 m/s	Velocity of flow
α_1	26.70°	26.70°	Guide blade angle
V_{w1}	8.33 m/s	8.33 m/s	Inlet whirling velocity
β_1	44.73°	76.57°	Inlet blade angle
V_{w2}	4.62 m/s	1.97 m/s	Outlet whirling velocity
β_2	13.70°	24.25°	Outlet blade angle

The 2D geometry of the runner blade profile is using Airfoil NACA 2412. Figure 2 shows the expanse blade profile which contains the outer and inner blades using Autodesk AutoCAD software. Figure 3 displays the 3D models of each runner blade that will be simulated for strength test using Autodesk Inventor stress analysis and flow test using CFD SolidWorks flow simulation.

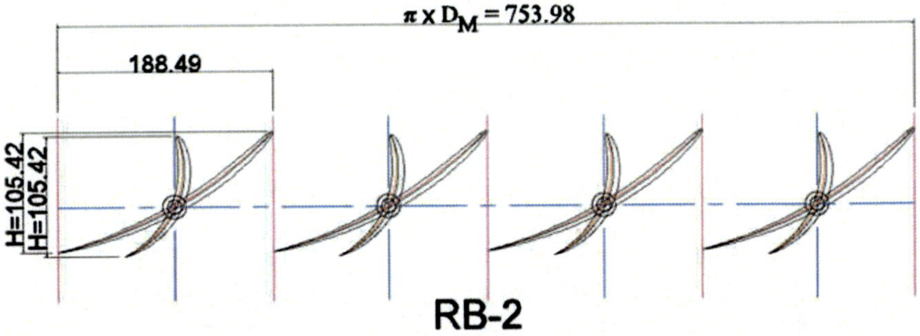

Fig. 2 2D geometry runner blade variant 2 (standard blade)

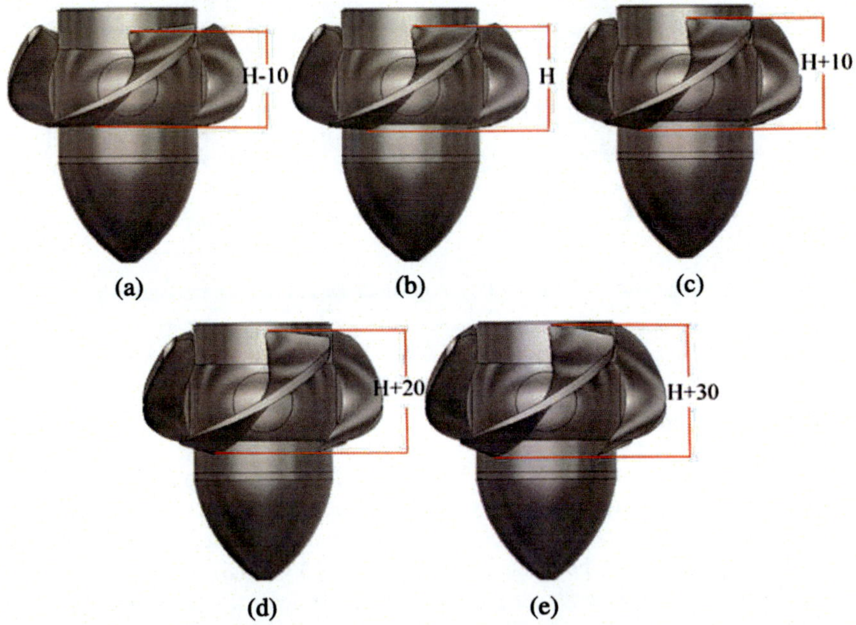

Fig. 3 3D models of runner blade: **(a)** RB-1 **(b)** RB-2 **(c)** RB-3 **(d)** RB-4 **(e)** RB-5

3 Results and Discussion

Results and discussion of the simulations that have been carried out can be seen as follows.

3.1 Momentary Stop Condition

Figures 4 and 5 show the results of CFD SolidWorks flow simulation for RB-3 in a momentary stop condition approach. From those figures, it shows that the pressure of the water entering the spiral casing has a red contour. When it goes to the runner blade, the contour becomes orange and the pressure is between 121,253.90 and 136,281.94 Pa, and after passing through the blades, it becomes green and the pressure decreases. Meanwhile, the velocity of the water entering the spiral case has a dark blue contour, and when it goes to the runner blade, it becomes light blue and the velocity is between 3.778 and 5.034 m/s.

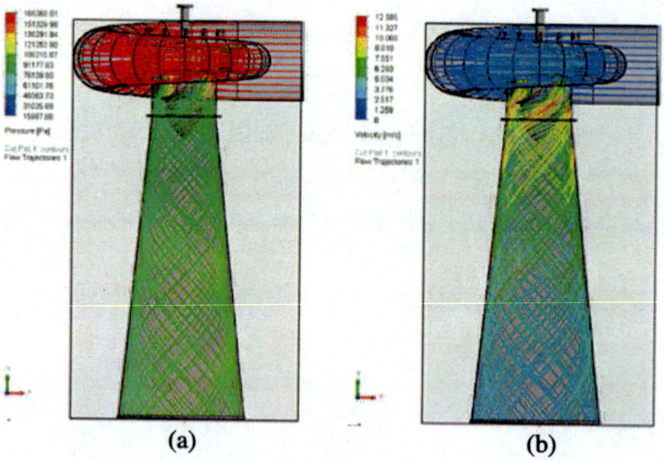

Fig. 4 Flow trajectories of momentary stop conditions in RB-3 **a** pressure **b** velocity

3.2 Rotating Condition

Figures 6 and 7 show the results of CFD SolidWorks flow simulation for RB-3 in rotating condition approach. From those figures, it shows that the pressure of the water entering the spiral casing has a red contour. When it goes to the runner blade, the contour becomes orange and the pressure is between 124,978.30 and 136,000.61 Pa, and after passing through the blades, it becomes green and the pressure decreases. Meanwhile, the velocity of the water entering the spiral case has a dark blue contour, and when it goes to the runner blade, it becomes light blue and the velocity is between 3.448 and 4.598 m/s.

3.3 Discussion

Figures 7 and 8 show the results of power and efficiency based on drag force and torque from the flow simulation of the four runner blade variants. These results are compared with the result of theoretical power, TURBNPRO, and a previous study by Abeykoon et al. [7]. From the two graphs above, it can be seen that the power in RB-1 to RB-3 variants tends to increase, while it tends to decrease from RB-3 to RB-5. It can also be seen that the power by torque is always above the power by drag force. Both power and efficiency of the RB-3 appear to be above of TURBNPRO and C. Abeykoon et al. Hence, the best runner blade variation that can be implemented on the Kaplan turbine is

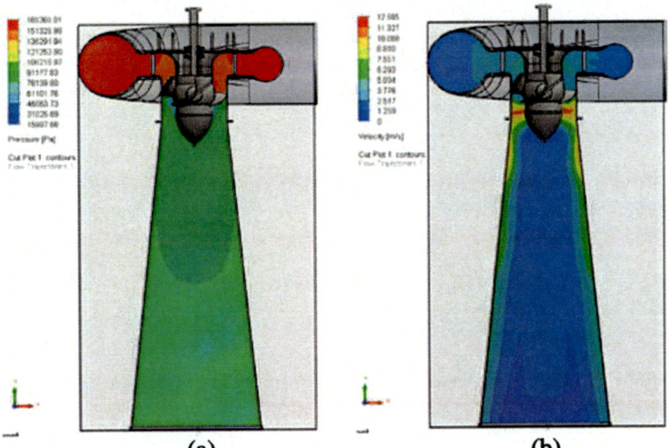

(a) (b)

Fig. 5 Cut plots of momentary stop conditions in RB-3 **a** Pressure **b** velocity

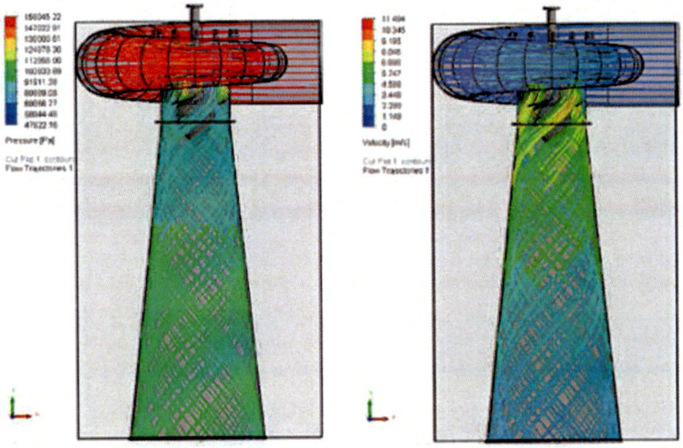

(a) (b)

Fig. 6 Flow trajectories of rotating conditions in RB-3 **a** pressure **b** velocity

the RB-3 variant. It generates better power when the turbine is at a momentary stop and rotating condition approach than other variations. Because in its height of inner blade, which is +5 mm, the water flow that passes through the runner and blade has greater pressure and velocity so as higher power and efficiency are obtained.

4 Conclusions

As implied in the results and discussions of simulation, it can be concluded that the height of the inner blade affects the power generated by the turbine in a momentary stop and rotating condition approach. In momentary stop condition approach, the power generated

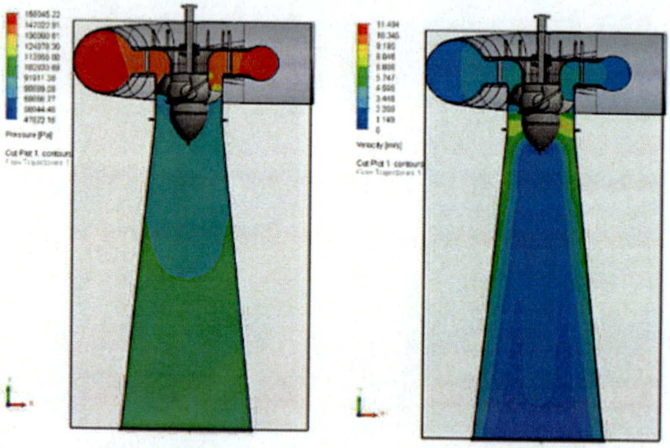

(a) (b)

Fig. 7 Cut plots of rotating conditions in RB-3 **a** pressure **b** velocity

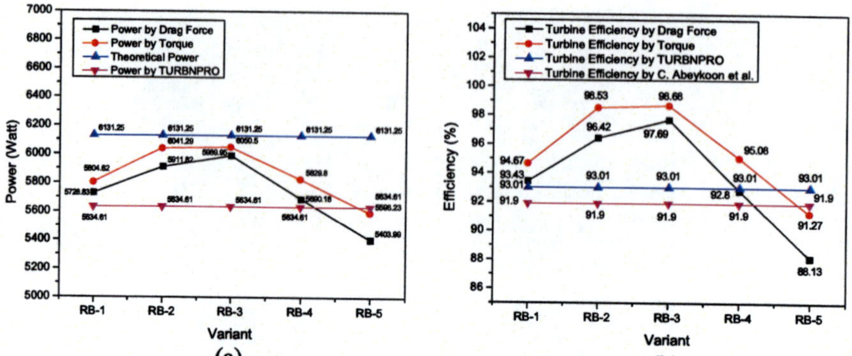

(a) (b)

Fig. 8 Performance of turbine **a** power **b** efficiency

by variant RB-1 is 5728.83 Watts, RB-2 is 5911.82 Watts, RB-3 is 5989.95 Watts, RB-4 is 5690.18 Watts, and RB-5 of 5403.99 Watts. In rotating condition approach, the power generated by variant RB-1 is 5804.62 Watts, RB-2 is 6041.29 Watts, RB-3 is 6050.5 Watts, RB-4 is 5829.80 Watts, and RB-5 is 5596.23 Watts. The highest power and efficiency were found in the RB-3 variant with the efficiency of 97.69% in a momentary condition approach and 98.68% in rotating condition approach.

Acknowledgements. The author would like to thank all mechanical engineering laboratory staff and our colleagues who support in this research.

References

1. Gorla SRS, Khan AA (2003) Turbomachinery design and theory. Marcel Dekker, New York
2. Penche C (1998) Layman's guidebook on how to develop a small hydro site. Politecnica de Madrid, Madrid
3. Nechleba M (1957) Hydraulic turbines their design and equipment. Artia Prague Printed, Czechoslovakia
4. Kovalev NN (1965) Hydroturbines design and construction (Gidroturbiny, Konstruktsii i Voprosy Proektirovaniya). Israel Program for Scientific Translations, Vol 64, No 11087, Jerusalem
5. Jawad LH (2019) Design and performance investigation of a hydraulic mini turbine based on renewable energy production system. J Babylon Eng Sci 27(3)
6. Permana E, Rudianto Y (2017) Design and velocity distribution of runner blade of kaplan turbine using CFD (computer fluid dynamic) for small hydroelectric power plant. IOP Conf Series Mat Sci Eng
7. Abeykoon C, Hantsch T (2017) Design and analysis of kaplan turbine runner wheel. In: Proceedings of the world congress on mechanical, chemical, and material engineering
8. Momin A, Dave N, Patel P, Panchal K (2017) Design and development of kaplan turbine runner blade. Int J Innov Res Sci Eng Technol (An ISO, vol. 3297, pp. 16519–16528)
9. JIS Handbook, Ferrous Materials & Metallurgy English Version ISBN 9784542137226 (2018)
10. De Siervo F, de Leva F (1978) Modern trends in selecting and designing kaplan turbines. Water Power Dam Constr 30(1):52–58

Investigation of the Runner Blade Arrangements on a 3-Blade Kaplan Turbine Against Turbine Power

Sirojuddin(✉), Alya Awanis Zahara, and Ragil Sukarno

State University of Jakarta, Jakarta 13220, Indonesia
sirojuddin@unj.ac.id

1 Introduction

One of the micro-hydropower (MHP) plant systems usually uses a Kaplan turbine to generate power. This plant utilizes a low head and water flow to strike the turbine blades [1]. The main component of the Kaplan turbine includes the spiral casing, guide vane, shaft, runner, and draft tube [2]. The turbine runner construction must consider the number of blades, the blade's thickness, and the arrangement of the runner blades because these have so much influence on the power and efficiency produced [3]. The runner blade has a very complicated design, so to get the outline runners, we have to go through airfoil NACA-2412 [4].

From the previous study, a Kaplan turbine with water discharge 0.35 m^3/s and head 2 m is predicted to produce 5.8 kW of power. The turbine was able to generate a voltage of 180 V, and hydraulic efficiency was about 70% [5]. The dimension of the blades inlet and outlet angles showed a major impact on the turbine's efficiency, where an increase of up to 42.03%. It would influence the tangential velocity difference, so the efficiency of the runner is from 50.98 to 93.01% [6]. The 3D model with SolidWorks and analysis on ANSYS 14 showed that Von Misses stress on the runner blade occurs at the connection between the hub and runner blade, and titanium alloys were the best material with a low weight where the maximum stress was 29.103 Mpa [7].

Furthermore, this study aims to investigate the arrangement of runner blade on Kaplan turbine 3-blade with inner blade height variation in −5, 0, +5, and +10 mm from standard blade height against the power generated by the turbine.

2 Materials and Method

The runner blade was designed for water discharge of 0.125 m^3/s, net head 5 m, and turbine theoretical power of 6.131 kW. The runner blade profile using NACA-2412 then process to 2D geometry with AutoCAD software. Next, the 3D models process in SolidWorks to get the optimum thickness and flow simulation result.

M. Kolhe et al. (eds.), *Recent Advances in Renewable Energy Systems*,
Lecture Notes in Electrical Engineering 876,
https://doi.org/10.1007/978-981-19-1581-9_11

2.1 Governing Equation

The preliminary design calculations were made by the following formula [8].

$$P_{\mathrm{TH}} = \rho \cdot g \cdot H_N \cdot Q \tag{1}$$

$$P_T = \rho \cdot g \cdot H_N \cdot Q \cdot \eta_T \tag{2}$$

$$N_s = \frac{N\sqrt{P_T}}{H^{1.25}} \tag{3}$$

$$k_u = 0.79 + 1.61 \times 10^{-3} \cdot N_s \tag{4}$$

$$D_M = \frac{k_u \cdot 60 \cdot \sqrt{2 \cdot g \cdot H_N}}{\pi \cdot N} \tag{5}$$

$$D_m = \left(0.25 + \frac{94.64}{N_s}\right) \cdot D_M \tag{6}$$

$$D_{\mathrm{cp}} = \sqrt{\frac{D_M^2 + D_m^2}{2}} \tag{7}$$

The description of the symbol can be see in Table 2. The power of turbine according to drag force and torque approach is calculated as follows in Table 1.

$$P_{\mathrm{df}} = z \cdot F_{\mathrm{dh}} \cdot \frac{\pi \cdot D_{\mathrm{cp}} \cdot N}{60} \tag{8}$$

$$P_{\mathrm{tq}} = T_y \frac{2 \cdot \pi \cdot N}{60} \tag{9}$$

$$\eta_T = \frac{P_{\mathrm{df}}}{P_{\mathrm{TH}}} \cdot 100\% \tag{10}$$

$$\eta_T = \frac{P_{\mathrm{tq}}}{P_{\mathrm{TH}}} \cdot 100\% \tag{11}$$

where P_{df} = power by drag force approach in momentary stop condition (Watt), P_{tq} = power by torque approach in rotation condition (Watt), η_T = turbine efficiency (%).

2.2 Parameter and Boundary Condition

The thickness of the blade is optimized according to Von Mises stress analysis using material JIS G3221–SFCM60S [9].

$$\sigma_{\mathrm{red}} = \sqrt{(\sigma_B + \sigma_T)^2 + \alpha(\tau + \tau_S)^2} \tag{12}$$

$$\sigma_{\mathrm{red}} \geq \sigma_a \quad \text{where } \sigma_a = \frac{\sigma_y}{\mathrm{SF}} \tag{13}$$

σ_{red} = reduction stress, σ_B = bending stres, σ_T = tension stress, τ = torsion stress, τ_s = shear stress, $\alpha = 3$ for Heuber Mises–Hencky, and $\alpha = 4$ for Tresca–Guest.

Table 1 Parameter and design data of turbine

Design parameter	Design dimensions	Description
z	3	Number of runner blades
Q	0.125 m^3/s	Water discharge
H_G	5.25 m	Gross head
H_N	5 m	Net head
η_T	95%	Estimated turbine efficiency
N	1000 rpm	Turbine speed
N_S	322.79 m-kW	Specific speed
P_{TH}	6,1312 kW	Theoretical turbine power
P_T	5,8246 kW	Actual turbine power
D_M	0.24 m	Runner diameter
D_m	0.14 m	Runner hub diameter
D_{cp}	0.19 m	Center point diameter
P	0.11036 MPa	Pressure
A	0.0289 m^2	Projected area of blade

Table 2 Boundary condition

Parameter	Input data
Turbulence model	K-epsilon
Analysis type	Internal
Fluid	Water (H_2O)
Density	1000 kg/m^3
Water discharge	0.125 m^3/s
Gravity	9.81 m/s^2 ($-Y$-axis)
Roughness	10 μm
Pressure	101,325 Pa
Temperature	293.2 K

2.3 Blade Profile

From Fig. 1, it can be calculated the dimension of each item using the formula in previous research [10], so the results are in Table 3.

The 2D geometry of the runner blade profile using NACA-2412 continued in AutoCAD software as shown in Fig. 2 variant of RB-2. After that the runner blade profile was made into a 3D model, and then, the thickness was optimized with Inventor software and to get the power generated using SolidWorks CFD flow simulation.

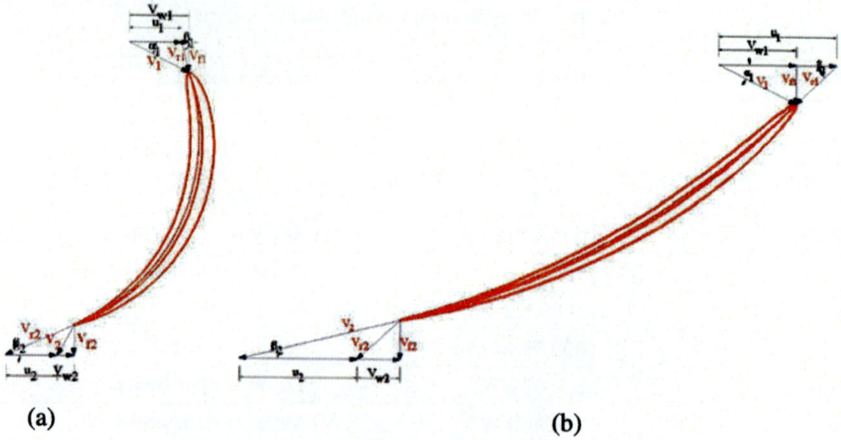

Fig. 1 Velocity triangle a inner blade, b outer blade

Table 3 Design data of turbine runner

Item	Outer	Inner	Description
u_1=u_2	12.56m/s	7.33m/s	Tangential velocity
$V_1 = V_2$	9.70 m/s	9.70 m/s	Absolute velocity
$V_{f1} = V_{f2}$	4.19 m/s	4.19 m/s	Velocity of flow
α_1	26.70°	26.70°	Guide blade angle
V_{w1}	8.33 m/s	8.33 m/s	Inlet whirling velocity
β_1	44.73°	76.57°	Inlet blade angle
V_{w2}	4.62 m/s	1.97 m/s	Outlet whirling velocity
β_2	13.70°	24.25°	Outlet blade angle

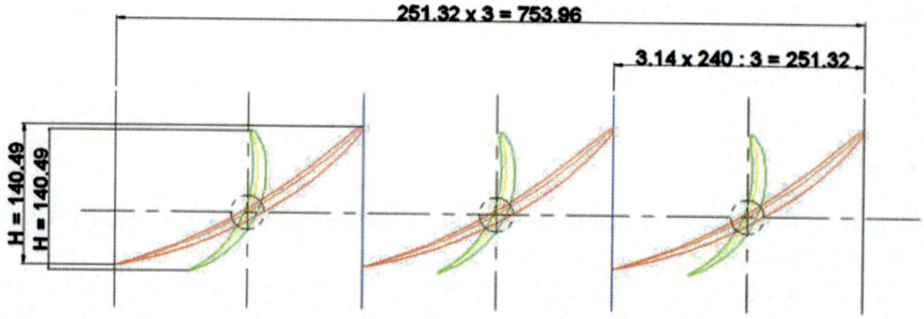

Fig. 2 2D geometry of runner blade RB-2 variant

This research analyzes 4 variations in the height of the inner blade in the upper and lower of the profile, that are RB-1 = −5 mm, RB-2 = 0 mm, RB-3 = +5 mm, and RB-4 = +10 mm from the height through the Y-axis of the standard blade as shown in Fig. 3.

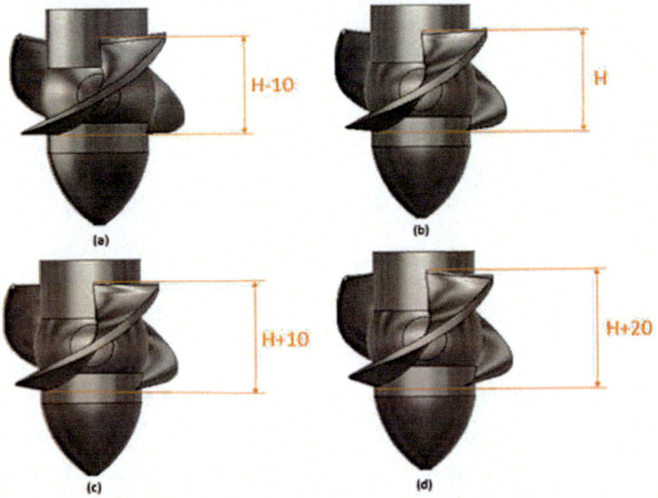

Fig. 3 Height of variant **a** RB-1, **b** RB-2, **c** RB-3, **d** RB-4

3 Result and Discussion

3.1 Momentary Stop Condition

From Fig. 4, the water pressure of RB-1 has a red contour where the pressure between 174,341.24 and 158,880.13 Pa enters the spiral case. Then, the water has a yellow contour between 143,419.02 and 127,957.91 Pa when it strikes the blade.

Meanwhile, the velocity of the water has a blue contour of 1.326–2.652 m/s when it enters the spiral case, and when it strikes the blade, it has a greenish contour of about 3.978–5.305 m/s, while the calculation results get 4.19 m/s of Vf1.

3.2 Rotation Condition

From Fig. 5, the water pressure of RB-1 has an orange contour between 158,880.13 Pa and 143,419.02 m/s when entering the spiral case. While the velocity of the water has a blue contour between 1.326 and 2.652 m/s when it enters the spiral case. Then, when it strikes the blade, it has a greenish contour of about 3.978–5.305 m/s.

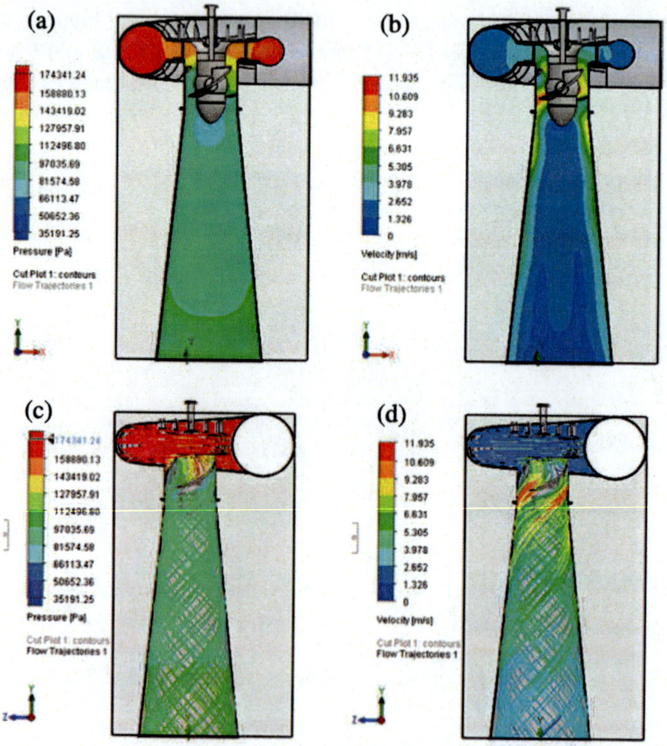

Fig. 4 Result from force RB-1 **a** pressure by cut plot, **b** velocity by cut plot, **c** pressure by flow trajectories, **d** velocity by flow trajectories

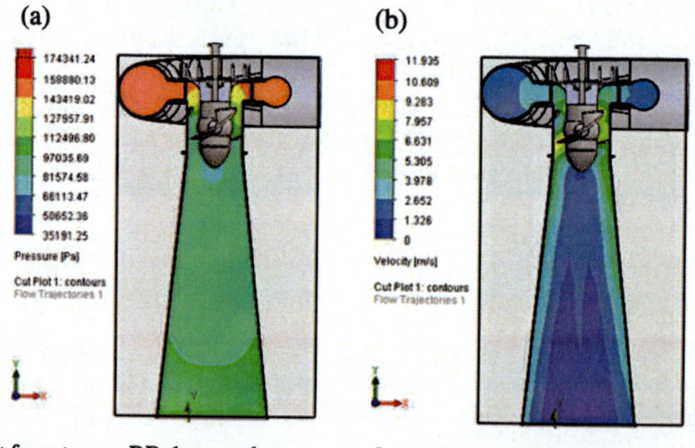

Fig. 5 Result from torque RB-1 **a** cut plot pressure, **b** cut plot velocity, **c** flow trajectories pressure, **d** flow trajectories velocity

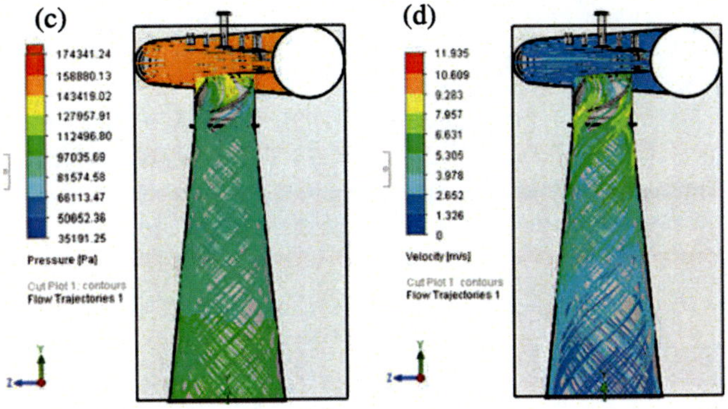

Fig. 5 continued

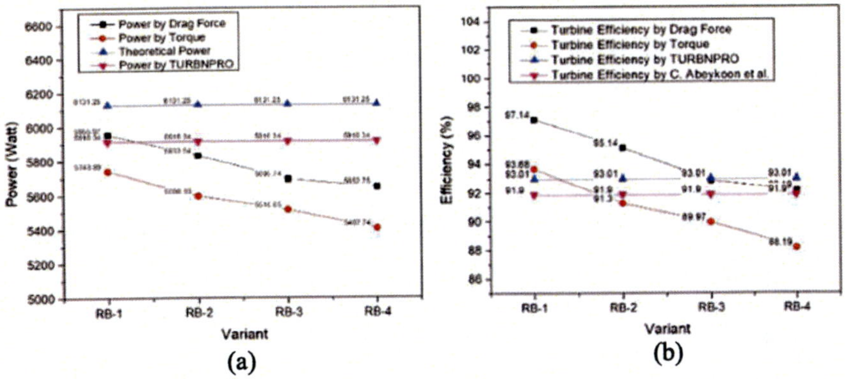

Fig. 6 Graph **a** power generated by the turbine, **b** efficiency generated by the turbine

3.3 Discussion

From the drag force and torque in the CFD simulation result, the power and efficiency generated by the turbine can be calculated using Formulas (8)– (11), after that validated by TURBNPRO and a previous study by C. Abeykoon et al. as shown in Fig. 6.

From Fig. 6a, can be seen that the theoretical power is 6131.25 W, and the power by TURBNPRO is 5916.34 W. From CFD simulation, the highest power by drag force and torque is variant of RB-1 with 5955.97 W and 5743.89 W, respectively.

In Fig. 6b, can be seen that turbine efficiency by TURBNPRO is 93.01%, while turbine efficiency by C. Abeykoon et al. is 91.9%. From CFD simulation, the highest power by drag force and torque is variant of RB-1 with 97.14% and 93.68%, respectively.

4 Conclusion

It was found that the height of the inner blade affects the power generated by the turbine in a momentary stop condition and rotation condition approach.

In a momentary stop condition, the power generated by a variant of RB-1 5955.97 W, RB-2 5833.54 W, RB-3 5695.74 W, and RB-4 5652.75 W. The higher power is a variant of RB-1 with maximum efficiency is 97.14%.

In a rotation condition, the power generated by a variant of RB-1 5743.89 W, RB-2 5598.33 W, RB-3 5516.65 W, and RB-4 5407.74 W. The higher power is also a variant of RB-1 with maximum efficiency is 93.68%.

Acknowledgements. The authors like to acknowledge assistance to all mechanical engineering laboratory staff and also my colleagues who have already supported this research.

References

1. Penche C, de Minas DI (1998) Layman's handbook on how to develop a small hydro site, 2nd edn. BMC Public Health 5(1)
2. Permana E, Rudianto YR (2017) Design and velocity distribution of runner blade of Kaplan turbine using CFD for small hydroelectric power plant. In: 1st annual applied science and engineering conference, material science and engineering, p 180
3. Abubakar M, Badshah S, Ahmad T, Rahman N (2014) Modelling and analysis of very low head Kaplan turbine runner blades for rural area of Punjab. Int J Sci Eng Res 5(7)
4. Nechleba M (1957) Hydraulic turbines: their design and equipment (Artia Prague Printed, Czechoslovakia)
5. Khan FU, Rahman W, Ahmad MM (2021) Modeling, simulation, and fabrication of micro Kaplan turbine, 10, pp 64–76
6. Abeykoon C, Hantsch T, Design and analysis of Kaplan turbine runner wheel. In: Proceedeedings of the 3rd worlds congress on mechanical, chemical, and material engineering (McM'17)
7. Ujwala MM, Chowdary PCK, Naik LS (2017) Int Refereed J Eng Sci 6(17)
8. de Siervo F, de Leva F (1978) Modern trends in selecting and designing Kaplan turbines. Water Power Dam Constr 30(1):52–58
9. JIS HANDBOOK, Ferrous Materials & metallurgy English version. ISBN 9784542137226
10. Gorla RSR, Khan AA (2003) Turbomachinery design and theory. Marcel Dekker, New York

Analysis Study of Performance and Reliability Impact in Boiler Through Differential Coal Calorific Value (Case Study: Pelabuhan Ratu Coal-Fired Power Plant)

Hendra Yudisaputro[1,2](✉), M. Nur Yuniarto[2], Yohanes[2], and Agus Wibawa[3]

[1] PT. Indonesia Power, Jakarta, Indonesia
hendra.yudisaputro@indonesiapower.co.id
[2] Department of Mechanical Engineering, Institut Teknologi Sepuluh Nopember, Surabaya, Indonesia
[3] PT. Pembangkitan Jawa Bali, Surabaya, Indonesia

1 Introduction

The utilization of low-rank coal for a coal-fired power plant in Indonesia has increased since 2007, especially in developing Fast Track Program phase one (FTP 1). In early 2020, as an Indonesian government-owned corporation, PT. Perusahaan Listrik Negara (Persero) decided to optimize fuel costs by changing the coal specifications of the Pelabuhan Ratu coal-fired power plant from 4500 to 3500–4200 kcal/kg. The program aims to reduce electricity generation costs, particularly fuel costs. Of course, this decision tends to negatively impact the plant performance and reliability because of several problems such as the coal characteristic under the combustion requirement, high water content, decreases in plant thermal efficiency, high oxygen content and tendency to spontaneously combustion [1]. In addition, the utilization of low-rank coal will limit the generator loads and capacity factors, and operational equipment costs higher than usual [2].

Several studies have determined the characteristics and impact of lignite coal on power plants' combustion and production costs. Bielowicz (2012), in his research on low-rank coal classification technology, concluded that during combustion, boiler losses for the evaporation process of coal moisture content cause an increase in fuel consumption and the workload of the main equipment in the boiler [3]. Tahmasebi, et al. (2016) showed their research on the effect of lignite coal moisture content on combustion patterns. High moisture content makes combustion delays of up to 83–160 ms at 10% and 20% water contents [4]. Besides, in a study conducted by Tian et al. (2016), it was shown that the utilization of lignite coal increases auxiliary power consumption for mill operation and combustion air fan systems [5].

Although several parameters of low-rank coal on the combustion process have been investigated in some literature, only a few studies discuss the impact of lignite combustion on equipment reliability and plant performance in actual cases. The objectives of this study are to find the changes in boiler and plant thermal efficiency, to know the equipment

M. Kolhe et al. (eds.), *Recent Advances in Renewable Energy Systems*,
Lecture Notes in Electrical Engineering 876,
https://doi.org/10.1007/978-981-19-1581-9_12

and component reliability, determine the suitable calorific value for optimizing the fuel cost, and set the maintenance strategy during utilization of low-rank coal.

2 Material and Methods

2.1 Coal Characteristic

Five different types of coal were selected for this study according to availability and research purpose shown in Table 1.

Table 1 Coal characteristic for the experimental testing

Parameter	Unit	Value				
Higher Heating Value (AR)	kcal/kg-f	3510	4080	4272	4338	4647
Total Moisture Content (AR)	wt%	45.0	36.9	35.3	34.0	27.5
Fixed Carbon Content (AR)	wt%	21.3	25.1	27.8	28.8	31.8
Volatile Matter (AR)	wt%	28.0	31.7	31.5	32.3	34.1
Ash Content (AR)	wt%	5.8	6.4	5.4	4.8	6.7

All coals are assumed to be homogeneous because there is no blending or mixing process. The distribution of coal is carried out directly from the barge to the boiler bunker. The coal sample is carried out at least five times at the coal feeder inlet, and coal quality is determined through proximate analysis using Bomb Calorimeter LECO AC-600 and LECO TGA-701, while ultimate analysis using LECO CHN 628 and LECO S 832.

2.2 Equipment Design Data Collection and Plant Operations

The operating parameters such as total airflow (t/h), coal flow (t/h) in each coal feeder, IDF motor current (Amperes), FDF motor current (Amperes), PAF motor current (Amperes), Mill motor current (Amperes), IDF damper opening (%), FDF damper opening (%), PAF damper opening (%) and Primary Air Header pressure (kPa). The purpose of this data collection is to compare the actual capability of the boiler at each load change to the design.

After the main boiler equipment characteristic against electrical generation is identified, then a simulation of load generation for various coal calorific values is carried out using the Gate Cycle 6.1.1 (GC 6.1.1.). This simulation aims to predict whether, with a specific calorific value, the unit load can produce maximum power by constraint coal consumption not more than 220 t/h.

2.3 Experimental Test

The plant condition during the experimental test is: (1) Target load is 340 MW with five different types of coal from 3500 to 4600 kcal/kg. (2) Sampling point for data collection based on ASME PTC 46-Overall Plant Performance. (3) It carried out 120 min for stabilization and ±120 min for test data retrieval. (4) Local data collection is taken every 15 min. (5) DCS data collection is taken every one minute. (6) Coal sampling is done every 30 min. Boiler efficiency calculated by heat loss method with formula,

$$\eta\,\text{Boiler}(100\%) = 100 - L_1 + L_2 + L_3 + L_4 + L_5 + L_6 + L_7 + L_8 \tag{1}$$

where L_1 is a dry gas, L_2 is moisture in hydrogen, L_3 is moisture in the fuel, L_4 is moisture in the air, L_5 is carbon monoxide, L_6 is radiation, L_7 is fly ash, L_8 is bottom ash.

Plant thermal efficiency is calculated with,

$$\eta\,\text{thermal} = (860 * 100)/\text{nphr} \tag{2}$$

where nphr is net plant heat rate, kcal/kWh which taken by multiply coal consumption, kg/h with coal calorific value, kcal/kg and divide with net power load, kWh.

2.4 Equipment Reliability Analysis

Identification of critical equipment on the main boiler equipment is determined by calculating the Asset Criticality Ranking (ACR). ACR is taken from multiplication between System Critically Ranking (SCR) and Equipment Criticality Ranking (ECR). The determination of the SCR value is carried out by comparing the critical level of the boiler system with other systems at the PLTU, while the ECR value is determined by comparing one equipment to another in one system based on the period when the use of low-calorie coal is started, namely, 2016–2020.

$$\text{SCR} = \sqrt{\frac{\left(S^2 + C^2 + E^2 + F^2 + \text{EFF}^2\right)}{5}} \tag{3}$$

$$\text{ECR} = \sqrt{\frac{\left(S^2 + C^2 + E^2 + F^2 + \text{EFF}^2\right)}{5}} \tag{4}$$

$$\text{ACR} = \text{SCR} * \text{ECR} \tag{5}$$

where (S) Safety; (C) Cost Maintenance; (E) Environmental; (F) Failure Frequency; (EFF) Efficiency Factor.

Then Reliability Block Diagram (RBD) simulations are carried out on AEROS software. Model validation on the reliability block diagram (RBD) is carried out by comparing the number of breakdowns in each component with five years, 2016–2020, or for 43,200 h. If the actual breakdown amount is under the simulation with a tolerance limit of 5%, the model is declared valid.

3 Results and Discussion

3.1 Plant Performance

Changes in the coal calorific value impact the boiler's dan plant thermal efficiency. Based on the results, boiler efficiency against various kinds of coal in Fig. 1. shows that high coal calorific value will increasing boiler efficiency, and vice versa. High coal calorific value has low moisture content and significantly increases boiler efficiency, such as 4647 kcal/kg coal, which has a low moisture content of 27.5%wt.

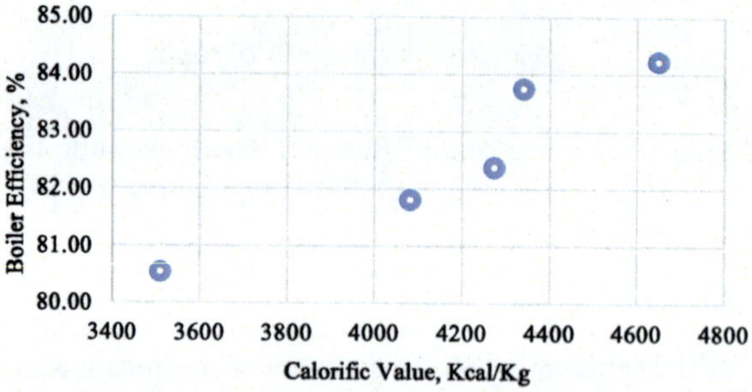

Fig. 1 Effect of coal calorific value on boiler efficiency

On the other hand, 3510 kcal/kg decreases boiler efficiency because it has high moisture at 45.0% wt. This phenomenon occurred due to the moisture content in coal obstructs the combustion process so that the boiler needs additional energy to separate water from the coal.

This experimental test also proves the truth of Li et al. [4], who conduct a study by varying coal moisture content on a laboratory scale. According to the study, the moisture content is the main reason for degradation in boiler efficiency. Some of the heat produced during the combustion process evaporates the coal moisture besides heating boiler pipes.

Figure 2 also shows that the thermal efficiency is significantly affected by the coal calorific value. High coal calorific value tends to increase the plant's thermal efficiency. This result is indicated by 4647 kcal/kg, where the thermal efficiency value increases to 32.60%. Likewise, with 3510 kcal/kg coal, the thermal efficiency significantly decreased to 30.12%. Based on the result, plant thermal efficiency aligns with boiler efficiency changes, although the turbine heat rate also makes some impact on it. However, the boiler is significant energy use in energy management where the coal is combusted, so that the contribution for change in plant performance is higher.

An increase in thermal efficiency cannot be used as an absolute consideration in determining the optimum heating value for the generation process. Therefore, it is necessary to calculate the fuel cost to determine which coal is suitable for economic production. Figure 3 shows that the optimum coal calorific value is between 4200 and 4400 kcal/kg with the lowest fuel cost of Rp398.81/kwh. During the utilization of coal with a calorific value of less than 4200 kcal/kg, there is an indication of an increase in fuel cost. This circumstance happened because the thermal plant efficiency going to decrease to 31.8%,

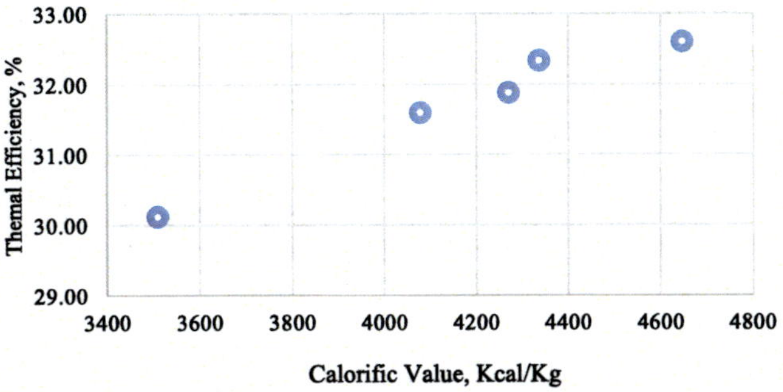

Fig. 2 Relationship between coal calorific value and plant thermal efficiency

which is indicated the coal consumption was a quite high event though the coal price during the condition was only Rp616.3/kg.

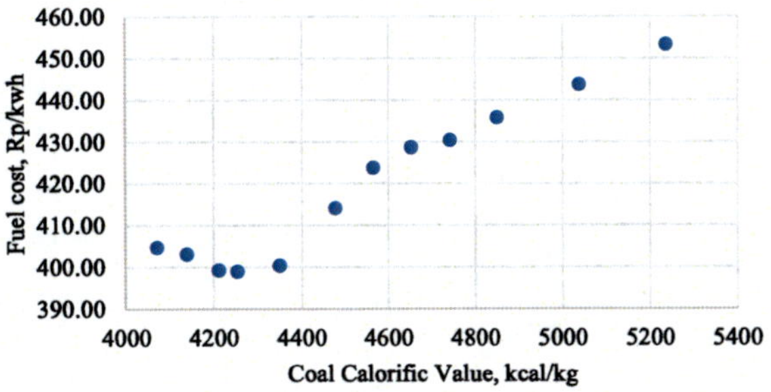

Fig. 3 Relationship between coal calorific value and plant thermal efficiency

The main cause of enhancement in fuel cost is the high price of coal. According to the calculation method from the Indonesian Minister of Energy and Mineral Resources, the coal price is formulated from moisture, sulfur, ash content and price reference of coal, hereinafter referring to HBA. High HBA, high calorific value and low moisture content will make high coal prices. On the other hand, it also shows that generator efficiency is not the leading indicator of whether a power plant is economical.

3.2 Plant Reliability

Based on ACR and SCR calculations and historical equipment failure data that cause production power loss, as shown in Fig. 4, mill equipment is the critical equipment affected by coal switching utilization.

Figure 4 shows the highest failure frequency due to the low calorific value found in the boiler fuel supply system from bunker to boiler. This fact happens because of an excess of coal consumption capacity from the utilization of low-rank coal, which causes significant damage to the mill and the coal feeder belt. Likewise, other disturbances regularly occur in the boiler fuel supply to bunker systems and slag and ash removal.

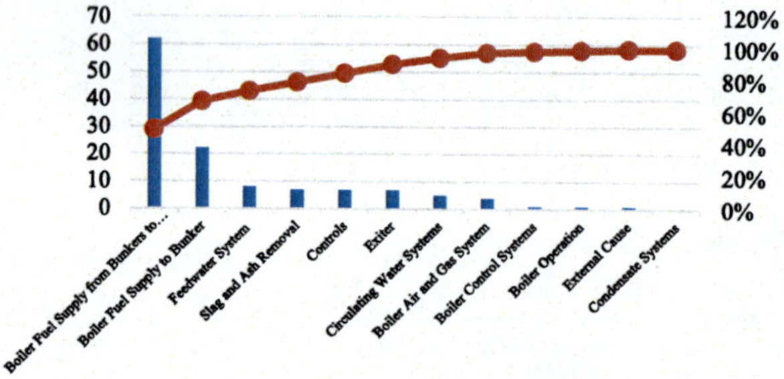

Fig. 4 Pareto of equipment failure

This analysis also shows the truth of the research results conducted by Thomas G. Woo (1979) [2].

The RBD model is simulated under conditions of 1000 simulation hours with a 1000 total data execution. The top five mill components that have the lowest reliability value $R(t) = 60\%$ are Coal Pipe, Damper, Lub. Oil Station, Pyrite Gate and Swing Valve, respectively, with breakdown time 530, 700, 1600, 1910 dan 2300 h, as shown in Fig. 5. The top five lowest availability of mill component also same with the lowest reliability by availability number 0.648, 0.74, 0.91, 0.922 dan 0.881, respectively.

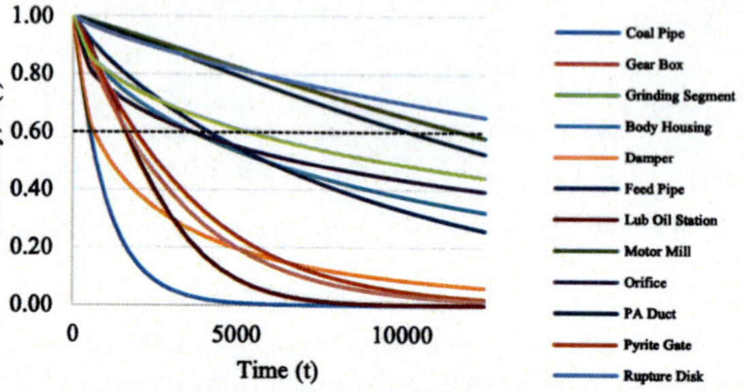

Fig. 5 Reliability of mill component

RBD simulation results from the mill component that is shown in Fig. 6 interpret that the coal pipe has a Mean Time Between Failure (MTBF) value of 530 h with an average operating time of 850 h the subsequently fails. It requires repairing time for 150 h. While the damper has an MTBF value of 70 h and an operating time of 670 h, it then fails, which requires a repair time of 330 h. At the same time, Lube oil, pyrite and swing valve components contribute to operating longer so that within 1000 h of operation, there is no indication of downtime.

4 Conclusion

The analysis shows that variations in coal affect some changes in boiler and plant thermal efficiency. While critical equipment is a mill with the lowest reliability and availability

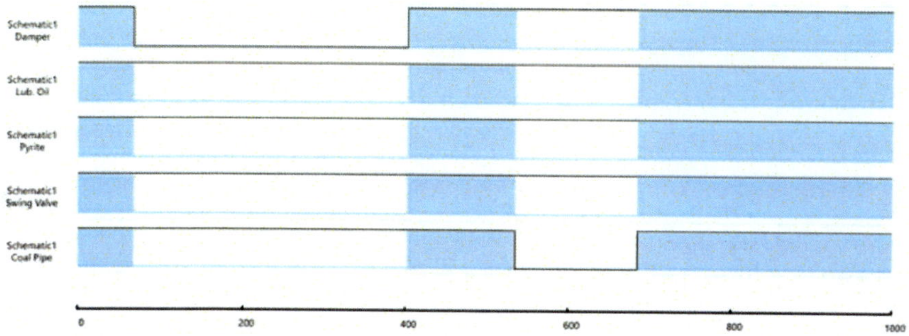

Fig. 6 A tristate plot of the five lowest reliability at mill components

component is Coal Pipe, Damper, Lub. Oil Station, Pyrite Gate and Swing Valve. The optimum coal calorific value for economic production cost is 4200–4400 kcal/kg. Then, a maintenance strategy has been carried out on critical equipment to reduce the negative impact of coal switching when reliability $R(t) = 60\%$.

References

1. Ma L, Fang Q, Yin C, Wang H, Zhang C, Chen G (2019) A novel corner-fired boiler system of improved efficiency and coal flexibility and reduced NOx emissions. Appl Energy. https://doi.org/10.1016/j.apenergy.2019.01.084
2. Woo TG (1979) Lignite fuel and power-plant availability. IEEE Trans Reliab R-28(4):279–282. https://doi.org/10.1109/tr.1979.5220603
3. Bielowicz B (2012) A new technological classification of low-rank coal on the basis of Polish deposits. Fuel 96:497–510. https://doi.org/10.1016/j.fuel.2011.12.066
4. Tahmasebi A, Zheng H, Yu J (2016) The influences of moisture on particle ignition behavior of Chinese and Indonesian lignite coals in hot airflow. Fuel Process Technol 153:149–155. https://doi.org/10.1016/j.fuproc.2016.07.017
5. Tian ZF, Witt PJ, Schwarz MP, Yang W (2012) Combustion of predried brown coal in a tangentially fired furnace under different operating conditions. Energy Fuels 26(2):1044–1053. https://doi.org/10.1021/ef2014887

Effect of the Oxide Scale on Tube Boiler Remaining Life of a 600 MW Coal Power Plant

Diki Purwadi(✉), Suwarno, and Vivien S. Djanali

Department of Mechanical Engineering, Institut Teknologi Sepuluh Nopember, Kampus Keputih-Sukolilo, Jl. Arif Rahman Hakim, Surabaya 60111, Indonesia
dikipurwadi@gmail.com

1 Introduction

Boilers are the primary system for converting combustion energy from coal into steam energy in power plants. The boiler requires optimal maintenance to minimize blackouts caused by boiler damage such as boiler tube leaks. Several damages on mechanism that often occurs is overheating. In creep regimes such as in superheaters and reheaters, the oxide layer on the internal tube will affect the heat which can cause overheating and affect the life of the tube. Tubes are damaged at high temperatures which often occurs due to the presence of an oxide layer on the tube surface. So that knowledge about time, temperature, and the nature of the oxidizing environment have a correlation with the morphology of the oxide, can provide an alternative technique to analyze the failure of the operation result and can be used to determine the remaining life of the tube and avoid tube failure according to Koshy [1]. Hamzah et al. [2] stated that one of the mechanisms of damage that occurs in tube boilers is due to an overheated oxide layer. The damage due to the oxide layer often occurs in areas exposed to high temperatures such as reheat and superheaters.

The tube operating temperature in the area increases during the operating life due to the vapor side oxide layer which affects the cooling of the tube. As the tube temperature increases, the rate of formation of the inner scale also increases. As the scale increases continuously, the temperature of the material and its cycle continue, getting higher and higher every year. Therefore, a method is needed to increase the metal temperature continuously as a result of the oxide layer being formed to estimate the residual high-temperature pipe operation in the creep area. One of the previous attempts to describe the scale of vapor side oxide growth in low-alloy ferritic steels (1–3% Cr) as a function of time and temperature with the results showing a linear variation of the logarithmic oxide layer thickness and penetration depth (metal lost from oxidation) with parameters Larson-Miller through equations according to *B. B. JIIA, S. N. Ojha and B. K. Misra* [3].

The growth of the oxide layer is influenced by geometry and heat transfer parameters such as steam temperature, a mass flow rate of steam, flue gas temperature, and convection coefficient on the outer surface of the tube. The effect of these parameters can be simulated to determine the effects that occur due to the thickness of the oxide layer and changes in temperature according to Purbolaksono et al. [4]. The results of

M. Kolhe et al. (eds.), *Recent Advances in Renewable Energy Systems*,
Lecture Notes in Electrical Engineering 876,
https://doi.org/10.1007/978-981-19-1581-9_13

measurements and tests on tubes that failed in the previous period need to be simulated as part of the analysis and validation regarding the impact of variations in oxides and temperatures that occur in the boiler and its relationship to the remaining life of the boiler tube. In the present work, a study on the effect of the internal tube oxide layer on the remaining life of the tube is presented by using the secondary superheater tube (SSH) as the model case.

PC Boiler with a load of 600 MW is a Steam Power Plant with the fuel used is coal with medium–high calories 4600–5000 kcal/kg (HHV/High Heating Value) with an average consumption of 7500 tons/day at a full load of 600 MW. In 2018–2020, Steam Power Plant with a load of 600 MW, there have been 3 times boiler leaks caused by overheating. And here are macro tube photos in 2018, and 2020 (Fig. 1).

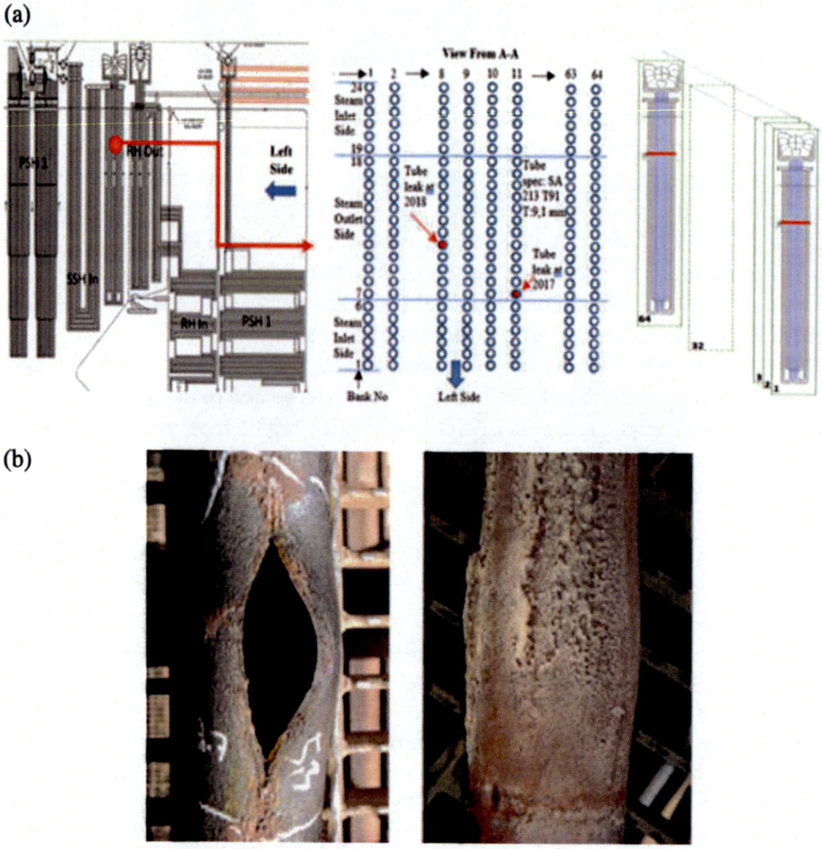

Fig. 1 **a** Boiler arrangement showing the location of the SSH and leakage location, **b** visual of the tube leak in 2018 (Tube No 7 & 4)

An overheating tube can be caused by the presence of scaling in the form of oxide scale with a thickness of >300 μ which can affect the rate of heat transfer from the flue gas to the tube wall which then does not transmit the heat transfer to the steam fluid due to the insulating oxide scale. So that the metal tube experiences excessive heat and exacerbated by the reduction in thickness due to corrosion will accelerate the occurrence of creep.

The tube boiler during the operation period is measured for thickness and also measurements of oxide scale in situ. Analysis was also carried out by taking metal temperature data to determine the temperature on the tube, while the measurements were made for tubes that had temperatures of 560 °C, 570 °C, and above 575 °C (Table 1).

Table 1 Design and geometry of the SSH out tube

Design SSH outlet	
Panel	64 panel
Tube	24 tube
Tube 1–6 and tube 19–24	SA 213 T22/ 50.8 OD/9.1 mm design thickness
Tube 7–18	SA 213 T91/ 50.8 OD/11.2 mm design thickness
Elbow lower SSH outlet	SA 213 T12/ 50.8 OD/5.6 mm design thickness

2 Method

The method used in the first stage is collecting data related to oxide growth with non-destructive methods and actual in situ measurements, the second is identifying the failure mechanism and taking samples from the failed tube by conducting laboratory tests including visual test, chemical composition test, hardness test, microstructure test, deposit test, and creep test and the last one uses the measurement results from laboratory tests and in situ measurements to simulate the prediction of remaining life due to the oxide layer, in this case, the secondary superheater (SSH out) case is taken.

The analysis begins with data collection including drawing, maintenance and operation history to determine the remaining life of the superheater out boiler area. The analysis of tube failure due to the oxide scale uses an empirical formula that connects the thickness of the oxide scale with the Larson-Miller parameter according to Rehn et al. [5] and other parameters, where the data used are test results data and data when overheating occurs in the tube boiler superheater out area. Data from 2017 to 2020, there have been 3 leaks caused by overheating due to the presence of oxide scale in the boiler tube which disrupts cooling in the pipes. The measurement of the scale thickness was carried out when the unit was stopped, and it was found that the boiler pipe in the superheater out area had an oxide scale >300 μ.

The analysis begins with data collection including drawing, maintenance and operation history to determine the remaining life of the superheater out boiler area. Boiler design, maintenance history, and operation history. The measurement point is in the SSH Out area on panels No. 50–60 on tubes 7–12.

In this study to approach the prediction of oxide scale growth and steam side oxide scale formation for ferritic steels with 1–3% chromium which is correlated with Larsen-Miller parameters as reported by Rehn et al. [5] used the following approach:

$$\mathrm{Log}\left(\frac{X}{0.0254}\right) = 0.00022P - 7.25 \quad (1)$$

where X is the thickness of the scale in mm. In the Larsen-Miller method [5], time and temperature are related to each other by the following equation:

$$\left(\frac{9}{5}\right)T + 492(C + \mathrm{LOG}t) = P \quad (2)$$

where P is the Larsen-Miller parameter; T is the temperature in degrees Celsius; t is the operating time in hours and C is a constant.

Meanwhile, according to *B. B. JIIA, S. N. Ojha and B. K. Misra* [3] used the following approach:

$$\mathrm{Log}10X = 7.1438 + 2.1761 \times 10 - 4T(20 + \log 10t) \quad (3)$$

where:

X = steam side oxide scale thickness in mils (10^{-4} inch)

T = temperature (°F + 460), and

t = operating hours

The way for calculating the stress using Eq. 3 according to Suwarno et al. [6], approached with simulations and algorithms in determining the creep and Larsen-Miller relationship calculated based on the following formula:

$$\sigma = P(\mathrm{Do} - t)2t \quad (4)$$

where P is the working pressure, Do is the outside diameter measured from the sample, t is the thickness.

Simulation of oxide growth can be carried out by several methods, for example, by using the "Larson Miller-like method" [2] using the basic kinetic equation where the oxide growth is proportional to the root of time and a constant that depends on the material and operating regime [3].

$$d = \sqrt{2} \cdot \mathrm{Kp} \cdot t \quad (5)$$

where d is the thickness growth of the oxide (μm), K is the rate constant of oxidation (μm²/h) dan t is operating time. So, by using this equation it can be predicted the thickness of the oxide in the operating time t [3].

This reference will be used for the remaining life analysis approach due to the oxide scale, where the algorithm is used in accordance with actual conditions so that the predictions are closer to the actual. The measurement results are then varied with operating parameters, in situ measurements and laboratory measurements to obtain a prediction of remaining life on the tube due to oxide scale.

3 Result

During the boiler tube operation, NDT thickness and oxide scale measurements were carried out in situ during inspection periods. An example of data of the oxide thickness is presented in Fig. 2 which shows that the thickness of the oxide layer depends on the tube number (position). The oxide scale generally can be up to 600 μm which is above common best practice for oxide scale thickness.

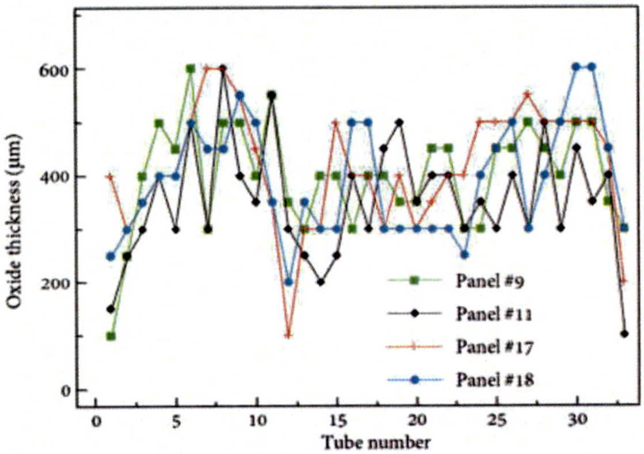

Fig. 2 The thickness of oxide layer from NDT test on selected tube and panel. The lines connected the points is given for eye guidance

Metal temperature data was also taken to determine the temperature on the tube, while the measurements were carried out for tubes that had temperatures of 560 °C, 570 °C, and can be above 575 °C depending on the location of the tubes (Table 2).

Table 2 Chemical composition of SA213-T91, comparing tube with 200,000 h and a new tube

Element	Fe	C	Si	Mn	P	S	Cr	Mo
Tube SA213-T91 (in used)	86.3	0.07	0.34	0.38	0.02	0.03	8.43	0.91
Tube SA213- T91 (new tube)	86.6	0.07	0.36	0.4	0.02	0	8.54	0.93

The results of the chemical composition test in Table 3 show that it belongs to the SA 213 T91 category. Chemical composition test results show no deviation in chemical composition.

In Fig. 3, The results of the micro test show that there have been carbide spheroidization. The SEM test shows that the sample has an oxide scale in the 2018–2020 range of 0.6–0.95 mm, while in 2017 the oxide scale value was 0.6 mm, in 2018 was 0.7 mm and in 2019 was 0.8–0.9 mm. This shows that the oxide scale grows where the tube temperature has been overheated compared to the design conditions. The SEM microstructure

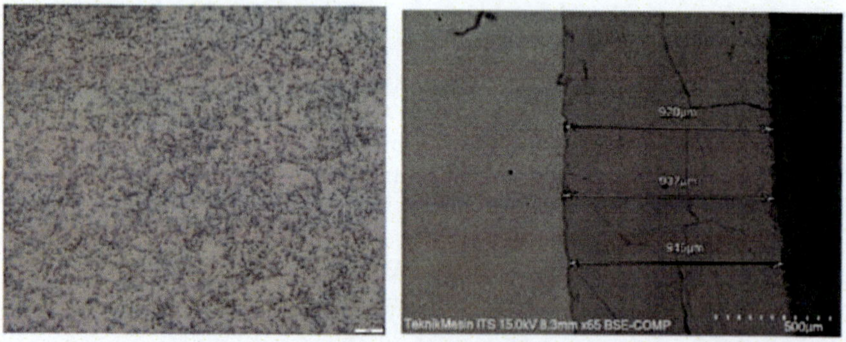

Fig. 3 Microstructure of the tube from the sampling indicate that the microstructure degradation in the form of spheroidization

also shows the presence of conditions where severe spheroidization has occurred. Micro voids are also seen located in the area close to the failures.

Furthermore, the simulation of the tube with several conditions/cases used in the remaining life modeling is carried out. The remaining life calculations in hours (h) and remaining life years, on the SH tube can be seen that thinning alone is not enough to cause creep failure, but the condition of the degraded material, which is also proven by creep testing, causes this tube has failed. So, it can be concluded that SSH failed due to creep.

Measurements on boiler pipes that operate more than 200,000 h show the effect of mass flow tube on tube thickness which has an impact on reducing tube lifetime. Based on the test (Figs. 4 and 5), it is known that temperature also affects the lifetime of the tube where temperature also has an impact on the growth of oxide scale, it can be seen in the simulation that if the tube only undergoes thinning, then the age of the tube exceeds 200,000 h of operation. However, if there is a combination of thinning and overheating due to the presence of oxide, the life will be shorter and the tube life will be shorter than 200,000. If the material is degraded due to creep, the probability of failure will be greater. Estimated growth of oxide scale on tube T91 and using empirical formula approach to determine the growth of oxide due to increase in temperature and time and its effect on boiler tube lifetime. Oxide growth is influenced by tube geometry and heat transfer parameters such as steam temperature, steam mass flow rate, exhaust gas temperature, and convection coefficient on the outer surface of the tube.

4 Conclusion

The results from several cases show that thinning alone is not enough to cause creep failure, but the condition of the degraded material, which is also proven by creep testing, causes this tube to fail, from several cases of pipe failure it is known that thinning causes residual lifetime the remaining is 18%, while if it is overheated the residual life is 7% and if it is degraded the remaining life is 3%. The operating temperature has a significant effect on the growth of oxide thickness and the growth will be faster if the operating

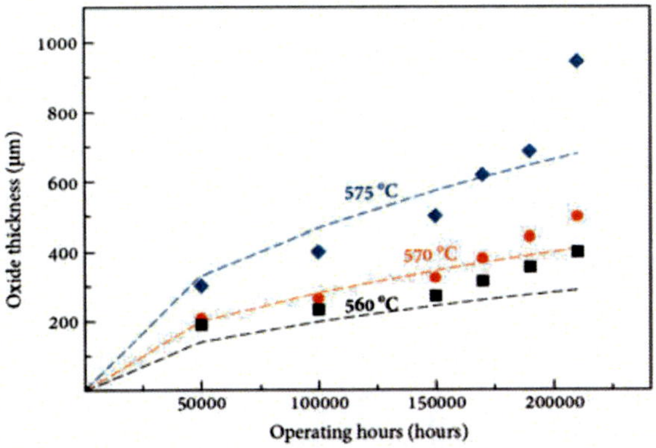

Fig. 4 The oxide growth as function of the tube operating hours as modeled and measured taken from the NDT test. The dash line is modeled, and mark points is measured by the NDT

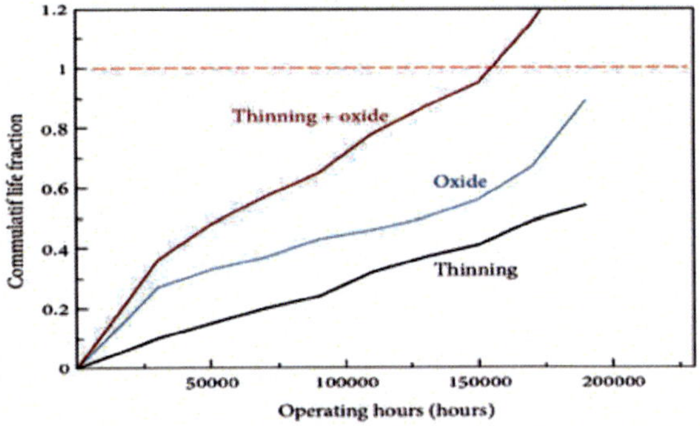

Fig. 5 Various scenario of tube life based on effect of oxide and thinning of the tube

temperature is above the design temperature. In tubes aged >200,000 h, the estimated oxide layer formed at an operating temperature of 560 °C is about 200–300 μm, while the actual is about 300–400 μm. The operating temperature of 570 °C is about 300–400 μm, while the actual is around 400–500 μm and at operating temperatures >575 °C is about 600–700 μm, while the actual is around 700–900 μm. Temperature and service hours also affect the lifetime of the tube which also affects the growth of the oxide layer, it can be seen in the simulation that if the tube only undergoes thinning, then the age of the tube exceeds 200,000 h of operation. Based on measurements on tubes operating >200,000 h the oxide scale can grow to >900 μm followed by the annual thinning rate of approx. 0.02–0.03%, while the growth rate of oxide is between 0.03–0.06% with an increase in fracture probability of 0.04–0.08%

References

1. Koshy M (2015) Super heater tube analysis for oxide scale growth at various operating conditions. Department of Mechanical Engineering, Srinivas Institute of Technology, Mangaluru, Karnataka, India
2. Hamzah MZ, Yeo WH, Fry AT, Inayat-Hussain JI, Ramesh S, Purbolaksono J (2013) Estimation of oxide scale growth and temperature increase of high (9–12%) chromium martensitic steels of superheater tubes. Eng Fail Anal 35(2013):380–386. https://doi.org/10.1016/j.engfailanal.2013.03.014
3. JHA BB, Ojha' SN, Misra BK (2008) Residual life estimation of high temperature tubings based on oxide scale thickness measurement, Institute of Minerals and Materials Technology, Bhubaneswar-751015, India 'Dept. of Met. Engineering, Institute of Technology, Banaras Hindu University, Varanasi-221 005
4. Purbolaksono J, Khinani A, Rashid AZ, Ali AA, Nordin NF (2009) Prediction of oxide scale growth in superheater and reheater tubes. Department of Mechanical Engineering, Universiti Tenaga Nasional, Km 7 Jalan Kajang-Puchong, Kajang 43009, Selangor, Malaysia b TNB Research Sdn Bhd, No. 1 Lorong Air Hitam, Kajang 43000, Selangor, Malaysia
5. Rehn JIM, Apblett Jr WR, Stringer J (1981) Controlling steamside oxide exfoliation in utility boiler superheaters and reheaters. Mater Perform, 27–31
6. Suwarno et al (2020) Laporan pengujian mekanikal properties boiler tube unit 6 dan 7 PLTU Suralaya, PT. ITS Tekno Sains, Institut Teknologi Sepuluh Nopember Surabaya

Numerical Study of the Generator Lubricant Cooler Air-Side Flow to Increase the Reliability of GTG#1.3 PLTGU Muara Karang

Aris Kurniawan[1](✉) and Sutardi[2]

[1] Department of Mechanical Engineering, FTIRS-ITS Collaboration Program With PT Perusahaan Listrik Negara, Jakarta, Indonesia
kurniawanaris85@gmail.com

[2] Department of Mechanical Engineering, FTIRS-ITS, Surabaya, Indonesia

1 Introduction

Muara Karang Block 1 Gas and Steam Power Plant (PLTGU) started commercial operations in 1992. Muara Karang PLTGU block 1 has a total capacity of 506 MW, consisting of a 3 × 107 MW PLTG (GE MS9001E) plus a steam turbine with a capacity of 185 MW. Muara Karang PLTGU block 1 is capable of operating with gas fuel or HSD. The main equipment used in the PLTGU includes a gas turbine, HRSG (heat recovery and steam generator) and a steam turbine and generator.

On October 12, 2012, the generator was replaced on a gas turbine 1.3, this replacement has several differences with other generator units, especially in the generator bearing lubricating cooling system which is separate from the gas turbine cooling system. The generator uses 2 bearings as a pedestal, during operation, the bearings are lubricated to maintain their performance. The generator lubricant cooling cycle is a closed system which is recirculated using 1 pump AC lubricant and the other standby, backed up by 1 DC lubricant pump when AC power loss. The generator lubricant circulates through the air-cooled heat exchanger (ACHE) cooling system then lubricates the generator bearings and finally returns to the grease tank. The following illustration of the generator lubricant cooling cycle is shown in Fig. 1.

Operation with 1 ACHE 2 fan has a low level of reliability because GTG#1.3 has tripped 5 times since the replacement of the generator with a different type in 2012. The trip on GTG#1.3 is due to the inlet lubricant temperature of the generator >320 K. From the damage data, it is compiled fishbone diagram to find out the main problem is shown in Fig. 2.

Many studies are related to heat exchangers. In paper [1], Re-Design Cooler Lubricants on Gas Turbines with Thermodynamics and Heat Transfer Analysis. In paper [2], Impact of ambient air temperature and heat load variation on the performance of air-cooled heat exchangers in propane cycles in LNG plants–Analytical approach. Numerical analysis of a cross-flow compact heat exchanger for vehicle applications [3]. In paper [4], Numerical Study of Cooling Performance Improvement on Lube Oil Cooler Gas Turbine Arranged in Series and Parallel with Variation of Lube Oil Flow Capacity.

M. Kolhe et al. (eds.), *Recent Advances in Renewable Energy Systems*,
Lecture Notes in Electrical Engineering 876,
https://doi.org/10.1007/978-981-19-1581-9_14

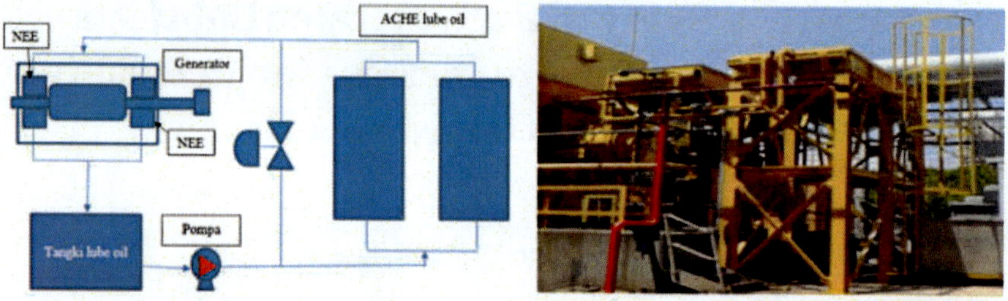

(a) (b)

Fig. 1 **a** Lubricant circulation system in GTG#1.3 generator, **b** actual condition of ACHE

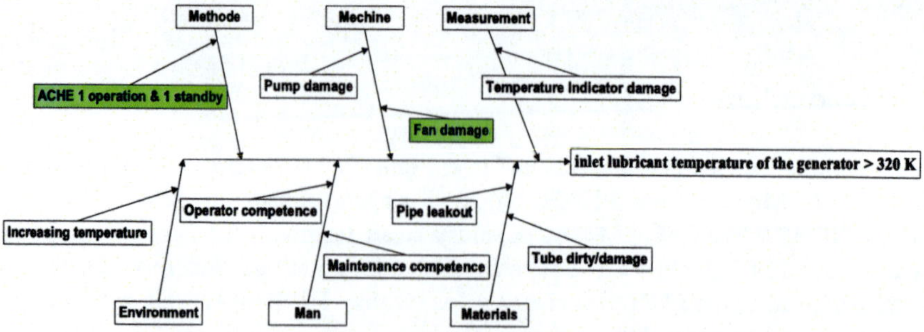

Fig. 2 Fishbone diagram the inlet lubricant temperature of the generator > 320 K

In paper [3], Numerical analysis of a cross-flow compact heat exchanger for vehicle applications.

2 Basic Theory

The heat exchanger is a device used to transfer heat between two or more fluids that have a temperature difference. The high-temperature fluid will move to the low-temperature fluid. ACHE is a cross-flow heat exchanger type that has a cross-flow direction of hot and cold fluids.

On the external flow side, calculations are carried out so that the heat transfer coefficient value is obtained which will later be entered into the simulation of the internal flow of air-cooled heat exchanger lubricants in single, series, and parallel tube conditions. The heat transfer coefficient Calculation Steps are as follows [5]:

(a) Velocity maximum ($V_{\max}$)

$$V_{\max} = \frac{S_T}{2(S_D - D)} V \tag{1}$$

Annotation:

V_{max} maximum cooling air velocity (m/s^2)
S_T distance between tubes perpendicular to flow (m)
S_D tube diagonal distance (m)
D tube diameter (m)
V fluid velocity toward the tube bank (m/s)

(b) Reynolds number ($R^m_{eD_{max}}$)

$$R^m_{eD_{max}} = \frac{V_{max} D}{v} \tag{2}$$

Annotation:

V kinematic viscosity

(c) Nusselt number ($\overline{N_{uD}}$)

$$\overline{N_{uD}} = C_1 R^m_{eD_{max}} P_r^{0.36} \left(\frac{\Pr}{\Pr s} \right)^{1/4} \tag{3}$$

Annotation:

Pr Prandtl number

(d) Heat transfer coefficient ($\overline{h}$)

$$\overline{h} = \overline{\mathrm{Nu}_D} \frac{k}{D} \tag{4}$$

Annotation:

K thermal conductivity (W/m K)

3 Methodology

3.1 ACHE Specification

The technical specifications of the GTG#1.3 generator lubricant air-cooled heat exchanger are as follows [6]:

Number of tubes: 152.

Outer Diameter tube: 16.4 mm.

Number of fans: 2 pieces.

Tube material: Aluminum Gold Fin Alloy (Al 98.7%) A1100H14.

3.2 Lubricant Properties

the lubricant used in the GTG#1.3 generator bearing is mobile dte oil light iso vg 32 and the mass flow of lubricant that flows is 3.29 kg/s. mechanical properties of the lubricant at a temperature of 343 K included in the CFD simulation can be seen in Table 1.

Table 1 Lubricant mechanical properties

Parameter		Input
Material (lubricant)	Density	836.15 kg/m^3
	Cp	2429 J/kg K
	Thermal Conductivity	0.12 W/m K
	Viscosity	0.031 kg/m s

3.3 Computational Fluid Dynamics

The CFD simulation used boundary conditions and controls residual can be seen in Table 2.

Table 2 Input parameter

Parameter		Input
Model	Energy equation	On
	Viscous	Realizable k-ℇ,enhance wall
Boundary condition	Inlet pelumas (mass flow)	mass flow inlet = 3.29 kg/s; Thermal = 343 K
	Tube	Material = Aluminum
		Convection ACHE single 2 fan = 307.353 W/m^{2}K
		Convection ACHE parallel and series 4 fan = 341.079 W/m^{2}K
		Convection ACHE parallels and series 3 fan = 325.466 W/m^{2}K
Controls	Residual	X velocity = 1e−04; Y velocity = 1e−04; Z velocity = 1e−04
		Continuity = 1e−04; Energy = 1e−05

Figure 3 shows the operating domain of ACHE 1 which will be compared with the simulation results with the actual operating data.

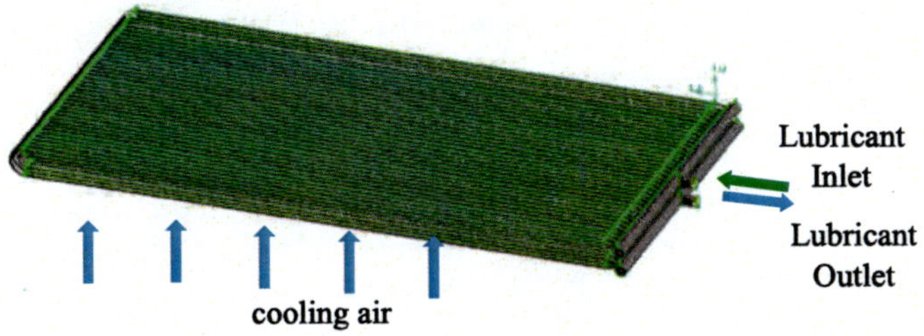

Fig. 3 Operation domain of ACHE single operation 2 fans

4 Result and Discussion

4.1 Validation

Simulation according to actual conditions is carried out first to get the correct modeling system. The simulation is carried out with a single ACHE operating model, namely, 1 ACHE passed by generator lubricant with 2 operating fans. The validation process is carried out by comparing the simulated ACHE outlet lubricant temperature with the actual ACHE outlet lubricant temperature. The comparison of the simulated ACHE outlet lubricant temperature with the ACHE outlet lubricant actual temperature can be seen in Table 3.

Table 3 ACHE outlet lubricant temperature comparison

Item	Source	Average temperature (K)	Error (%)
ACHE outlet lubricant temperature	Simulation	317.47	0.148
ACHE outlet lubricant temperature	Experiment	317	

4.2 Grid Independency Test

Grid independency test is done by comparing the simulated ACHE outlet lubricant temperature with the actual ACHE outlet lubricant temperature. The results of the grid independence test can be seen in Table 4.

4.3 Temperature Distribution

The following are the results of the CFD simulation with various variations. There are 5 types of ACHE variations that are ACHE single 2 fans (see Fig. 4), ACHE parallel 4 fans (see Fig. 5) and ACHE series 4 fans (see Fig. 6).

Based on the picture above, quantitative data is made to facilitate analysis, as shown in Table 5. The table shows the operation of the ACHE series 4 fan has the lowest lubricating temperature. ACHE operation which produces the lowest lubricant outlet

Table 4 ACHE outlet lubricant temperature simulation results for grid independence test

Mesh	Simulation temperature	Experimental temperature	Unit	Relative error (%)
180,372	323.48	317	K	–
589,153	317.10	317	K	1.97
696,300	317.31	317	K	0.07
972,404	317.47	317	K	0.05

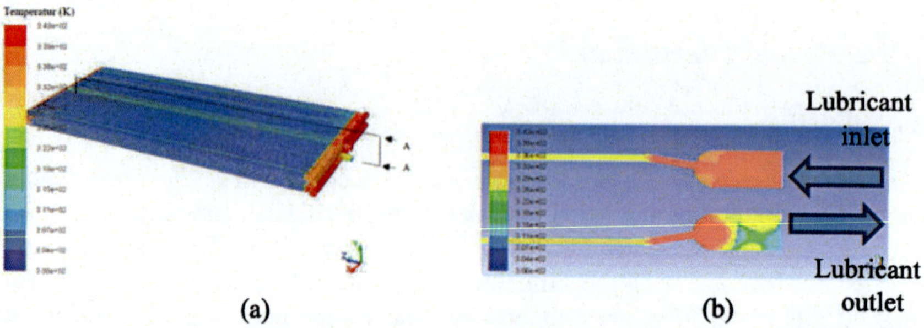

Fig. 4 **a** ACHE series 2 fan lubricant temperature contour **b**. Iso-surface lubricant temperature contour $x = -5$ (section A-A Fig. 2) inlet and outlet of lubricant

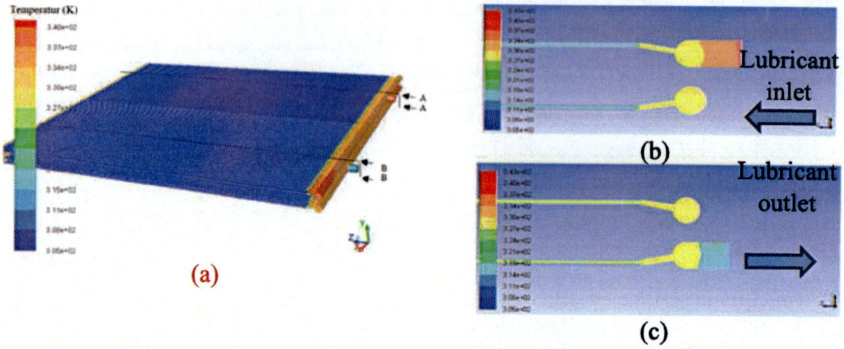

Fig. 5 **a** ACHE parallel 4 fan lubricant temperature contour **b**. Iso-surface lubricant temperature contour (section A-A) inlet of lubricant **c**. Iso-surface lubricant temperature contour (section B-B) outlet of lubricant

temperature with the ACHE 4 fan operating series model, which is 312.02 K. If one of the fans fails to operate it is still within safe limits because the ACHE lubricant outlet temperature is 315.31 K because it is far from the lubricant outlet temperature limit of 320 K.

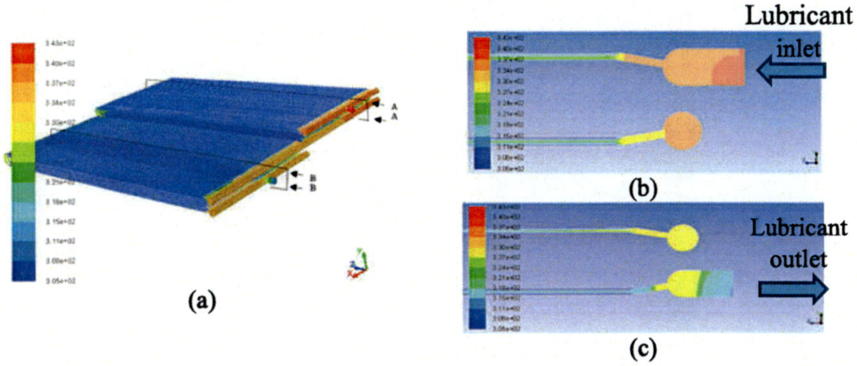

Fig. 6 **a** ACHE series 4 fan lubricant temperature contour **b**. Iso-surface lubricant temperature contour (section A-A) inlet of lubricant **c**. Iso-surface lubricant temperature contour (section B-B) outlet of lubricant

Table 5 Quantitative results of CFD on various operating models of ACHE

Simulation	Operating Model	ACHE inlet lubricant temperature (K)	ACHE outlet lubricant temperature (K)	ACHE outlet temperature limit (K)
1	ACHE single operasi 2 fan	343	317.47	320
2	ACHE seri operasi 4 fan	343	312.02	
3	ACHE seri operasi 3 fan	343	315.31	
4	ACHE parallel operasi 4 fan	343	314.12	
5	ACHE parallel operasi 3 fan	343	315.94	

4.4 Pressure Distribution

The following are the results of the cfd simulation with ACHE single 2 fans (see Fig. 7).

In actual conditions, ACHE single 2 fan has ACHE inlet lubricating pressure of 2.6 bar (260,000 Pa) and ACHE outlet lubricant pressure of 2 bar (20,000 Pa). Based on CFD simulations with various operating models ACHE still meets the minimum pressure limit for lubricant to enter the bearing generator, for more details can be seen in Table 6.

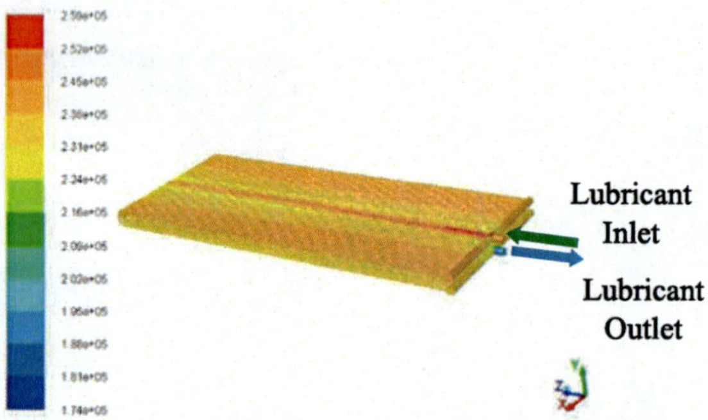

Fig. 7 ACHE lubricant pressure contours in ACHE single 2 fans

Table 6 ACHE lubricant pressure in various operating models

Simulation	Operating Model	ACHE inlet lubricating pressure (Pa)	ACHE outlet lubricant average pressure (Pa)	ACHE pressure limit (Pa)
1	ACHE single operasi 2 fan	260,000	198,252	150,000
2	ACHE seri operasi 4 fan	260,000	182,351	
3	ACHE seri operasi 3 fan	260,000	181,630	
4	ACHE parallel operasi 4 fan	260,000	169,341	
5	ACHE parallel operasi 3 fan	260,000	165,441	

4.5 Cooling Capacity

Cooling capacity is a measure of the ability of the cooling system to dissipate heat, which in this study is ACHE's ability to remove heat contained in the lubricant. After performing the CFD simulation, the cooling capacity is obtained in various ACHE operating configurations, this can be seen in Table 7.

Based on Table 7, it can be seen that the ACHE 4 fan operation series produces the highest cooling capacity compared to other operating patterns, this is due to a larger cooling surface area, more coolant flow and gradual cooling.

4.6 Operating Costs

The costs required to operate ACHE based on the number of fans operating for 1 year can be seen in Table 8.

Table 7 Cooling capacity of ACHE in various operations

Simulasi	Model Pengoperasian	Cooling capacity (kW)
1	ACHE single operasi 2 fan	204.02
2	ACHE seri operasi 4 fan	247.57
3	ACHE seri operasi 3 fan	221.28
4	ACHE parallel operasi 4 fan	230.79
5	ACHE parallel operasi 3 fan	216.25

Table 8 ACHE operating costs based on the number of fans operating

Simulation	Operating model	Operating cost per year
1	ACHE seri operasi 4 fan	Rp189,741,600.00
2	ACHE seri operasi 3 fan	Rp142,306,200.00
3	ACHE parallel operasi 4 fan	Rp189,741,600.00
4	ACHE parallel operasi 3 fan	Rp142,306,200.00
5	ACHE single operasi 2 fan	Rp94,870,800.00

In terms of operating costs, the ACHE series 4 fan has the largest value, but the operating value can be less significant when compared to the reliability obtained with the ACHE series 4 fan operation.

5 Conclusion

The simulation results of Computational Fluid Dynamics on single, parallel and series ACHE arrangements, it is obtained: ACHE single operation 2 fans have the lowest cooling capacity so that the ACHE outlet lubricant temperature is highest but the resulting pressure drop is the smallest. ACHE parallel operation of 4 fans has a larger cooling capacity than ACHE parallel operation of 3 fans because of the larger cooling mass flow. However, ACHE parallel operation of 4 fans and 3 fans has almost the same pressure drop. ACHE operation series 4 fans have the largest cooling capacity compared to various ACHE operating models with a large pressure drop but for ACHE lubricant outlet connections it is still within safe limits. The most optimal operating method is ACHE installed series 4.

References

1. Rosady SDN, Dwiyantoro BA (2014) Re-design of cooler lubricants in gas turbines with thermodynamics and heat transfer analysis, Sepuluh Nopember Institute of Technology, Surabaya

2. Fahmy MFM, Nabih HI (2016) Impact of ambient air temperature and heat load variation on the performance of air-cooled heat exchangers in propane cycles in LNG plants—Analytical approach, Elsevier, Kairo
3. Carluccio E, Starace G, Ficarella A, Laforgia D (2005) Numerical analysis of a cross-flow compact heat exchanger for vehicle applications. Elsevier, Appl Thermal Eng 25:1995–2013
4. Wibowo AK, Dwiyantoro BA (2014) Numerical study of cooling performance improvement on lube oil cooler gas turbine arranged in series and parallel with variation of lube oil flow capacity, Institut Teknologi Sepuluh Nopember (ITS), Surabaya
5. Incropera FP, Dewitt DP, Bergman TL, Lavine AS (2007) Fundamentals of heat and mass transfer, Seven. Willey, New York
6. Incropera FP, Dewitt DP, Bergman TL, Lavine AS (2012) Operating and maintenance manual ACHE, Denco Lubrication ltd, Hereford

Optimization of Coal Blending with Backpropagation Neural Networks (BPNN) and Genetic Algorithms (GA) in Tangential In-Furnace Blending Boilers

Mohamad Kurnadi(✉), Sutikno, and M. Khoirul Effendi

Department of Mechanical Engineering, Institut Teknologi Sepuluh Nopember, Surabaya 60111, Indonesia
moku05@gmail.com

1 Introduction

There is a power plant built by PT. PLN in Phase 1 Fast Track Program (FTP 1) with a capacity of 3 × 315 MW. The power plant was built to support the program to see energy using low-rank coal (LRC) so that it can reduce the cost of supplying electricity. The fuel design used in the power plant is low range coal (LRC) with a design calorific value of 4200 kcal/kg and maximum consumption of 220 tons/hour [1]. Although by design the power plant uses LRC, in fact, electricity production cannot reach its maximum if only using LRC with that's design calorific value. Therefore, it is hoped that to obtain maximum electricity production, a solution is needed to obtain coal that is suitable for power plant. Coal has a very important role in determining the production and combustion characteristics of a power plant. Although in principle each power plant has a coal design with a certain quality, the availability of coal that can be used by the power plant is of various sources and qualities.

Coal blending is the process of mixing high-quality coal and low-quality coal to obtain medium-quality coal. The coal blending process is an alternative in overcoming the problems that are often faced by a power plant that gets coal that does not meet specifications [2]. In fact, the availability of coal at a power plant consists of various types of coal from various sources and suppliers so that they have the same quality. Variations in coal quality will affect the production, efficiency, and emissions parameters of a power plant.

Previous research has been carried out by Nia [3], the effect of the furnace blending method with the composition of two types of coal on the tangential boiler combustion process. The comparison of the composition of medium rank coal (MRC) and low-rank coal (LRC) used is 50:50, by feeding LRC coal to two burner sets and MRC coal to the other two burner sets. The research methodology is using simulations in that the results of feeding MRC coal to burners A and B (lower burner levels) and LRC coal at burner levels C and D (upper burner levels) show the smallest heating area and more efficient combustion. In this study, variations of MRC and LRC coal were

M. Kolhe et al. (eds.), *Recent Advances in Renewable Energy Systems*,
Lecture Notes in Electrical Engineering 876,
https://doi.org/10.1007/978-981-19-1581-9_15

carried out as well as the effect of coal composition on coal burners to produce output such as load, combustion efficiency in the boiler, flue gas temperature, NO_x, SO_x, and optimal unburned carbon with the backpropagation neural network and genetic algorithm method. With this research, it is expected to get the optimal composition and understand the effect of coal blending so that it is expected to get a better performance which results in maximum electricity production and low exhaust emissions (Fig. 1).

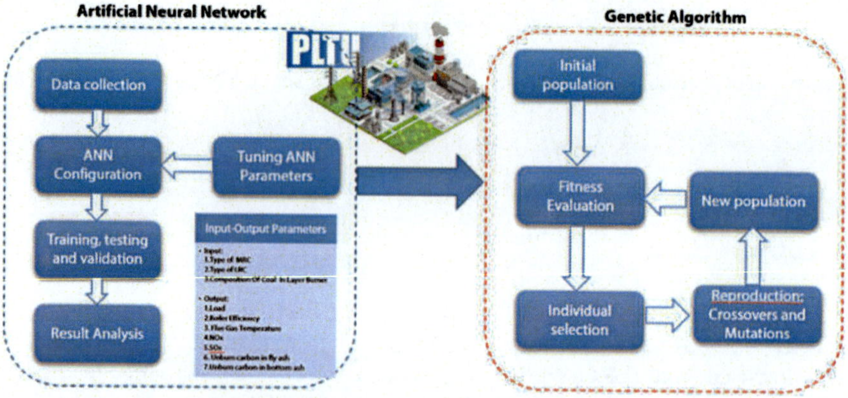

Fig. 1 Modeling scheme with backpropagation neural networks and genetic algorithms

2 Methodology

In this research, we use backpropagation neural networks and genetic algorithms. The backpropagation neural network used in this research is to obtain a mathematical model that connects the input with the output which then becomes a fitness function for the genetic algorithm to find the optimal output value (Fig. 2).

Artificial neural networks is a classification algorithm whose working principle mimics human biological neural networks [4]. The process in the artificial neural network starts from the input each of which has a weight and then ummed by a propagation function (summing function). The result will then be compared with the threshold value by the activation function. The picture above is an illustration of a simple neural network model with a neuron that processes n inputs ($x1$, $x2$, ..., xn) which has a weight of w ($w1$, $w2$, ..., wn) with a bias weight of b, an activation function f and an output y, the mathematical function of the below model is [5]:

$$y = b + \sum_{i=1}^{n} x_i w_i \tag{1}$$

One of the most widely used artificial neural network models is the backpropagation neural network (BPNN). This method is an excellent method in dealing with complex pattern recognition [5] such as beam angle optimization in intensity-modulated radiotherapy treatment (IMRT) planning [6], used in end milling to predict the surface roughness

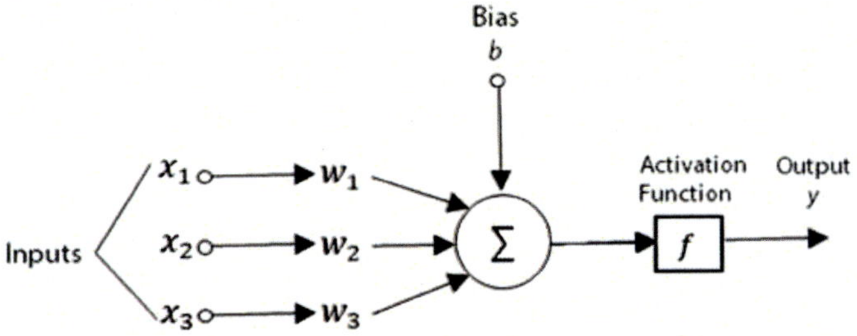

Fig. 2 Simple model of backpropagation neural network

of the Ti-6AI-4 V alloy material [7], predicting the quality of carbon fiber reinforced polymer (CFRP) end milling process [8], find the correlation between parameter inputs and the design responses on Airless Tires Design [9]. Genetic algorithm is a way to find a suitable and optimal solution to a problem, following the mechanisms of natural genetics and natural selection [10].

Data collection in this study is based on actual data on the Power Generation Performance Test which is carried out regularly every month. Due to the purpose of this research is to find the optimization of coal blending MRC, LRC, and burner layer composition in the tangential boiler, the data requirements in this study are only limited to variations of MRC, LRC, and layer composition as input variations as well as flue gas temperature, NO_x content, SO_x, and unburned carbon are used as the output response of the input variation (Table 1).

Table 1 Design of experiment

Input parameter	Level 1	Level 2	Level 3
MRC	MRC1	MRC2	MRC3
LRC	LRC1	LRC2	LRC3
Composition of coal in burner layer	1st composition	2nd composition	–

where:

MRC 1	Coal supplied by BA company
MRC 2	Coal supplied by Dizamatra company
MRC 3	Coal supplied by PLNBB company
LRC 1	Coal supplied by Arutmin company
LRC 2	Coal supplied by EEI company
LRC 3	Coal supplied by PLNBB company
Burner Layer Composition 1	Composition with MRC on the bottom layer
Burner Layer Composition 2	Composition with MRC on the top layer (Fig. 3).

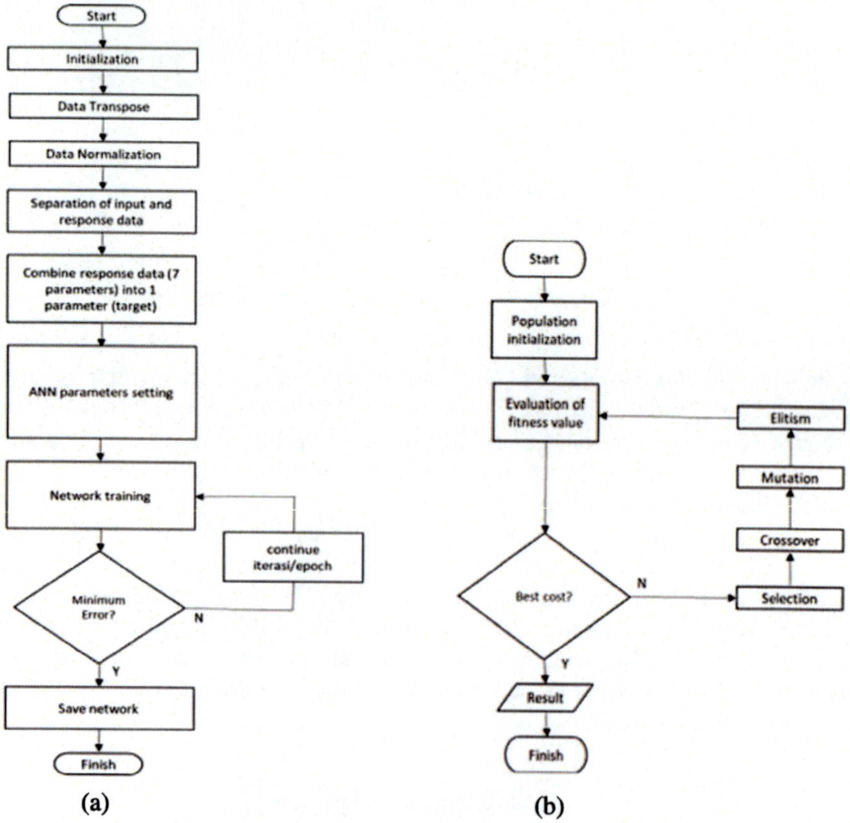

Fig. 3 **a** Flowchart of BPNN, **b** flowchart of genetic algorithm

3 Result and Discussion

Based on data collection, problem limits and conformity with the design of the experiment table, the data was obtained as shown in Table 2. In order for the data to be entered into the modeling using Matlab software, data processing must first be carried out by coding and normalization. Normalization of this data aims so that all data have the same range of values. In this study, data normalization was carried out with a data range of −1 to 1 using the equation:

$$x' = \frac{x - \min(x)}{\max(x) - \min(x)}(\text{new}_{\max} - \text{new}_{\min}) + \text{new}_{\min} \tag{2}$$

where:

x' normalized value
x initial value before normalization

Table 2 Variations of input and response targets for coal blending

NO	Input			Response						
	Type of MRC	Type of LRC	Composition of coal in layer burner	Load (MW)	Boiler EFF (%)	Flue gas temp (°C)	NO_X (mg/Nm^3)	SO_X (mg/Nm3)	Carbon in fly ash (wt %)	Carbon in bottom ash (wt %)
1	BA	ARUTMIN	LRC-LRC-MRC-MRC	298	84.12	156.57	175.03	499.94	0.53	0.52
2	BA	ARUTMIN	LRC-LRC-MRC-MRC	290	83.03	172.50	105.00	328.00	0.23	0.93
3	BA	ARUTMIN	MRC-MRC-LRC-LRC	255	84.99	180.98	142.25	62.75	9.01	5.91
4	BA	ARUTMIN	MRC-MRC-LRC-LRC	295	85.87	178.15	126.25	256.50	2.56	7.85
5	BA	ARUTMIN	MRC-MRC-LRC-LRC	292	83.41	155.07	189.43	498.65	0.34	0.69
6	BA	ARUTMIN	MRC-MRC-LRC-LRC	286	81.52	134.84	135.78	426.48	0.16	1.03
7	BA	EEI	LRC-LRC-MRC-MRC	294	82.15	168.26	156.42	362.00	0.11	0.90
8	BA	EEI	LRC-LRC-MRC-MRC	296	81.31	135.65	159.91	178.63	0.31	0.82
9	BA	EEI	LRC-LRC-MRC-MRC	293	82.32	173.57	156.42	362.00	0.14	1.28
10	BA	EEI	LRC-LRC-MRC-MRC	296	82.06	137.72	222.38	405.66	0.22	1.11
11	BA	EEI	LRC-LRC-MRC-MRC	295	82.59	164.71	132.62	604.17	0.22	1.11
12	BA	EEI	MRC-MRC-LRC-LRC	290	79.73	172.50	105.00	328.00	0.23	0.93
13	BA	PLNBB LRC	LRC-LRC-MRC-MRC	298	87.78	154.71	255.51	716.03	0.14	0.98
14	BA	PLNBB LRC	LRC-LRC-MRC-MRC	301	83.29	158.56	224.73	765.32	0.54	0.47

(continued)

Table 2 (*continued*)

NO	Input			Response						
	Type of MRC	Type of LRC	Composition of coal in layer burner	Load (MW)	Boiler EFF (%)	Flue gas temp (°C)	NO_X (mg/Nm^3)	SO_X (mg/Nm3)	Carbon in fly ash (wt %)	Carbon in bottom ash (wt %)
15	BA	PLNBB LRC	LRC-LRC-MRC-MRC	292	85.48	157.33	255.28	490.13	0.25	0.11
16	BA	PLNBB LRC	LRC-LRC-MRC-MRC	291	85.95	157.73	272.88	496.60	0.30	0.24
17	BA	PLNBB LRC	LRC-LRC-MRC-MRC	292	84.06	168.89	187.92	111.97	0.74	0.71
18	BA	PLNBB LRC	MRC-MRC-LRC-LRC	300	85.47	165.86	107.60	603.17	0.33	0.82
19	BA	PLNBB LRC	MRC-MRC-LRC-LRC	290	85.96	154.13	183.42	694.70	0.75	0.15
20	BA	PLNBB LRC	MRC-MRC-LRC-LRC	287	83.63	168.59	269.92	574.81	0.14	0.98
21	BA	PLNBB LRC	MRC-MRC-LRC-LRC	289	83.46	172.51	246.89	277.85	0.74	0.71
22	DIZAMATRA	EEI	LRC-LRC-MRC-MRC	290	81.49	161.10	248.59	449.23	1.45	0.81
23	DIZAMATRA	PLNBB LRC	LRC-LRC-MRC-MRC	298	83.60	154.71	341.00	410.85	0.40	1.43
24	DIZAMATRA	PLNBB LRC	LRC-LRC-MRC-MRC	295	82.83	157.55	210.77	263.41	0.41	0.81
25	DIZAMATRA	PLNBB LRC	LRC-LRC-MRC-MRC	299	83.06	155.17	275.32	358.94	0.68	1.88
26	DIZAMATRA	PLNBB LRC	LRC-LRC-MRC-MRC	291	83.11	160.41	201.88	183.77	0.25	0.11
27	PLNBB MRC	EEI	LRC-LRC-MRC-MRC	286	82.32	125.07	161.21	381.99	0.20	0.98
28	PLNBB MRC	PLNBB LRC	LRC-LRC-MRC-MRC	291	82.77	160.35	254.50	254.19	0.10	0.16

(*continued*)

Table 2 (*continued*)

NO	Input			Response						
	Type of MRC	Type of LRC	Composition of coal in layer burner	Load (MW)	Boiler EFF (%)	Flue gas temp (°C)	NO_X (mg/Nm^3)	SO_X (mg/Nm3)	Carbon in fly ash (wt %)	Carbon in bottom ash (wt %)
29	PLNBB MRC	PLNBB LRC	LRC-LRC-MRC-MRC	291	83.51	155.10	256.32	137.64	0.11	0.41

$\min(x)$ smallest value of data
$\max(x)$ largest value of data
new_{max} maximum limit of normalization (1)
new_{min} minimum limit of normalization (−1).

The output target in this study consists of seven variables, namely, load, boiler efficiency, flue gas temperature, NO_x, SO_x, unburn carbon in fly ash and bottom ash and to facilitate the neural network in making mathematical functions that connect the input and target in network training, it is necessary to combine all the outputs into one variable. The formulation of the incorporation of the target network used in this study is as follows:

$$\begin{aligned}\text{Target} = {} & (-0.25 \times \text{Gross Load}) - (0.25 \times \text{Efficiency}) \\ & + (0.1 \times \text{Flue Gas Temperature}) + (0.1 \times \text{NOx}) \\ & + (0.1\text{SOx}) + (0.1 \times \text{Unburn carbon in fly ash}) \\ & + (0.1 \times \text{Unburn carbon in bottom ash})\end{aligned} \tag{3}$$

In this study, experiments were carried out by performing various numbers of neurons, hidden layers, and activation functions to get the smallest error value. The mean square error (MSE) value obtained in the study is 0.013077 which is obtained from the network model with a total of 4 neurons in each hidden layer, five (5) hidden layers and the activation function used is tansig. MSE is a method used to measure the error rate in a forecasting model whose calculations follow the formula:

$$\text{MSE} = \frac{\sum_{i=0}^{n}(y_i - Y_i)^2}{N} \tag{4}$$

where:

y_i actual value
Y_i prediction value
N number of data.

The correlation coefficient (R) of the network formed as a whole is 0.53456. The correlation coefficient is the level or measure of the strength of the linear relationship between the variables X and Y which in this study indicates the relationship between the predicted value of the backpropagation neural network and the given target value. The correlation coefficient obtained in this study is considered not good enough (good if $R = 1$) which indicates that the predicted value of the resulting network is not good enough. Several things that can cause this are the lack of precise BPNN parameters or the data used in the training is not much and good (Table 3).

As the input problems in the backpropagation neural network model, the input problems used in the genetic algorithm are variations in the type of coal MRC, LRC , and burner layer composition which are used to produce optimal output. The optimization function used in the genetic algorithm is an optimization function with one goal that is an optimal value which in this study is a combination of output variations in the form

Table 3 Neural network inputs and targets

NO	Input			Target
	Type of MRC	Type of LRC	Composition of coal in layer burner	
1	−1	−1	−1	−0.42,336
2	−1	−1	−1	−0.31,611
3	−1	−1	1	0.25,474
4	−1	−1	1	−0.29,789
5	−1	−1	1	−0.30.741
6	−1	−1	1	−0.25,844
7	−1	0	−1	−0.27,033
8	−1	0	−1	−0.40.335
9	−1	0	−1	−0.24,054
10	−1	0	−1	−0.31,951
11	−1	0	−1	−0.26,456
12	−1	0	1	−0.11,114
13	−1	1	−1	−0.52,447
14	−1	1	−1	−0.28,068

(*continued*)

Table 3 (*continued*)

NO	Input			Target
	Type of MRC	Type of LRC	Composition of coal in layer burner	
15	−1	1	−1	−0.39,154
16	−1	1	−1	−0.38,718
17	−1	1	−1	−0.40.023
18	−1	1	1	−0.52,019
19	−1	1	1	−0.40.145
20	−1	1	1	−0.12,546
21	−1	1	1	−0.2202
22	0	0	−1	−0.08,076
23	0	1	−1	−0.26,182
24	0	1	−1	−0.33,933
25	0	1	−1	−0.28,999
26	0	1	−1	−0.35,491
27	1	0	−1	−0.33,459
28	1	1	−1	−0.27,145

(*continued*)

Table 3 (*continued*)

NO	Input			Target
	Type of MRC	Type of LRC	Composition of coal in layer burner	
29	1	1	−1	−0.36,113

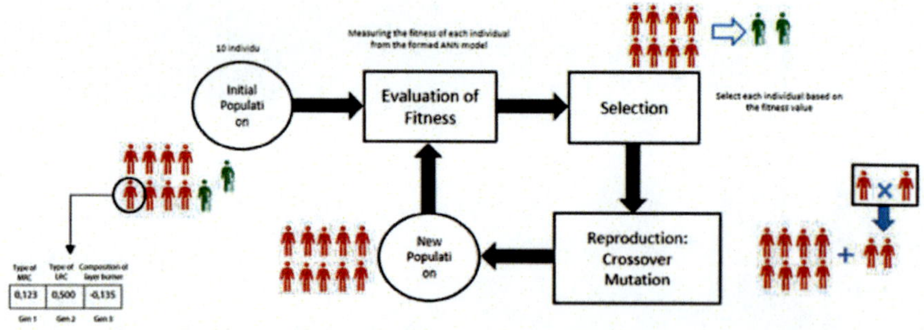

Fig. 4 Overview of the coal blending optimization process with genetic algorithms

of load, boiler efficiency, flue gas temperature, NO_x, SO_x, unburn carbon in fly ash, and bottom ash.

The initialization process in the coal blending problem is the process of generating the initial generation in a population. The individuals in this research model are random real numbers consisting of 3 genes that indicate the type of coal MRC, LRC, and their composition in the burner layer whose combination is limited to a lower limit of −1 and an upper limit of 1 (according to the normalized value of the backpropagation neural network input). The evaluation process for calculating the fitness value of each individual that has been randomly generated at the population initialization stage obtained from the backpropagation neural network. The optimization used in this study is to find the minimum value of the individual who has been raised so that the evaluation process is the process of finding the individual with the smallest value. The selection process is the process of selecting individuals with the best fitness values and conversely eliminating individuals who have low fitness values in a population. After that, the individual will be reproduced by crossover and mutation.

The optimal value generated from the two algorithm models is −0.39402 resulting from blending MRC coal types from the BA company, LRC coal from PLNBB LRC company and variations in the composition of the burner layer with MRC coal in the lower layer and LRC coal in the upper layer (LRC-LRC-MRC-MRC). Based on this optimal value, we can also predict the output by breaking down and denormalizing these values based on the weights and normalization formulas that have been set previously (Eqs. 2 and 3). The results of the denormalization of the optimal value of −0.39402 will produce a load of 280 MW, boiler efficiency of 84.15%, flue gas temperature of 151.92 ^{O}C, NO_x of 21.35 mg/Nm^3, SO_x of 400.19 mg/Nm^3, unburned carbon in fly and bottom ash of 4.38 and 3.83%wt (Figs. 4 and 5).

4 Conclusion

The results show that artificial neural network is a machine learning whose learning is supervised by input and output data provided so that the mathematical function of both can be known. In training the backpropagation neural network with five hidden layers, the number of neurons per layer is 4 and the type of transfer function "tansig"

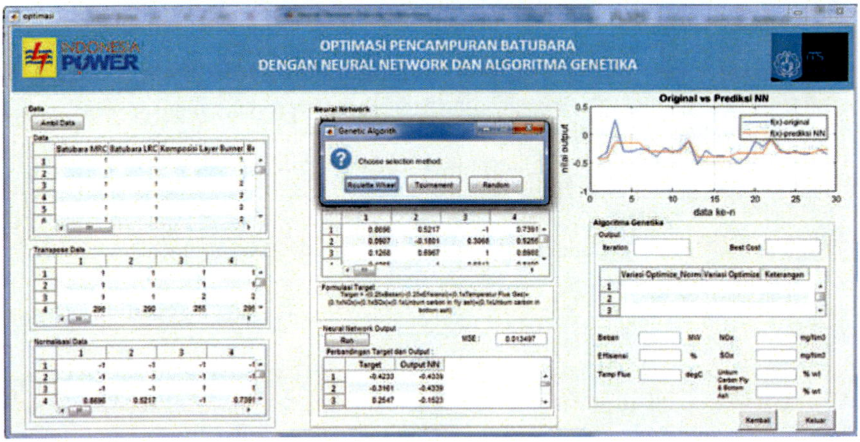

Fig. 5 Coal blending optimization system interface with backpropagation neural network and genetic algorithms

result the network with the MSE is 0.0130177 and correlation coefficient 0.53456 which indicates that the network obtained is not good enough. This is because the data used is incomplete, abundant, less accurate, and precise. In this study, the mathematical function generated by the backpropagation neural network is used as a fitness function in the genetic algorithm. Based on the results obtained in this optimization system, it was found that the coal blending of MRC 1 (BA company), LRC 3 (PLNBB company) and the layer burner composition 1 (composition of MRC in the lower burner layer and LRC in the upper burner layer) produce optimal output (value −0.39402) which is predicted to produce a load of 280 MW, boiler efficiency of 84.15%, flue gas temperature 151.92 ^{O}C, NO_x 21.35 mg/Nm^3, SO_x 400.19 mg/Nm^3, unburned carbon in fly and bottom ash 4.38 and 3.83%wt.

References

1. Dongfang electric corporation: boiler operation manual (2011)
2. Suprapto S (2009) Blending coal for power generation. Research and Development Center for Mineral and Coal Technology (tekMIRA), Bandung
3. Ariningtyas N (2014) Numerical study of the effect of feeding configuration of two types of coal with in-furnace blending method on combustion process in tangential boilers. Department of Mechanical Engineering, Faculty of Industrial Technology, Sepuluh Nopember Institute of Technology (ITS), Surabaya
4. Badrul M (2016) Optimization of neural networks with genetic algorithms for predicting post-conflict local election results. Bina Insani ICT Journal. 3(1):229–242
5. Agustin M, Prahasto T (2012) The use of backpropagation neural networks for selection of new student admissions in the computer engineering department at Sriwijaya state polytechnic. Jurnal Sistem Informasi Bisnis 02
6. Dias J et al (2013) A genetic algorithm with neural network fitness function evaluation for IMRT beam angle optimization. Springer, Berlin

7. Soepangkat BOP et al (2018) Multi-objective optimization in drilling Kevlar Fiber reinforced polymer using grey fuzzy analysis and backpropagation neural network–genetic algorithm (BPNN–GA) approaches. Int J Precis Eng Manufact
8. Effendi MK et al (2019) The combined methodology of backpropagation neural network with genetic algorithm to optimize delamination factor and surface roughness in end-milling of carbon fiber reinforced polymer composites. AIP Conf Proc 2187:030006
9. Pramono AS, Effendi MK (2019) Optimization in airless tires design using backpropagation neural network (BPNN) and genetic algorithm (GA) approaches. AIP Conf Proc 2187:050001
10. Kusumadewi S (2004) Building an artificial neural network using matlab and excel link. Graha Ilmu, Yogyakarta

Exergy Analysis in Gas Turbine Power Plant with Different Offline Compressor Washing Methods

Arif Budianto(✉) and D. Bambang Arip

Department of Mechanical Engineering, Institut Teknologi Sepuluh Nopember, Surabaya 60111, Indonesia
arifbudi.abdullah@gmail.com

1 Introduction

Starting in early 2019, Unit 1 Blok 2 Muara Karang Combine Cycle Power Plant made a few changes to the offline compressor washing method which is usually in previous years. In the previous washing method, when carrying out the offline compressor washing process, the air extraction valve was opened, so the water used for washing the compressor would be removed after stage #6 and stage #11 of the compressor. Meanwhile, in the advance blade washing method, when performing the offline compressor washing process, the air extraction valve is closed so that the water used for the compressor washing process is not removed through the air extraction valve but instead go to the last stage of the compressor and then released through in the combustor chamber drain. Figure 1 is the air extraction design in Unit 1 Blok 2 Muara Karang Combine Cycle Power Plant. Klaus Burn et al. conducted an experiment by doing compressor washing using various levels of water and detergent purity. The results showed that the dirt removed from the blades will be deposited in the final stage when the cleaning liquid evaporates. Compressor fouling results in reduced power output and overall engine efficiency [1]. For example, 1% reduction in the efficiency of an axial compressor can account for a 1.5% increase in heat rate for a given power output [2]. Saravomoto performs fouling modeling on compressors comparing three engines. The modeling uses a simulation of the impact of the fouling model (1: 0.75) on compressor outlet pressure (CDP), power output, mass flow rate and thermal efficiency. In this study, the effect of applying different offline compressor washing methods will be studied more deeply using exergy analysis. The use of exergy principles is very important in understanding thermal processes. This analysis makes it possible to measure inefficient sources [3]. The main part of the application of exergy analysis is the design and optimization of thermal and chemical systems [4]. Tsatsaronis defines exergy as the maximum work that can be achieved from an energy source under conditions prevailing in a particular environment [5]. The exergy of a steady stream of a substance is equal to the maximum amount of work that can be obtained if the current is brought from its initial state to the dead state by a process in which the current only interacts with the environment [6]. In the dead state, the mechanical, thermal and chemical equilibrium conditions between the system and

M. Kolhe et al. (eds.), *Recent Advances in Renewable Energy Systems*,
Lecture Notes in Electrical Engineering 876,
https://doi.org/10.1007/978-981-19-1581-9_16

the environment are met [7]. Thamir K. Ibrahim et al. built a gas turbine model using the first and second laws of thermodynamics. The components of the system include an air compressor, combustion chamber and gas turbine. Each of these components plays a role in the efficiency of gas turbine power plants. The lower temperature of the compressor intake air will further increase the efficiency of the system. In addition, changing the combustion chamber will change the air–fuel ratio to the better condition, so that it can increase the gas turbine inlet temperature higher than before [8].

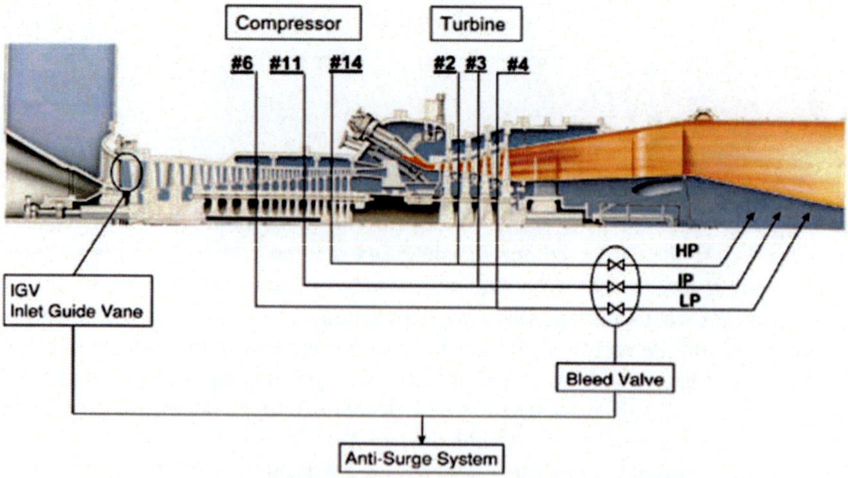

Fig. 1 Air extraction configuration of Gas Turbine M701F

2 Methodology

2.1 Energy Analysis

The energy analysis used in this study is in accordance with the Brayton cycle rules used in gas turbines where in this cycle the calculation of the gas turbine system will involve energy entering and leaving the system. The equations in this study are shown as below [9].

$$\dot{W}_c = \dot{m}_a(h_2 - h_1) \tag{1}$$

h_2 and h_1 indicate the enthalpy leaving and entering the compressor, while W_C is the work of the compressor and m_a is the flow rate of air entering the compressor [9].

$$\dot{m}_a \cdot h_2 + \dot{m}_f \cdot \text{LHV} = \dot{m}_g \cdot h_3 \tag{2}$$

$$\dot{m}_g = \dot{m}_a + \dot{m}_f \tag{3}$$

Lower heating value (LHV) is a property of fuel that varies depending on the specifications of the fuel used. In this study, the fuel used is natural gas [9].

$$\dot{W}_T = \dot{m}_g(h_3 - h_4) \tag{4}$$

h_3 and h_4 indicate the enthalpy leaving and entering the turbine gas, while W_T is the work produced by the turbine gas, and m_g is the flue gas flow rate.

2.2 Exergy Analysis

Exergy is the maximum state of a system to produce work which is limited by the dead state. The exergy equation is an analysis of the first and second laws of thermodynamics. Exergy can occur due to the nature of irreversibility in the thermal system. In addition, exergy can also be used to analyze losses thermodynamically in a thermal system.

Physical exergy. Physical exergy occurs due to differences in temperature and pressure in the observed thermal system with the dead state of the surrounding environment. The physical exergy of a thermal system can be calculated using the following equation [9]:

$$\dot{E}_x = \dot{m}_a[(h - h_0 - T_0(S - S_0 - R * LN(P/P_0))] \tag{5}$$

Chemical exergy. Chemical exergy is an exergy component associated with the difference in the chemical composition of a system with that of the environment. To calculate the chemical exergy flow in the fuel (CaHb), the following equation can be used [10].

$$\dot{E}_{\text{fuel}} = \dot{m}_f \cdot \text{LHV}(1.033 + 0.0169b/a - 0.0698/a) \tag{6}$$

2.2.1 Exergy Destruction

The rate of exergy destruction in each component can be calculated by calculating the rate of exergy in and out of the equipment [8]. Basically, the exergy flow in each equipment will decrease after experiencing a process. The exergy breakdown rate can generally be defined in Eq. (7):

$$\dot{E}_{x,\text{in}} + \dot{E}_{x,\text{out}} = \dot{I}_{\text{dest}} \tag{7}$$

Air Compressor:

$$\dot{E}_{x1} + \dot{W}_C = \dot{E}_{x2} + \dot{I}_{\text{AC}} \tag{8}$$

Combustion Chamber:

$$\dot{E}_{x2} + \dot{E}_f = \dot{E}_{x3} + \dot{I}_{\text{CC}} \tag{9}$$

Gas Turbine:

$$\dot{E}_{x3} = \dot{E}_{x4} + \dot{W}_T + \dot{I}_{\text{GT}} \tag{10}$$

2.3 Research Methodology

The research begins by collecting data on the power plant at the same time. The data were taken in two different conditions, after the old method and the new method offline compressor washing. Then the data is processed using thermodynamic equations and using Gate cycle software. The Gate cycle simulation results are validated by comparing the Gate cycle data and the data from manual calculations. After Gate cycle modeling is validated, several operating parameters are varied on the Gate cycle to see the effect of washing on the compressor. Then the data from Gate cycle modeling is analyzed to see the comparison of the old compressor washing method with the new method. The gas turbine modeling scheme in this study is shown in Fig. 2.

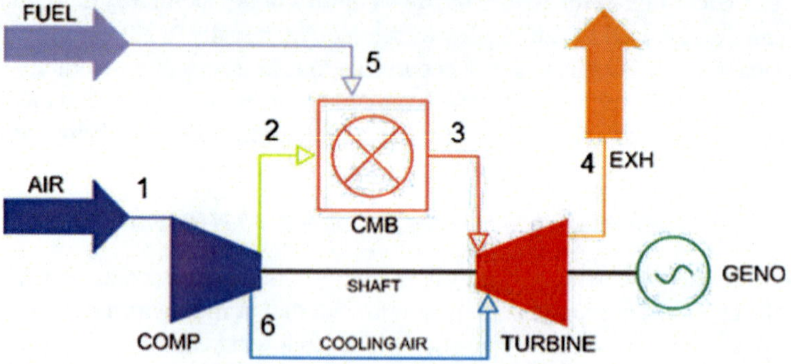

Fig. 2 Modeling scheme on gate cycle

3 Result and Discussion

The data for this study as mention before were taken from Muara Karang Combine Cycle Power Plant, Indonesia [11] (Table 1).

Tables 2 and 3, show that the exergy destruction rate on the compressor with the new compressor washing method has a smaller value. This is because with the new compressor washing method, the compressor efficiency will increase so that the exergy destruction rate of compressor will be smaller.

Tables 4 and 5, show that the exergy destruction rate in the combustion chamber with the new compressor washing method has a smaller value. This is because with the new compressor washing method, the fuel rate will be smaller so that the difference between the rate of heat entering the combustion chamber and the exergy rate of the combustion chamber will be smaller.

Tables 6 and 7, show that the rate of exergy destruction in gas turbines with the new compressor washing method has a smaller value. This is because with the new compressor washing method, the turbine inlet temperature will be smaller so that the power produced by the gas turbine and the exergy rate of the gas turbine will be smaller which results in a smaller exergy destruction rate.

Table 1 Base rating Mitsubishi M701F3 Blok 2 Muara Karang

Parameter	Unit	GT 1	GT 2
Power output	kW	253,036	253,257
Heat rate	J/kW hr	9,443,720	9,445,000
Ambient temperature	°C	29.1	29.1
Inlet pressure	m Bar	1010	1010
Exhaust flow	kg/hr	2,233,552	2,280,000
Exhaust temperature	°C	606.7	608.9
Inlet pressure drop	m Bar	2.94	2.94
Outlet pressure drop	m Bar	8.6	8.25
Fuel temperature	°C	42	42
Fuel LHV	kJ/kg	48,150	48,150

Table 2 The exergy destruction rate of compressor with the old method

T_in, °C	Exergy destruction rate, kW		
	50% load	75% load	100% load
25	14,749	16,013	18,455
27	14,852	16,124	18,582
29	14,954	16,232	18,707
31	15,055	16,340	18,834

Table 3 The exergy destruction rate of compressor with the new method

T_in, °C	Exergy destruction rate, kW		
	50% load	75% load	100% load
25	14,488	15,331	17,092
27	14,589	15,438	17,208
29	14,690	15,543	17,325
31	14,789	15,646	17,442

Figure 3 shows that at 100% loading has the highest gas turbine efficiency value. While the lowest gas turbine efficiency value is at 50% loading.

Table 4 The exergy destruction rate of combustor with the old method

T_in, °C	Exergy destruction rate, kW		
	50% load	75% load	100% load
25	145,321	176,354	209,846
27	145,863	176,992	210,565
29	146,394	177,612	211,326
31	146,926	178,230	212,088

Table 5 The exergy destruction rate of combustor with the new method

T_in, °C	Exergy destruction rate, kW		
	50% load	75% load	100% load
25	145,015	175,795	209,450
27	145,545	176,421	210,194
29	146,074	177,035	210,938
31	146,603	177,647	211,685

Table 6 Exergy destruction rate of the gas turbine with the old method

T_in, °C	Exergy destruction rate, kW		
	50% load	75% load	100% load
25	16,108	20,635	26,287
27	16,211	20,766	26,454
29	16,315	20,897	26,621
31	16,418	21,028	26,788

Table 7 Exergy destruction rate of the gas turbine with the new method

T_in, °C	Exergy destruction rate, kW		
	50% load	75% load	100% load
25	16,050	20,505	26,300
27	16,152	20,635	26,468
29	16,256	20,765	26,636
31	16,359	20,896	26,802

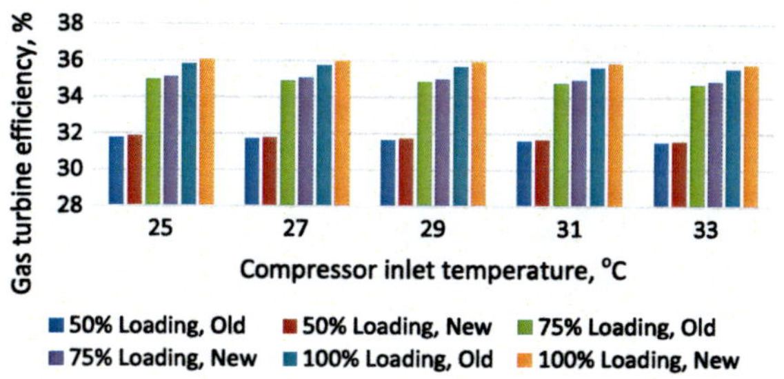

Fig. 3 Gas turbine efficiency on variations of compressor inlet temperature

4 Conclusion

The results show that there is an increase in compressor efficiency in the new compressor washing method by 0.27% at 50% loading, 0.6% at 75% loading and 1.21% at 100% loading. This increase in compressor efficiency has an impact on increasing gas turbine efficiency by 0.25% at 50% loading, 0.34% at 75% loading and 1.37% at 100% loading. In addition, the results of the study indicate that increasing inlet temperature of the compressor, the exergy destruction of the compressor, combustion chamber and gas turbine will increase.

References

1. Brun K, Grimley TA, Foiles WC, Kurz R (2015) Experimental evaluation of the effectiveness of online water washing gas turbine. ASME J Eng Gas Turb Power (137):042605
2. Scott JN (1977) Improving Turbo compressor efficiency via performance analysis techniques. ASME Paper (77):53
3. Seddigh F, Saravanamuttoo HIH (1991) A proposed method for assessing the susceptibility of axial compressors to fouling. ASME J Eng Gas Turb Power (113):595
4. Moran MJ, Sciubba E (1994) Exergy analysis: principles and practice. ASME J Eng Gas Turb Power (116):285
5. Tsatsaronis G (1993) Thermoeconomic analysis and optimisation of energy systems. Prog Energy Combus Sci (19):227
6. Kotas TJ, The exergy method of thermal plant analysis. Butterworths, London, UK
7. Bejan A, Tsatsaronis G, Kotas TJ, Moran MJ (1996) Thermal design and optimization. Wiley, Toronto, Canada
8. Ibrahim TK et al (2017) Thermal performance of gas turbine power plant based on exergy analysis. Appl Thermal Eng (115):977
9. Moran MJ, Shapiro HN (2006) Fundamentals of engineering thermodynamics, 5th edn. Wiley
10. Samosir WL et al (2015) Analisis exergy pada combustion chamber pembangkit listrik tenaga gas uap (PLTGU) teluk lembu 30 mw. Jom F. Teknik Universitas Riau, vol 2
11. Mitsubishi Heavy Industries, LTD (2000) GT Compressor Blade Washing Unit

Swappable Battery Innovation as a Drone Frame Structure with Purpose to Increasing the Flight Time Duration

Muhammad Haekal Shafi[1], Valiant Tirta Amarta[1], Ferdina Ramadhansyah[2], Puguh Pambudi[1], and Alief Wikarta[1](✉)

[1] Department of Mechanical Engineering, Institut Teknologi Sepuluh Nopember, Surabaya 60111, Indonesia
wikarta@me.its.ac.id

[2] Department of Industrial Mechanical Engineering, Institut Teknologi Sepuluh Nopember, Surabaya 60111, Indonesia

1 Introduction

In this era of the industrial revolution 4.0, technological developments have entered various sectors ranging from agriculture, infrastructure, authorities (police and army), and other sectors. In supporting the development of existing technology, currently, the use of drones or unmanned aerial vehicles (UAV) has been widely used. The use of drones is very helpful in various sectors. For example, drones have been used as a land-based mapping tool by the Center for Research and Development of Agricultural Land Resources (BBSDLP) [1], land-based mapping carried out in Indonesia has progressed quite rapidly. The drone also takes part in infrastructures monitoring [2] and air patrol by the authorities [3]. However, in the use of drones, there is one very vital drawback, namely the flight time is quite short for one battery [4].

In the development of flight time on drones, several solutions have been developed and implemented to increase flight time and battery charging time. One example of the solution is the battery of graphene developed by California lithium battery, this type of battery is very fast in charging, larger capacity with the same size, lightweight, and is environmentally friendly [4]. Another example is the drone made by the Impossible Aero Company [5], where this drone is able to fly for a long time. This drone utilizes a battery as the main structure of the frame, but the impossible drone has drawbacks, namely in terms of battery charging which is still long, and the battery cannot be swapped.

Drones use their energy to fly by generating thrust and lift. The related aspects of drone energy consumption, namely: payload weight, battery weight, drone frame weight, airspeed, and flight distance [6].

To increase the mileage, battery capacity can be added, but keep in mind that the addition of battery capacity also increases the weight of the battery which causes an increase in energy consumption [7].

Therefore, comes a new idea of swappable battery innovation as a drone frame structure with the purpose of increasing the flight time duration. Swappable battery

M. Kolhe et al. (eds.), *Recent Advances in Renewable Energy Systems*,
Lecture Notes in Electrical Engineering 876,
https://doi.org/10.1007/978-981-19-1581-9_17

packs on drones that use batteries as frames so can increase the number of batteries installed on drones which are indirectly expected to increase battery capacity and drone flight time. The concept of a swappable battery has a purpose so that the battery is easy to replace when it runs out.

Based on research conducted by Chan and Kam [8], the energy requirement of the drone is calculated by adding up the power requirement of the motor with the existing power loss then multiplied by the flying time of the drone. For the energy that the battery can supply in joules, it is calculated by multiplying the battery voltage by the battery capacity (Ah) and 3600.

Based on research conducted by Hwang et al. [9], the energy requirement of the drone is obtained by multiplying the required thrust force by the desired speed. Then, for the energy that can be supplied by a battery, it is influenced by the current issued by the battery, the change in voltage that occurs, and the rate of change in voltage that occurs in the battery.

2 Methodology

In this research, to add more battery energy capacity, 16 41,200 Li-ion battery cells are used to power the drone. Those batteries then being assembled with a 4s4p configuration so that the battery pack can meet the required specification of the motor about 14.4 V. The battery configuration also provides 236.16 Wh of energy, three times more than a regular drone battery. The battery pack is separated into four modules with a 2s2p configuration of each module. Each battery module is placed on each arm of the drone. From the explanation above, the drone illustration can be seen in Figs. 1 and 2 and the detailed specification is shown in Table 1.

Fig. 1 3D illustration swappable battery innovation and complete design

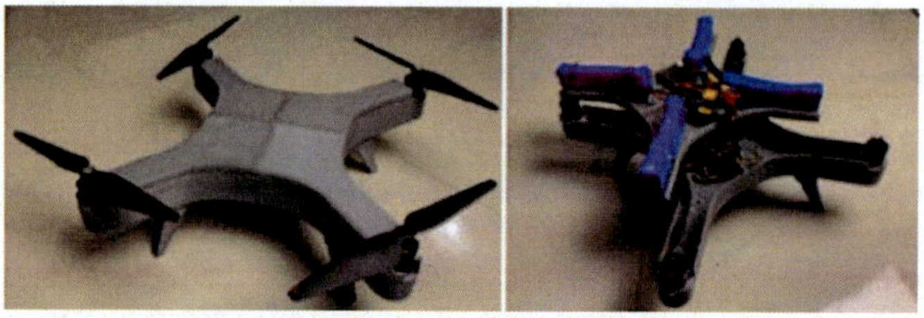

Fig. 2 Drone prototype

Table 1 Drone specification

Drone specification	
Number of rotor	4
Vehicle weight	2022 g
Frontal area	0,245 m^2
Propeller radius	130 mm
Number of rotor	4
Nominal capacity (C_0)	16,800 mAh
Rated discharge time (t_0)	22.4 min
Fully charged voltage (V_0)	16.8 V
Standard voltage (V_s)	14.4 V
Weight (W_b)	800 g

Figure 3 Flight duration test is being held by two methods, theoretical measurement and actual measurement. Theoretical flight duration measurement is done based on the research conducted by Hwang et al. [9]. The first step for the theoretical measurement is calculating the required power with the following equation

$$P_{re,\text{rotor}} = TxU_i + DxU \tag{1}$$

T is the thrust required to lift the drone,U_i is the induced speed, D is the drag force, and U is the forward or horizontal speed of the drone. In this experiment, the value of U is equal to the wind speed so that the drone can be assumed to only hover and not move. D is a function of drone drag coefficient (C_d), air density (ρ), drone frontal area, and U.

$$D = \frac{1}{2}\rho C_d A_f U^2 \tag{2}$$

U_i is a function of U, T, disk area (A), air density (ρ), the inclination angle of the drone when the drone is moving forward (θ), and U_i itself. The value of U_i is obtained by iteration process.

$$U_i = \frac{T}{2\rho A} x \frac{1}{\left(\sqrt{(U.\cos\theta)^2 + (U.\sin\theta + U_i)^2}\right)} \tag{3}$$

$$\theta = \tan^{-1}\frac{D}{W} \tag{4}$$

θ is a function of D and the total weight of the drone (W). Disk area (A) for the multirotor drone is a function of a number of rotor (N) and propeller radius (r).

$$A = N\pi r^2 \tag{5}$$

After the power required is obtained, the required battery power can be found by dividing the value of the power required from the drone with the drone motor efficiency.

Based on the manufacturer specification [10], the average efficiency with the variation of the throttle is 0.77.

$$P_{re} = \frac{P_{re,\text{rotor}}}{\eta} \tag{6}$$

To calculate the battery consumption in this drone, iteration calculation is used with the flowchart calculation as shown in Fig. 3.

The first step from the process is determining the time interval (Δt) as much as j (= 0, 1, 2, …). After that, in each iteration step, the voltage drop which is a function of fully charged voltage (V_0), nominal capacity (C_0), and the slope of the linear curve representing a battery voltage drop (k) is calculated.

$$V_{j+1} = V_0 - k(C_0 - C_j) \tag{7}$$

The value of k is a function of the fully charged voltage (V_0), standard voltage (V_s), nominal capacity (C_0), and a fraction of the nominal capacity (λ).

$$k = \frac{V_0 - V_s}{\lambda C_0} \tag{8}$$

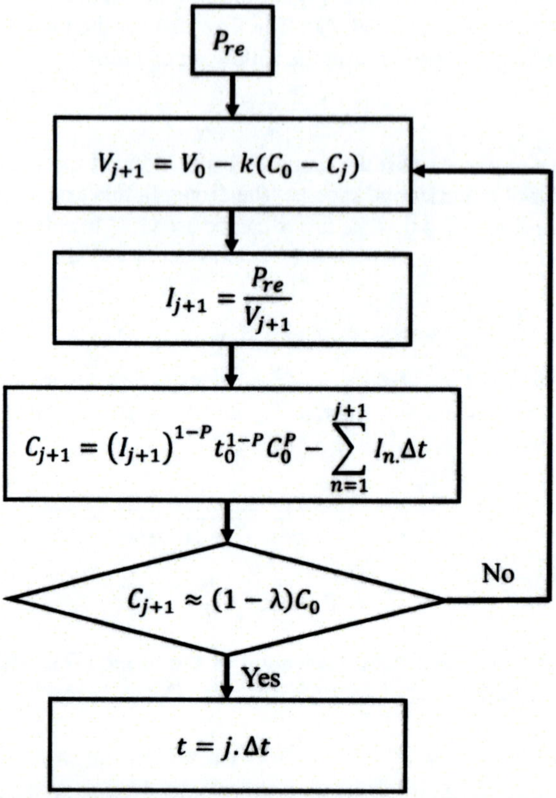

Fig. 3 Theoretical flight time calculation flowchart

After the voltage drop is obtained, the current required for the drone is calculated to meet the drone's power needs.

$$I_{j+1} = \frac{P_{re}}{V_{j+1}} \tag{9}$$

Then, the battery capacity after the time span Δt can be calculated with the following equation.

$$C_{j+1} = \left(I_{j+1}\right)^{1-P} t_0^{1-P} C_0^P - \sum_{n=1}^{j+1} I_n . \Delta t \tag{10}$$

Iteration is being done until the value of the battery capacity after the time span is equal to the minimum battery capacity. The minimum battery capacity can be calculated with the following equation.

$$C_{j+1} \approx (1 - \lambda)C_0 \tag{11}$$

For the actual measurement, to test the flight duration, the battery is fully charged at 16.8 V, then the drone is set to hover 5 m above the ground. The flight duration was, then measured until the battery reach variations of voltage, 13.6, 12.8, and 12 V. The flight duration measurement starts during the drone takeoff and stops when the drone land.

3 Results and Discussion

The methodology in this section would be explaining about the results that we got from the test. It has been achieved two types of flight duration, theoretical flight duration and actual flight duration. Both results are compared with an experiment conducted by Hwang et al. to see the relative error from two methods theoretical and actual tests.

3.1 Theoretical Flight Duration

In this subchapter, the results of theoretical methods of flight duration measurement are discussed. In this theoretical calculation, the value of Cd of the drone is assumed to be 0.81 [9], then the efficiency value of the drone's motor is 0.77 from the manufacturer's specification data. Meanwhile, the horizontal speed value of the drone (U) is assumed to be the same as the wind speed at the test site, Surabaya City, Indonesia, on August 16, 2021, at 18 km/h [11]. From these data, the theoretical flight duration value is obtained as follows (Table 2).

3.2 Actual Flight Duration

The measurement of flight duration starts when the drone takes off until the drone lands again. The throttle used during the actual test is 60%. The test was carried out in moderately windy conditions [11], so that these conditions can affect the results of

Table 2 Theoretical flight duration with the variation of end voltage

End voltage	Theoretical flight duration (min.)
13.6	24.25
12.8	29.25
12	36.67

drone testing carried out. The bigger the wind, the heavier the drone will be to move and balance while flying.

The results of the actual flight duration measurement are shown in Table 3. For test result with an end voltage of 12 V, it has been tested but the thrust generated by the motor cannot fly the drone properly after the voltage drops below 12.8 V. This happens because as the voltage decreases, the RPM of the motor will also decrease so that the power-to-weight ratio for flying the drone is not appropriate.

Table 3 Actual flight duration with the variation of end voltage

End voltage	Actual flight duration (min.)
13.6	25
12.8	30.96
12	–

3.3 Comparison

This subchapter discusses the comparison of the results of the tests that have been carried out between theoretical and actual to find out the relative errors that occurred in this study. After obtaining the error value in the testing of this research, it is then compared with the tests conducted by Hwang et al. with the aim of comparing the accuracy of the theoretical calculation equations in the study compared to the actual test results.

From Table 4, it can be seen that the value of the actual test results that we carried out was greater than the theoretical value. Meanwhile, in the research of Hwang et al., the actual value of flight duration is smaller than the theoretical value. This difference in value can occur due to the wind which can increase the payload when pushing the drone down and reduce the payload when pushing the drone up. The average error value in the study conducted by Hwang et al. is 2.41%, while from the tests, the error value is 4.26%. The error value of our measurements is still below 5% so the test can be said to be valid.

Table 5 shows the comparison of flight time duration between a drone with conventional batteries and a drone with swappable batteries as the frame structure. Both drones have the same dimensions and motor specifications. The conventional drone is shown in

Table 4 Comparison of test results with Hwang et al

Results	Fully charged voltage (V_0)	End of discharge voltage	Flight velocity (m/s)	Flight duration (min)		Error (%)
				Theoretical	Actual	
Experiment	16.8	12.8	5	29.25	30.96	5.52
	16.8	13.6	5	24.25	25	3
Hwang et al	24.5	22.2	0	22.24	22.15	0.4
	24.5	22.2	1.4	23.5	23.11	1.69
	24.5	22.2	12	23.7	22.47	5.47

Fig. 4 above. On drones with swappable batteries, the battery capacity is increased up to four times. This is because the placement of the battery takes advantage of the dimensions of the drone and becomes the structure of the drone frame itself so that the space on the drone can be maximized. However, due to the addition of this battery, the weight of the drone doubled. In terms of flight time duration, although the battery capacity has increased by four times, the actual flight time duration has only increased to 2.4 times. This happens because the drone gets heavier with the same motor specifications so that the energy consumed is also getting bigger and causes the battery to run out easily.

Table 5 Comparison between conventional drones and swappable battery drones specification

Results	Fully charged voltage (V_0)	Total weight (kg)	Nominal capacity (C_0, mAh)	Theoretical flight time (min.)	Actual flight time (min.)
Conventional	16.8	1.002	4200	13.16	12.8
Swappable	16.8	2.002	16,800	29.25	30.96

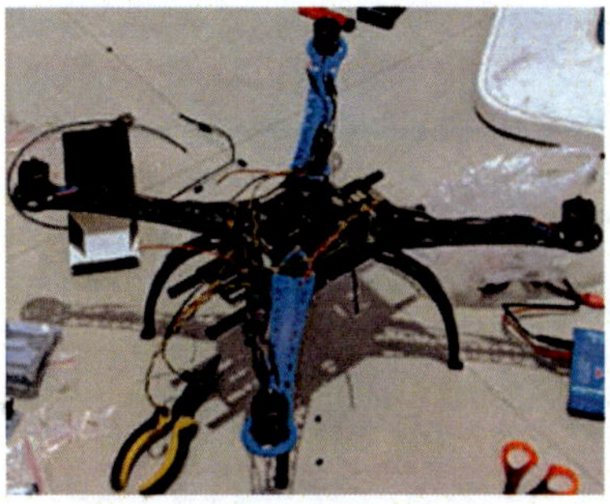

Fig. 4 Conventional drone

4 Conclusion and Recommendation

4.1 Conclusion

Based on the research that has been tested by two methods of theoretical and actual testing. This research has some conclusions which described as follows:

1. The value of flight time duration of a drone can be estimated through battery voltage, the thrust required, drone dimensions, and drone speed.
2. The use of swappable battery innovation as a drone frame structure can increase the battery capacity of a drone up to four times and flight time duration by 2.4 times.

4.2 Recommendation

The results above show that the addition of battery capacity is not comparable to the flight time duration because the flight time duration only increased by half of the battery addition. It happens because the battery addition also increases the total weight of the drone, adding more power required to fly. Thus, further study is needed to calculate the optimum battery addition and the flight time duration.

References

1. BBSDLP Newspage. http://bbsdlp.litbang.pertanian.go.id/ind/index.php/layanan-mainmenu-65/info-aktual-2/896-drone-teknologi-pembantu-pemetaan-terbaru, last accessed 9 Jan 2021
2. Buchari E, Octaviansyah D, Chairuddin MT, Saribu LD (2018) Penggunaan drone untuk mendapatkan data kecelakaan Lalu Lintas. J Indonesia Road Saf 1(3):147–156
3. Kementerian PUPR Newspage. https://binamarga.pu.go.id/index.php/berita/pengolahan-foto-drone-untuk-visualisasi-pemodelan-3d, last accessed 9 Jan 2021
4. Kardasz P, Doskocz J, Hejduk M, Wiejkut P, Zarzycki H (2016) Drones and possibilities of their using. J Civil Environ Eng 6(3):1–7
5. Vayu Aerospace Homepage. http://www.vayuaerospace.com/, last accessed 17 July 2021
6. Zhang J, Campbell JF, Sweeney DC II, Hupman AC (2020) Energy consumption models for delivery drones: a comparison and assessment. Transp Res Part D: Transp Environ 90:102668
7. Stolaroff JK, Samaras C, O'Neill ER, Lubers A, Mitchell AS, Ceperley D (2018) Energy use and life cycle greenhouse gas emissions of drones for commercial package delivery. Nat Commun 9(1):1–13
8. Chan CW, Kam TY (2020) A procedure for power consumption estimation of multi-rotor unmanned aerial vehicle. J Phys: Conf Ser 1509(1):012015
9. Hwang MH, Cha HR, Jung SY (2018) Practical endurance estimation for minimizing energy consumption of multirotor unmanned aerial vehicles. Energies 11(9):2221
10. T-motor (2018) Airgear 450 specification. T-motor web resource. https://store.tmotor.com/goods.php?id=726 Accessed 07 Aug 2021
11. BMKG Weather Forecast. https://www.bmkg.go.id/cuaca/prakiraan-cuaca.bmkg?Kota=Surabaya&AreaID=501306&Prov=35, last accessed 16 Aug 2021

Changes the Governor Valve Operation Mode to Improve Efficiency HP Turbine

Aripin Gandi Marbun(✉)

Department of Mechanical Engineering, Institut Teknologi Sepuluh Nopember (ITS), Surabaya 60111, Indonesia
arifinmarbun84@gmail.com

1 Introduction

One of the main equipment of a steam power plant is a turbine. The turbine itself is equipment that functions to produce mechanical energy both potential and kinetic up to 3000 rpm. Furthermore, it will be converted into electrical energy using a generator and an excitation system. Problems that generally occur in steam turbines are corrosive turbine blades, cracks in the blades, vibration and a decrease in turbine efficiency. The damage to the equipment is usually caused by the non-standard quality of the steam used and the selection of materials and improper installation. Meanwhile, the decrease in efficiency can be caused by equipment damage, improper operation of the turbine, and quality of water use. Based on the efficiency report of the steam power plant 2 × 200 MW unit 1 in June 2020, it is known that at a load of 196 MW, the efficiency value of the turbine HP is 79.96%, where there is a deviation from the baseline efficiency value of 6.57% [1]. This is certainly detrimental to the performance of the unit and needs to be improved so that the turbine efficiency value is closer to the baseline (see Table 1).

Table 1 Performance test unit 1 results for June 12, 2020

Parameter	Unit	Base Line	Actual	Deviation
Gross power output	MW	196	196	0.00
Heat loss method	Kcal/kwh	2625.11	2968.71	−343.61
Turbine heat rate (HHV basis)	Kcal/kwh	1991.40	2083.00	−91.60
Boiler efficiency (HHV basis)	%	83.48	77.14	6.35
HP turbine efficiency	%	**86.53**	**79.96**	**6.57**
IP turbine efficiency	%	93.59	88.42	5.17
LP turbine efficiency	%	69.15	70.81	−1.66

Given the deviation of the HP turbine efficiency from the baseline with the current performance test results, it is necessary to find a way to increase turbine efficiency. One

M. Kolhe et al. (eds.), *Recent Advances in Renewable Energy Systems*,
Lecture Notes in Electrical Engineering 876,
https://doi.org/10.1007/978-981-19-1581-9_18

way to increase turbine efficiency is to optimize the operating system by changing the parameters used. It is hoped that with this optimal operating system, the losses in the HP turbine will be smaller so that it has an impact on increasing the efficiency of the HP turbine. The test conditions during the performance test unit 1 are:

a. The unit operates with manual control and the operating turbine is single mode.
b. The desuperheater is normally and shoot blowing during data collection.
c. Gross energy production are taken from recordings on the local kWh meter in the metering room while coal consumption is taken from counter readings in the DCS central control room, and coal samples are taken from the operating tapping coal feeder.
d. The fly ash sample was taken at the air pre-heater outlet using an isokinetic coal sampling kit, while the bottom ash was taken at the SSC.
e. Sample testing or analysis is carried out at an external laboratory.

Implementation of performance tests as material for efficiency reports is carried out once a month. The goal is to know the performance of each equipment and provides data as a recommendation for improving equipment efficiency.

2 Method

These steam power plants use steam as a medium to turn turbines. The specifications contained in the turbine are as follows Table 2 [2]:

The steam turbine is a prime mover that converts the potential energy of steam into kinetic energy and then converts it into mechanical energy in the form of turbine shaft rotation. The rotation of the turbine shaft is generated due to a steady flow of steam entering the steam nozzle. The steam enters the steam jet, here the steam velocity is increased, and some of the kinetic energy of the steam is sent to the turbine blades which causes the turbine blades to rotate. The speed of the rotation turbine in this steam power plant is 3000 rpm, the higher the consumer load, the greater the need for steam to maintain turbine rotation. The size of the load is very influential on the steam that will be produced, if the load is high enough, then the amount of steam needed is also large and vice versa.

The regulation of the amount of steam that enters the turbine is carried out by the governor valve (GV) which works automatically. The turbine at this plant has three levels such as high pressure (HP) turbine, intermediate pressure (IP) turbine, and low pressure (LP) turbine, see Fig. 1. The working principle of a steam turbine is that it starts from superheated steam coming from a boiler that has high temperature and high pressure and flows into the HP turbine. Inside the turbine, there are fixed blades and moving blades which have a shape in such a way that it will expand the steam. The steam energy received by the turbine blades is used to drive the turbine shaft.

In the HP turbine, energy absorption occurs which causes the temperature and pressure of the steam to decrease. The steam coming out from the HP turbine is then forwarded back to the boiler to be reheated to increase the temperature and enthalpy of the steam fluid. After being reheated, the steam flows to the IP turbine which will drive the

Table 2 Turbine technical specification

Description	Parameter
Model	N220-12.75/535/535
Type	Subcritical pressure, intermediate reheat, three-casing, two steam discharge condensed turbines
Admitted main steam pressure/temperature	12.75 MPa/535 °C
Reheated steam pressure/temperature	2.616 MPa/535 °C
Rated Power:	200 MW
Cooling water flow at the rated condition:	36,300 t/h
Feedwater temp:	251.3 °C
Steam flow at rated power:	660 t/h
LP casing min steam discharge:	128 t/h
The number of stages:	The number of stages:
• HP casing:	• 1 governing stage + 11 pressure stages
• IP casing:	• 10 pressure stages
• LP casing:	• 2 × 5 pressure stages

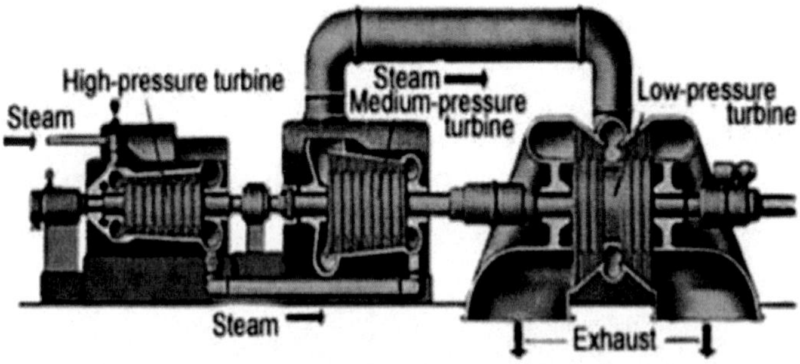

Fig. 1 Steam turbine components [3]

blades of the intermediate pressure turbine and LP turbine, so that the motion of these blades will strengthen the turbine shaft movement. After turning the turbine, the steam is condensed in the condenser to be converted back into the water. The reference power cycle and the corresponding temperature–entropy (T-s) diagram is illustrated in Fig. 2. It consists of an idealized Rankine cycle in which a working fluid produces power by flowing in a loop made of various pieces of equipment [4].

The first law of thermodynamics reads the change in energy in a closed system, equal to the amount of heat energy entering the system to the surrounding environment. The HP turbine efficiency calculation does not require a steam flow rate variable, but only uses pressure and temperature measurements that enter the HP turbine equipment.

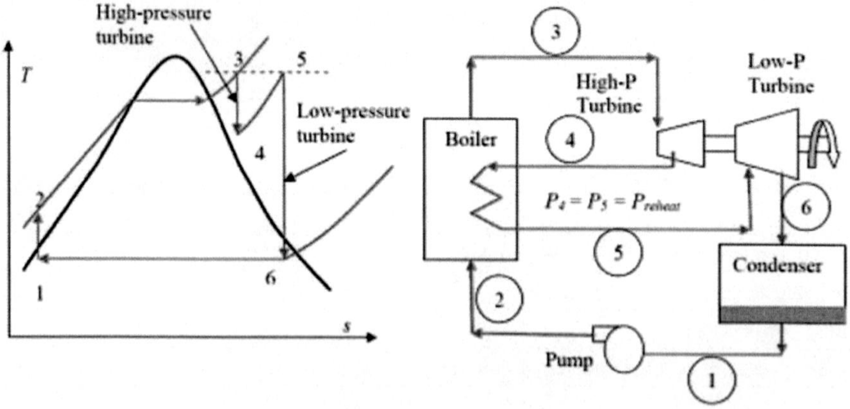

Fig. 2 Rankine cycle at steam power plant

Based on the calculation method using the first law of thermodynamics, HP steam turbine efficiency can be determined by the equation which is

$$\eta_{HP} = \frac{h_{\text{in}} - h_{\text{out}}}{\Delta h_s} \tag{1}$$

h_{in} enthalpy superheated steam on the inlet side,
h_{out} enthalpy superheated steam on the outlet side, and
Δh_s isentropic drop enthalpy.

The HP turbine efficiency calculation standard above is in accordance with the EPRI Heat Rate Improvement Reference Manual TR-109546. The steam expansion process in the HP turbine is described as follows Fig. 3:

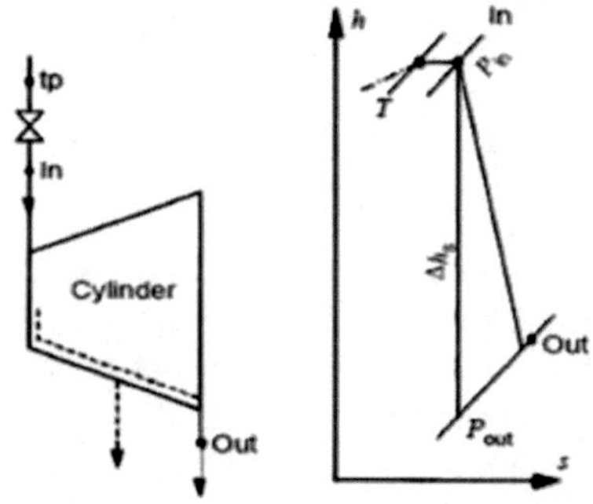

Fig. 3 Steam expansion on HP turbine [3]

Average generation for percent in heat rate change = 0.18%/1.0% efficiency decrease that cost per USD/day = \$381.77. The effect on heat rate due to HP Turbine efficiency deviation must be calculated from the computer heat balance program or calculated manually [3].

There are two kinds of operation modes for the GV in this steam power plant, there are single mode and sequence mode.

2.1 Single Mode

Single mode is operating mode with the % opening of GV numbers one until four are the same, according needs the amount of main steam flow and the set point of the main steam pressure as shown in figure below (see Fig. 4).

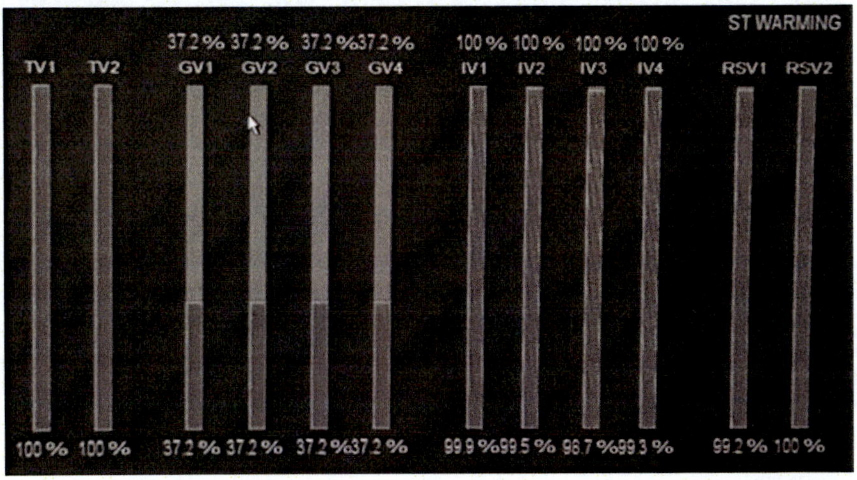

Fig. 4 Single-mode display on DCS

From Fig. 4, we can be seen that the opening of the GV one until four is the same at the position of 37.2% with a load of 165 MW. So, there are still losses in GV because all GV positions are throttled.

2.2 Sequence Mode

Sequence mode is GV operating mode that is not the same as one another, where GV 1 is first attempted to open larger than the other GV or possible there is a fully closed. As in the full load, GV 1 and 2 are fully open, GV 3 is 45% and GV 4 is 10% (see Fig. 5).

While at a load of 100 MW (load 50%) the opening of GV 1 and 2 is 75.1%, GV 3 is 17.9% and GV 4 is 0%. This causes the distribution of the main steam that enters the HP turbine is not the same. Following Fig. 5 of the sequence mode at a load of 100 MW. From the sequence mode display above, it can be confirmed that GV number 4 has no losses because the position is fully closed, so the losses that occur are only in GV numbers 1, 2 and 3. This certainly can increase the efficiency of the HP turbine because it has reduced losses in GV.

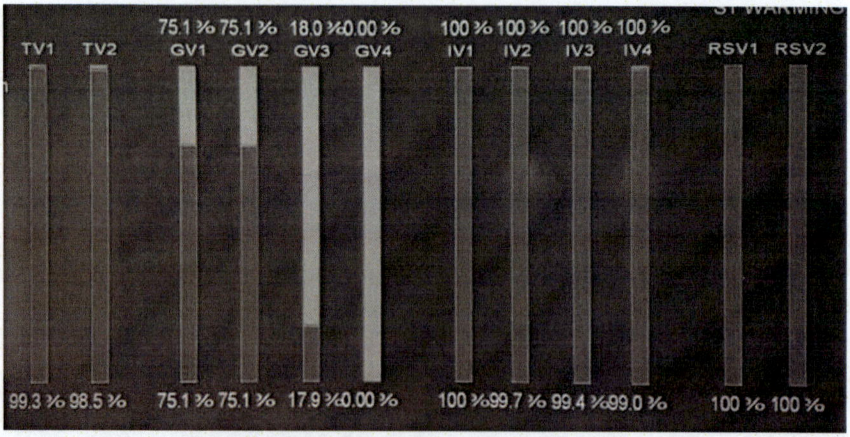

Fig. 5 Sequence mode display on DCS

3 Analysis/Results and Discussion

Starting from the initial operation when commissioning unit one in 2014, the turbine valve operating mode was carried out in single mode. The data was taken for single mode every 15 min on June 12, 2020, as follows (Table 3).

The following is a single-mode parameter data diagram for HP turbine efficiency and opening of GV numbers 1, 2, 3, and 4 at a load of 125, 130, 145, 165, 184, and 196 MW carried out on June 12, 2020 (see Fig. 6).

From the data above, we can see that the greater the unit load, the higher the efficiency of the HP turbine, as well as the opening of GV numbers 1, 2, 3, and 4.

3.1 Sequence Mode Data

On August 10, 2020, the experiment was conducted to change the turbine operating pattern from single mode to sequence mode. From the experimental results, after steady conditions, parameter data related to changes in the operating mode are taken in Table 4.

The following is a diagram of the sequence mode parameter data for HP Turbine efficiency and the opening of GV numbers 1, 2, 3, and 4 at a load of 125, 130, 145, 165, 184, and 196 MW which was carried out on August 10, 2020.

From Fig. 7, we can be seen that GV numbers 1, 2, 3, and 4 are not always the same in the opening. For GV 1 and 2, the average is full open, while GV 4 tends to be very small, even almost fully closed. This of course results in reduced losses on the GV.

3.2 Data Processing

From the single mode and sequence mode parameter data that have been taken, the HP turbine efficiency data is collected in both modes adjusted for the same load of 196 MW, resulting in the calculation results for single mode are as follows formula (1) as follows in Table 5.

Table 3 Single-mode operation parameter data

Parameter	10:00	10:15	10:30	10:45	11:00	11:15	11:30	11:45	Mean
MS temp. (°C)	532.72	535.14	532.70	532.70	532.72	535.15	531.50	535.15	533.47
Cold reheat steam temp. (°C)	331.99	333.75	332.87	332.87	332.87	334.59	331.95	333.71	333.07
Hot reheat steam temp. (°C)	534.72	534.72	534.72	534.72	535.21	536.66	533.8	536.7	535.16
MS pressure (MPa)	10.85	10.88	10.88	10.76	10.72	10.82	10.78	10.88	10.82
Cold reheat (MPa) Steam pressure	2.26	2.27	2.27	2.25	2.25	2.26	2.26	2.27	2.26
Hot reheat (MPa) Steam pressure (MPa)	2.136	2.148	2.148	2.124	2.124	2.136	2.136	2.148	2.13

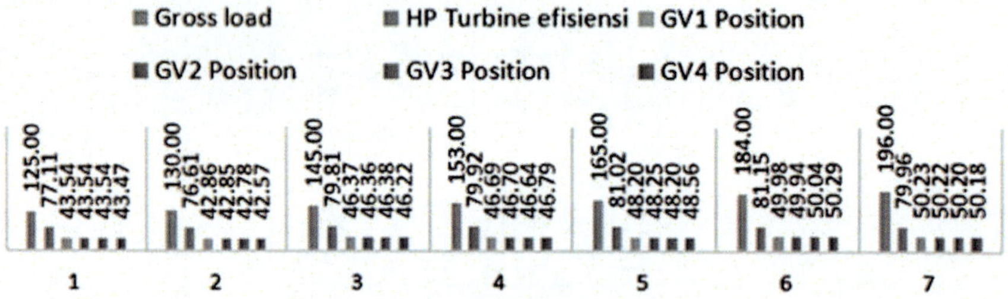

Fig. 6 Single-mode parameter graph

Table 4 Sequence mode operation parameter data

Parameter	10:00	10:15	10:30	10:45	11:00	11:15	11:30	11:45	Mean
MS temp. (°C)	531.74	532.84	532.84	530.42	534.22	530.52	531.70	529.26	531.69
Cold reheat steam temp. (°C)	335.01	335.12	335.12	334.19	336.92	334.28	335.16	332.57	334.79
Hot reheat steam temp. (°C)	536.47	537.9	537.9	536.47	537.88	536.47	537.88	536.45	537.17
MS pressure (MPa)	11.35	11.19	11.22	10.81	10.99	10.97	11.24	11.09	11.11
Cold reheat (MPa) Steam pressure	2.63	2.59	2.59	2.51	2.54	2.56	2.61	2.57	2.58
Hot Reheat (MPa) Steam pressure (MPa)	2.50	2.47	2.47	2.38	2.42	2.43	2.48	2.44	2.45

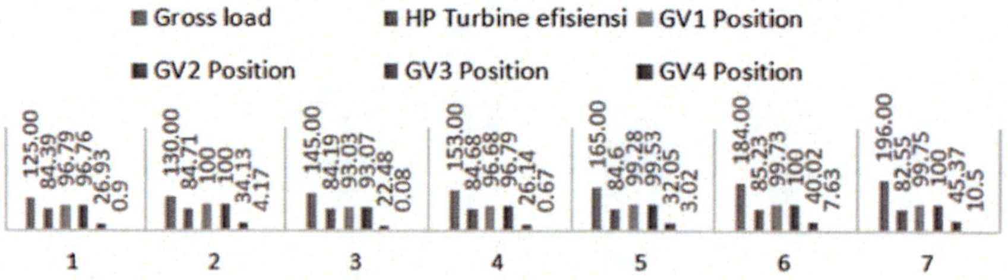

Fig. 7 Sequence mode parameter graph

4 Conclusion

From the comparison of the data for Tables 4 and 5, the efficiency value of the HP turbine single mode is 79.96%, while the efficiency value of the HP turbine sequence mode is 82.55%, resulting in an increase in the efficiency of the HP turbine by 2.59%. From

Table 5 HP turbine efficiency single mode and sequence mode

Parameter	Symbol	Unit	Single mode	Sequence mode
Main steam enthalpy	*hms (in)*	kJ/kg	3,448.29	3,443.88
Cold reheat enthalpy	*hcrh (out)*	kJ/kg	3,090.22	3,090.01
Main steam entropy	*Sms*	kJ/kg°C	6.646	6.641
MS in—cold reheat enthalpy	*hms (in)–hcrh (out)*	kJ/kg	358.07	353.87
MS entropy and MS temperature enthalpy	hshp	kJ/kg	3,000.47	3,015.21
Delta enthalpy	Δh_s	kJ/kg	447.82	428.67
HP turbine Eff	η	%	79.96	82.55

the August 2020, efficiency report, it was also found that there was an improvement in NPHR unit one during the change in operating mode until 25.15 kcal/kwh.

References

1. Hutabarat D (2020) Efficiency guide for GV turbine mode sequence Pangkalan Susu steam power plant. Kohesi J
2. Guangdong Power Engineering Corporation (2014) Operation and maintenance training program turbine operation part pangkalan susu steam power plant. Guangdong China
3. Sudhakar R (Sep 2019) Dimensional analysis and lower half control valve seats replacement in high-pressure Turbine. IJERT
4. Sarr J-AR, Mathieu-Potvin F (Aug 2016) Increasing thermal efficiency of Rankine cycles by using refrigeration cycles: a theoretical analysis. Canada
5. Tsou J (1998) Heat rate improvement reference manual. Electric Power Research Institute, Inc., Pleasant Hill

Energy Absorption Analysis on Crash-Module Shape and Configuration of Medium-Speed Train

Achmad Syaifudin[1(✉)], Agus Windharto[2], Andri Setiawan[3], and Abdul Rochman Farid[4]

[1] Department of Mechanical Engineering, Institut Teknologi Sepuluh Nopember, Surabaya, Indonesia
saifudin@me.its.ac.id
[2] Department of Industrial Design, Institut Teknologi Sepuluh Nopember, Surabaya, Indonesia
[3] Graduate Program of Supply Chain Management, Institut Teknologi Sepuluh Nopember, Surabaya, Indonesia
[4] Division of Product Supports, Indonesian Railway Company Ltd. (PT. INKA), Surabaya, Indonesia

1 Introduction

PT Industri Kereta Api (railway industry limited stock company, known as PT INKA), as the first fully integrated rolling stock manufacturer in Southeast Asia, is developing a semi-fast train that will be operated on Java Island to increase people's interest in using public transportation. With a maximum speed up 160 km/h, this semi-fast train is expected to shorten the mileage from Jakarta to Surabaya. Thus, it can attract many plane passengers on short–medium flights to switch to trains. Trains as a means of mass transportation should protect passengers from high potential accidents. According to data released by the Indonesian Transportation Safety Committee, there have been five train accidents from 2014 to 2015 caused by collisions [1]. To reduce the risk of human victims in train collisions, railway companies should implement a passive safety system in the application of crashworthiness technology, which could be expected able to minimize the consequences of accidents.

According to SNI 8826 (national standard in Indonesia) and EN 15227, crashworthiness is a numerical analysis carried out to analyze the ability of the vehicle structures to absorb energy in the event of a collision. This standard describes the level of crashworthiness that will reduce the consequences related to the safety of passengers and crew of railway facilities due to train accidents when active safety measures are not sufficient. This standard also provides a level of protection by addressing the consequences of the types of collisions that cause fatal injuries to passengers and crew on railroads [2, 3]. For medium- and high-speed trains, an active safety system needs to be coupled with a passive safety system to increase the safety of passengers if a collision cannot be avoided. The train's active safety system is useful for reducing the speed of the train when a

M. Kolhe et al. (eds.), *Recent Advances in Renewable Energy Systems*,
Lecture Notes in Electrical Engineering 876,
https://doi.org/10.1007/978-981-19-1581-9_19

potential collision arises suddenly. When the speed has reached a safe level for collision, the passive safety system greatly helps the train reduce the risk of passenger death. Two types of passive safety systems often called crash energy management (CEM) should be adopted in a medium- and high-speed train.

CEM is a technique that enhances the crashworthiness of vehicles by controlling the load's flow into the body. This process involves taking advantage of the energy absorbed by the other components of the vehicle. CEM has been widely used in the design of passenger rail cars. Conventional car bodies can support high loads but are not designed to withstand impacts [4]. CEM is represented by the form of a train structure design that is having a role to reduce excessive deformation and a collision damper/absorber module that is installed on the front or rear end of the carriage [5–10].

As a stage of developing medium-speed trains and compliance to the standards of SNI 8826 and EN 15227, this study is carried out to determine the effect of variations in the cross-sectional shape and configuration of the impact absorber module considering total deformation and collision pulses. Moreover, the safety level of the medium-speed train during a collision will be discussed as well.

2 Methods

In this study, a Diesel Multiple Unit (DMU) medium-speed trains made from an interior–exterior facet model as shown in Fig. 1 is used to generate a finite element model. Two element types can be selected for the simulation, i.e., solid or surface element. To reduce numerical iteration, the surface element (shell element) is chosen to represent the train's model. Three crash-module shapes are being investigated in this study: circular, square, and hexagonal cross-sections. This cross-sectional variation of the crash-module shape had been proposed by Shah et al. and Muhammad and Akbar [9, 11]. All crash-module shapes are having similar volumes to normalize the result of absorbed energy.

Fig. 1 Exterior model of medium-speed train: **a** Isometric view and **b** Side view

The design of the passive safety structure installed on the medium-speed train model can be shown in Fig. 2. This passive safety structure according to CEM consists of three components: (i) an upper frame/pillar to protect the roof of the driver's cabin in the event of an accident, (ii) a rigid cabin front wall to protect the driver in the event of a collision from the front of the train, and (iii) crash-module absorber to eliminate impact energy in the event of an accident [12]. Figure 2a depicts the configuration of the crash module installed on the crash wall. In this study, crash module combined with crash-box was proposed to investigate the influence on collision pulse and to seek the overriding potential, as shown in Fig. 2b. The impact simulation through an explicit solver in ANSYS Workbench 19.0 was carried out to determine the crash-module shape that is able to absorb the largest impact energy after an active safety system lowers the train's speed.

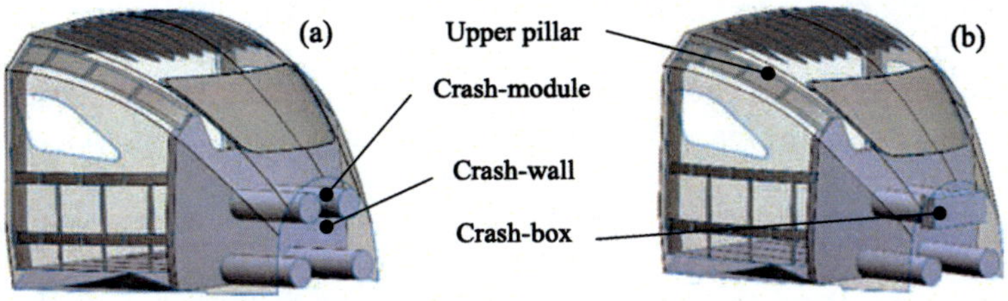

Fig. 2 Front-end passive safety model: **a** Configuration 1, without crash-box, and **b** Configuration 2, with crash-box

The materials used for the train structures are aluminum 6005A and mild steel AISI 1018. Aluminum 6005A is used in underframe components, extruded aluminum, carriage shells, and impact-absorbing modules. Mild steel material is used inside frame components, crash-boxes, and roof frames. The glass components are laminated with polyvinyl butyral (PVB). The material properties are defined by the bilinear elastic–plastic isotropic, while the fracture and high-stress rate properties of the material under impact loading are defined by the Johnson–Cook material model. The material properties and the Johnson–Cook constants are summarized in Table 1.

Table 1 Material properties and material behavior used [13–15]

Properties	Al 6005A	AISI 1018	Laminated glass
Density	2700 kg/m^3	7870 kg/m^3	2530 kg/m^3
Young's modulus	69 GPa	205 GPa	70 GPa
Poisson ratio	0.33	0.29	0.22
Material behavior	Johnson–Cook	Johnson–Cook	Johnson Holmquist continuous
Material constant	A = 270 MPa, B = 569 MPa, N = 0.83, M = 1, TM = 605, C = 0.2, έo = 0.001	A = 532 Mpa, B = 229 MPa, N = 0.3, M = 1, TM = 1530, C = 0.0274, έo = 1	A = 1.06, B = 0.88, C = 0.02, M = 0.35, N = 1.1, Hugoniot Elastic Limit = 5.97 GPa, D = 0.05, D2 = 0.85, K = 39.65 GPa, K2 = 7.85GPa, K3 = −4.84 GPa

Model discretization was carried out using a quadrilateral element with an element size of 35 mm, and element uniformity was enforced throughout the geometry to avoid time step errors. The symmetric model advantage is applied to the simplified train's car body; thus, a total of 75,261 elements and 81,094 nodes are generated after the meshing process. As a stage of the verification process for the finite element model, kinetic energy resulting from the crashworthiness simulation of a medium-speed train with rigid walls was compared to that resulted from analytical calculations. The kinetic energy is the energy amount of the train's structure at the speed before the collision.

The loading conditions of the train are shown in Fig. 3, where the x-axis indicates the longitudinal direction, the y-axis indicates the vertical direction, and the z-axis indicates the lateral direction. The mascara (mask-of-car) of the medium-speed train as the main object of study is supposed to have a deformable stiffness, while the simplified car body remains rigid to prevent deformation. To enable the car body moves along with the mascara, the center of mass of the car body is coupled and connected to the node on the mascara. This simulation strategy was introduced by Watroba et al. to simplify the simulation model [16]. Then, the simplified train's car body is subjected to the actual weight of the car body and mascara on its center of gravity. Besides, the weight of 1.4 kN of two drivers on the mascara area is applied. Based on technical specifications of the medium-speed train, a trainset consisting of six motor cars and six trailer cars. After calculating the longitudinal load based on the configuration of the passengers in each car body, the required traction for acceleration conditions is 141.34 kN and for emergency deceleration conditions is −1.31 kN.

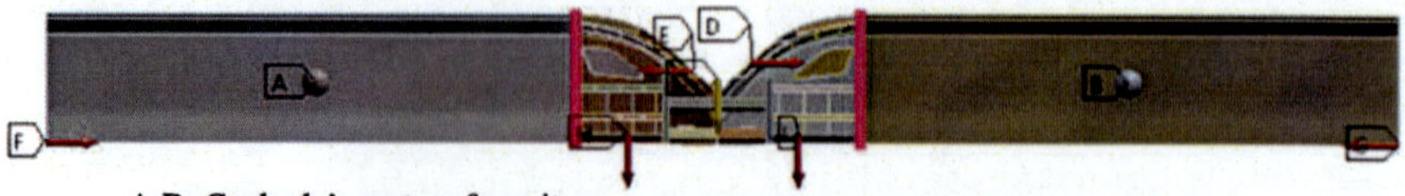

A-B: Car body's center of gravity
E-D: Drag force
F-G: Tensile coupler force
H-I : Driver's weight

Fig. 3 Boundary conditions applied

The crashworthiness test was conducted according to a collision scenario between two identical train units to obtain the total deformation and energy absorption capacity. The crash-module configuration will follow Fig. 2a with three variations of the cross-sectional area used: circular, hexagonal, and square types. Moreover, three-speed variations were given to observe the effect of the train's speed before the collision, i.e., 36 km/h, 45 km/h, and 54 km/h. By using air density (ρ) of 1.2 kg/m^3, frontal aerodynamic area of 8.738 m^2, and drag coefficient of 0.348, the air resistance at speed of 36 km/h is 0.15 kN, at speed of 45 km/h is 0.28 kN, and at speed of 54 km/h is 0.35 kN. As for emergency deceleration conditions, the compressive load at speed of 36 km/h is −0.55 kN, at speed of 45 km/h is −0.65 kN, and at speed of 54 km/h is −0.72 kN.

In terms of observation on overriding potential and collision pulse, a collision scenario between two identical train units is conducted at speed of 36 km/h subsequently after the optimum crash-module shape is obtained. After a crash occurred, both trains will move at the same speed. Overriding is measured from the difference between two collision points on two modules of two trains, which is called relative vertical distance.

Collision pulse depicts the motion of a vehicle during the collision phase, namely the deceleration of the vehicle at different time samples [17]. In the observation of overriding, the collision of two identical trains can be given an initial vertical offset of 40 mm [18]. There are two configurations of crash module and two variations of vertical offset differences applied. As a result, four various combinations will be analyzed, i.e., configuration 1 with zero initial vertical offsets for combination I, configuration 2 with zero initial vertical offsets for combination II, configuration 1 with 40 mm initial vertical offset for combination III, and configuration 2 with 40 mm initial vertical offset for combination IV. The collision pulse and relative vertical distance can be derived from the results of the simulation using Eqs. (1) and (2), respectively.

$$a = \left(\frac{v_2 - v_1}{t_2 - t_1}\right) / 9.8 m/s^2 \tag{1}$$

$$h_y = \left|\text{probe}_A - \text{probe}_B\right| \tag{2}$$

where a is the collision pulse in gravitational acceleration unit, v is the train's velocity before and after collision (m/s), t is the time difference, and h_y is the relative vertical distance between two identical trains.

3 Results and Discussion

3.1 Deformation of Mascara and Energy Absorption Capacity

The sample results of the crashworthiness simulation with variations speed for the circular cross-section and the corresponding energy balance are shown in Fig. 4. It illustrated how the crashworthiness simulation has run well, according to the associated energy balance. It displayed that the crash module has worked well in absorbing the impact energy generated from the collision and the laminated glass has shattered due to the strong impact of the collision.

All simulation results for crashworthiness are summarized in Table 2. It is indicated that crash-module shape with circular and hexagonal cross-section has a large deformation, much different if compared to square cross-section. As for the collision pulse, the simulation results showed that the circular and hexagonal cross-sectional shapes have good damping power, while the square cross-sectional shape has the opposite damping power, resulting in a large collision pulse. The impact energy-absorbing module with the circular cross-sectional shape has the highest impact energy absorption value compared to other shapes because it undergoes greater plastic deformation and receives a larger impact load compared to other cross-sectional shapes. These results are in accordance with several studies previously conducted by other researchers [5, 9–11, 19]. However, it also revealed that the square cross-sectional shape is suitable for the protective frame on the upper and lower sides of the driver desk because of its good rigidity to withstand axial impact loads. The good rigidity of the square cross-sectional shape is indicated by the low deformation yielded.

To validate the resulted kinetic energy, a crashworthiness test was carried out between the medium-speed train versus rigid walls with a speed of 10 km/h. It is necessary to

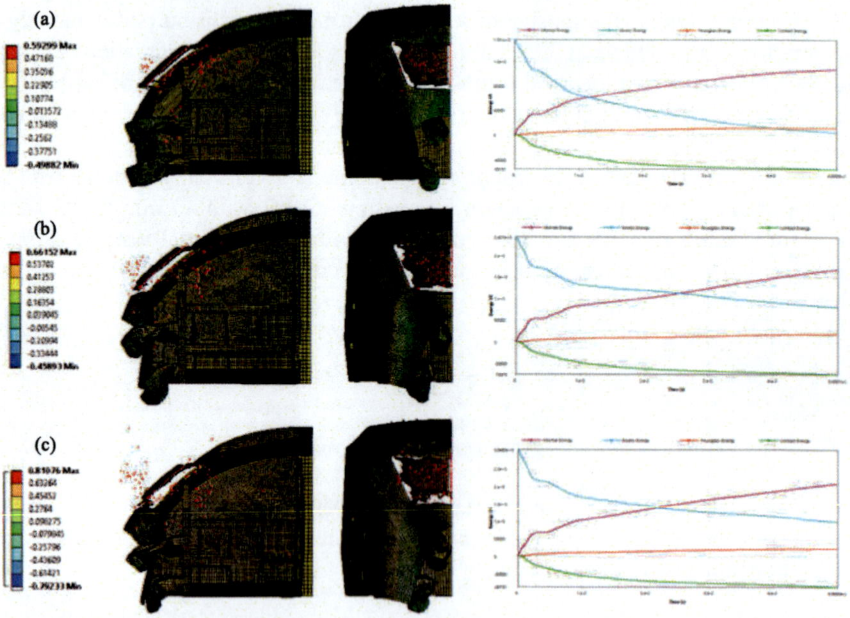

Fig. 4 Deformation on mascara using circular crash-module shape, side, and front view: **a** 36 km/h, **b** 45 km/h, and **c** 54 km/h

observe the energy graph of the medium-speed train without influence by other identical trains.

Figure 5 shows the energy summary graph of the circular cross-sectional shape with a collision speed of 10 km/h. The internal energy graph showed the amount of strain energy absorption capacity of the structures. The total kinetic energy absorbed by the structures is 120.82 kJ for the symmetric model or 241.64 kJ for the full model. The total mass of the medium-speed train's structure is 4,829 kg. Based on the formulation of kinetic energy with a speed of 10 km/h, the total kinetic energy of the train's structure is

Table 2 Deformation and collision pulse under variation of cross-section

Cross-section	Deformation (mm)			Collision pulse (*g*)		
	36 km/h	45 km/h	54 km/h	36 km/h	45 km/h	54 km/h
Circular	592.99 (22.80%)	661.50 (25.44%)	810.70 (31.10%)	3.45	4.14	5.58
Hexagonal	592.93 (22.80%)	655.60 (25.21%)	805.23 (30.85%)	4.65	5.48	6.21
Square	430.00 (16.00%)	506.21 (18.40%)	589.05 (21.24%)	15.20	16.37	17.89

241.45 kJ. This result validated the simulation model of crashworthiness by a difference of 0.07%.

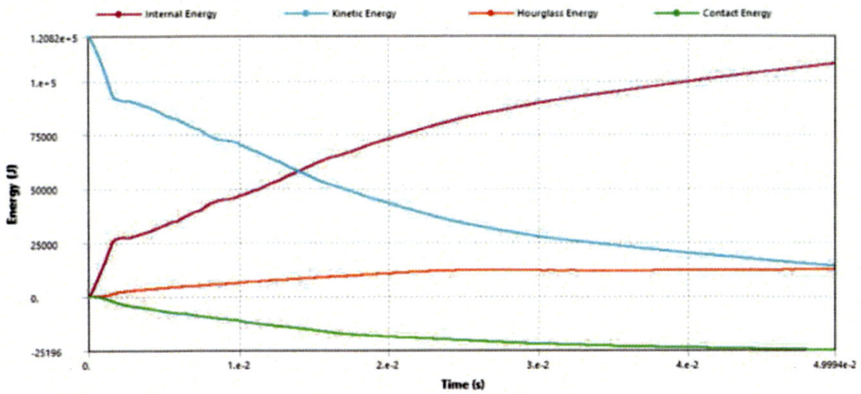

Fig. 5 Energy summary graph of medium-speed train vs rigid wall, with a speed of 10 km/h

To analyze the energy absorption capacity for various cross-sectional shapes, the energy summary graph for a speed of 10 km/h was compared. The energy absorption capacities for circular, hexagonal, and square cross-sections are 107.93 kJ, 98.85 kJ, and 75.22 kJ, respectively. It represented that the circular cross-section has the highest absorption of kinetic energy, in line with many other previous studies [9–11, 19].

3.2 Overriding and Collision Pulse

Following the analysis of crash-module shape, the chosen cross-sectional area was implemented in this simulation, i.e., crash-module shape with the circular cross-section. The relative vertical distance and corresponding collision pulse can be summarized in Table 4, which is obtained at 0.2 s, as the end-time of simulation according to the standards [2, 18].

Table 4 Values of relative vertical distance and collision pulse

Simulations	Relative vertical distance, h_y (mm)	Collison pulse, a (g)
I	13	3.45
II	14	4.03
III	40	0.95
IV	52	2.39

The results of the analysis indicated that for all simulation combinations, the collision of two identical trains at a speed of 36 km/h is still within the safety limit required by the standard, namely the maximum relative vertical distance is 100 mm and the maximum pulse collision is 5 g [18]. In addition, the initial vertical offset of up to 40 mm, which may occur due to pitching or when the train makes sudden braking is relatively safe for both trains. Thus, both crash-module configurations (with or without crash-box) are acceptable. However, it should be noted that the use of crash-box according to configuration 2 causes a larger collision pulse and an increase in the vertical relative distance that may also escalate the overriding potential. What is noteworthy from this simulation is that the emergence of an initial offset can reduce the resulting collision pulse, which is inversely proportional to the relative vertical distance that occurs after the collision.

4 Conclusions

Numerical studies on the crashworthiness of the medium-speed train equipped with various crash-module shapes and configurations have been carried out for varying speeds. The simulation results indicated that circular and hexagonal cross-sections can improve the safety of the train's structures and protect passengers in the event of a collision. According to the first crash test scenario based on the standards of SNI 8826 and EN 15227, the train will remain safe for the collision at a speed of 45 km/h. It is also recommended to use a square cross-section as the protective frame for the upper and lower side of mascara. From the overriding simulations, the addition of a crash-box connecting the crash-module shape is not recommended because it increases the potential for overriding.

Acknowledgements. This study was carried under the financial assistance of the Innovative-Productive Re-search grant (RISPRO-LPDP) project funded by the Education Fund Management Institute, Indonesia, with a contract number 39/LPDP/2019. The authors also gratefully acknowledge Made Abbie Y.P and Putu Pradiva for their hard struggle in conducting explicit numerical simulations.

References

1. Indonesian Transportation Safety Committee (KNKT): investigasi Kecelakaan Perkeretaapian—Media Release KNKT (2016)
2. Badan Standardisasi Nasional: SNI 8826:2019 Aplikasi Perkeretaapian—Crashworthiness pada sarana perkeretaapian. (2019)
3. European Committee for Standardization: the European Standard EN 15227 (2005) Railway applications—Crashworthiness requirements for railway vehicle bodies, pp 1–34
4. Tyrell D, Gordon J (2013) Crash energy management: an overview of federal railroad administration research. https://onlinepubs.trb.org/onlinepubs/trnews/trnews286CrashTest.pdf, https://doi.org/10.17226/22530
5. Setiawan R (2017) Michael Pamintori, dan: analisis crashworthiness Struktur Kereta Penumpang Indonesia, pp 191–195
6. Ambrósio JAC, Silva MPT (2004) Structural and biomechanical crashworthiness using multibody dynamics. Proc Inst Mech Eng, Part D: J Automobile Eng 218:629–645. https://doi.org/10.1243/0954407041166076

7. Ambrósio JAC, Pereira MFOS (1998) Flexible multibody dynamics with nonlinear deformations: vehicle dynamics and crashworthiness applications. Presented at the. https://doi.org/10.1007/978-3-662-03729-4_16
8. Ambrósio JAC, Pereira MS (1997) Multibody dynamic tools for crashworthiness and impact. In: Crashworthiness of transportation systems: structural impact and occupant protection. Springer Netherlands, Dordrecht, pp 475–521. https://doi.org/10.1007/978-94-011-5796-4_19
9. Muhammad IR, Akbar M (2019) Studi parametrik rancangan modul penyerap impak mekanisme deformasi plastis tipe axial crushing. J Online Mahasiswa (JOM) 6:1–7
10. Budi Pratiknyo Y (2017) Rachman Setiawan, dan: the overview of impact energy absorber module on plastic deformation mechanisms
11. Shah MKM, Ahmad N, Wani OI, Sahari J (2016) Study of crashworthiness behavior of thin-walled tube under axial loading by using computational mechanics. Int J Mater Metall Eng 10. https://doi.org/10.5281/zenodo.1130389
12. Matsika E, Ricci S, Mortimer P, Georgiev N, O'Neill C (2013) Rail vehicles, environment, safety, and security. Res Transp Econ 41:43–58. https://doi.org/10.1016/J.RETREC.2012.11.011
13. Peng Y, Ma W, Wang S, Wang K, Gao G (2019) Investigation of the fracture behaviors of windshield laminated glass used in high-speed trains. Compos Struct 207:29–40. https://doi.org/10.1016/J.COMPSTRUCT.2018.09.009
14. Pothnis JR, Perla Y, Arya H, Naik NK (2011) High strain rate tensile behavior of aluminum alloy 7075 T651 and IS 2062 mild steel. J Eng Mater Technol 133. https://doi.org/10.1115/1.4003113
15. Wang C, Cheng H, Su B (2012) Johnson–Cook material model and simulation experiment of aluminum alloy 6005A. Advan Mater Res 472–475:510–514. https://doi.org/10.4028/WWW.SCIENTIFIC.NET/AMR.472-475.510
16. Wątroba P, Pawlak M, Gąsiorek D (2018) Validation of the numerical model of impuls I electric multiple unit driver's cab. Springer Proc Math Stat 249:411–422. https://doi.org/10.1007/978-3-319-96601-4_37
17. Sequeira GJ, Brandmeier T, Joy G (2020) Evaluation and characterization of crash-pulses for head-on collisions with varying overlap crash scenarios. Transp Res Procedia. 48. https://doi.org/10.1016/j.trpro.2020.08.156
18. European Committee for Standardization (2005) The European standard EN 15227 (2005): railway applications—crashworthiness requirements for railway vehicle bodies
19. Velmurugan R, Muralikannan R (2009) Energy absorption characteristics of annealed steel tubes of various cross sections in static and dynamic loading. Latin Am J Solids Struct 6:385–412

Numerical Study Air Preheat Coil as Porous Medium to Analyse Flow Characteristics and Improve Productivity in Muara Karang Unit 4 Steam Power Plant

Rizal Mahendra Pratama and Tri Yogi Yuwono(✉)

Institut Teknologi Sepuluh Nopember, Surabaya, Indonesia
triyogi@me.its.ac.id

1 Introduction

Power plants convert the main energy from a fuel into electricity that is useful for society. There are many types of power plants, such as steam power plants, combine cycle power plants, diesel power plants that use fossil energy, and some power plants that use renewable energy. Muara Karang Unit 4 steam power plant is one of the power plants that supply electric energy in Java, especially in the load centre areas of DKI Jakarta. It was built in 1982 with residual oil. When operating with oil, the combustion airflow from force draft fan (FDF) will be preheated through a heat exchanger (HE) called air preheat coil (APC) to prevent sulphur deposition on the cold side of the air heater (AH), which can cause damage to AH elements. In 1995, a gasification project was carried out due to economic, environmental, and gas supply ability considerations. It was modified to burn both residual oil and natural gas. Nowadays, most of the operational process uses natural gas as the main fuel. Fuel oil is only used as a backup when the gas supply is interrupted or maintenance. When operating using natural gas, the APC, which is still attached to the air duct, is no longer needed and becomes airflow resistant. It will increase the pressure drop of the airflow supplied by the FDF, thus requiring greater FDF power to get the air requirements to the boiler. Removing modules of APC entirely or partially is expected to reduce the airflow resistance. It will automatically reduce the power required by the FDF and improves the productivity of the power plant.

Some literature studied modelling HE using computational fluid dynamic (CFD) commercial software. CFD is used for research on various types of HE to analyse several problems such as flow misdistribution, fouling, pressure drop, and thermal analysis [1]. Modelling HE as PM in CFD simulation is widely used for many purposes. It is because it can also simplify the methodology from complex geometry. Some literature used this method to simplify methodology, such as modelling finned plate HE to study hydrodynamic characteristics [2] or modelling an oil cooler of an aeroplane as a porous medium [3]. Even in the power plants industry, this approach is used to steam coil air heater (SCAH) and get results close to actual conditions without spending a lot of effort [4].

M. Kolhe et al. (eds.), *Recent Advances in Renewable Energy Systems*,
Lecture Notes in Electrical Engineering 876,
https://doi.org/10.1007/978-981-19-1581-9_20

This study will simulate airflow in APC if its tube module is removed entirely. It uses the PM approach in CFD commercial software to analyse APC flow and heat transfer characteristics with 172 MW load production and various module scenarios. Such scenarios variation aimed to reduce the pressure drop of the airflow supplied by the FDF. It will automatically reduce the self-energy usage by the FDF and improves productivity of the power plant.

2 Numerical Method

2.1 CFD Model and Simulation

The air supplied by FDF passes through the APC before entering the AH. The duct is mainly divided into three sections: inlet, APC, and outlet. APC consists of four modules assembled independently from each other. Each APC module consists of 153 circular finned tubes in a staggered configuration. A 3D geometry model based on actual geometry with the boundary condition for inlet defined as mass flow inlet and outlet as pressure outlet (see Fig. 1). The model meshing is a hexahedral subMap airflow direction on Z-axis.

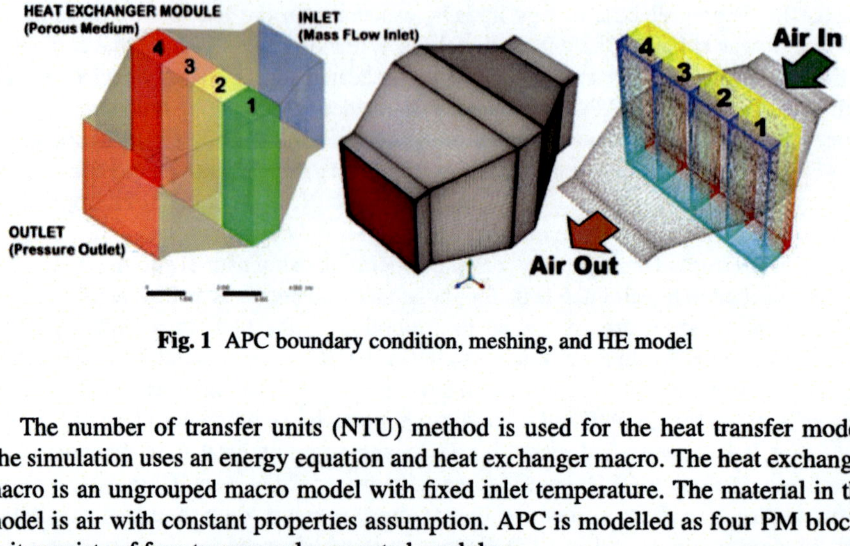

Fig. 1 APC boundary condition, meshing, and HE model

The number of transfer units (NTU) method is used for the heat transfer model. The simulation uses an energy equation and heat exchanger macro. The heat exchanger macro is an ungrouped macro model with fixed inlet temperature. The material in the model is air with constant properties assumption. APC is modelled as four PM blocks as it consists of four transversely mounted modules.

In this simulation, the proposed model domain is validated with the data measured during commissioning (see Table 1). The comparison results show a very good agreement, which gives a difference of 2.21% at the inlet pressure, 0.48% at outlet temperature, and 0.18% at pressure drop. So, it can be concluded that the modelling and simulation processes are valid and describe the actual conditions, which can then be applied to solve the problems posed in this study. Furthermore, in this simulation, grid independence test was also carried out for the five proposed grid sizes, and the results show that mesh A, B, and C have the values of inlet pressure, pressure drop, and outlet temperature still changing. However, for mesh C, D, and E, the values of the three parameters are relatively the same (see Table 2). Therefore, in this study, mesh C will be chosen for reasons of time and energy efficiency because it has a smaller number of meshes than D and E.

Table 1 Validation results

Parameter	Unit	References	Results	Error (%)	Description
Inlet press	Pa	7673.49	7507.44	2.21	Commissioning
Outlet temp	°C	68.50	68.83	0.48	Commissioning
Press. drop	Pa	1325.78	1323.44	0.18	Calculation

Table 2 Grid independency test results

Mesh	Size	Inlet press. (Pa)	Press drop (Pa)	Error press (%)	Outlet temp. (°C)	Error temp. (%)
Mesh A	792,000	7500.00	1316.00	0.74	69.10	0.88
Mesh B	1,017,280	7502.50	1318.50	0.55	69.05	0.80
Mesh C	1,270,720	7507.44	1323.44	0.18	68.83	0.48
Mesh D	1,415,120	7506.89	1322.89	0.22	68.75	0.37
Mesh E	1,559,520	7507.42	1323.42	0.18	68.93	0.63

Furthermore, in this study, two scenarios of APC conditions were carried out: the scenario of removing the APC module in which S0 for all modules was installed and S1 when all modules were removed. Simulation from both are including pressure drop, temperature, and power required by FDF. That will be compared to conclude which scenario gives the best performance.

Heat Exchanger Modelling. If APC is assumed to be a porous medium (PM) model, the first parameter used to define the PM model is porosity (γ). In this case, porosity is defined as the volume fraction of free space to the total volume of APC, as expressed in Eq. (1) below. After being calculated by Eq. (1), the APC studied has a porosity value of 0.98.

$$\gamma = \frac{\text{Total Volume } APC - (\text{Volume Tube} + \text{Volume Fin})}{\text{Total Volume } APC} \tag{1}$$

The viscous and inertial resistance contributes to the pressure drop along the APC. These factors are represented in the PM model by adding the source term (*Si*) in the momentum equation. This source term is shown in Eq. (2) below.

$$S_i = -\left(\frac{\mu}{\alpha} v_i + C_2 \frac{1}{2} \rho |v| v_i\right) \tag{2}$$

Inertial resistance becomes dominant as the velocity increases, while the viscous resistance is negligible in this study. Then, the pressure drop can be calculated by Eq. (3) below.

$$\Delta P = \frac{\rho V^2}{2} \times C_2 \times \Delta n_x \tag{3}$$

where C_2 is inertial resistance coefficient, v is inlet velocity, and Δn_x is PM thickness. C_2 could be defined using an empirical equation. The empirical equation to calculate the pressure drop for flow that passes through a row of a tube with several types and arrangements.

$$\Delta P = \frac{\rho u^2}{2} Eu \times z \tag{4}$$

where u is intertube velocity, Eu is Euler number, and z is the number of rows. Then, comparing Eqs. (3) and (4), with replacing u with v, the inertial resistance coefficient can be calculated by Eq. (5) below.

$$C_2 = \frac{\left(\frac{a}{a-1}\right)^2 \times Eu \times z}{\Delta n_x} \tag{5}$$

Eu for finned tubes with staggered arrangements is defined with Eq. (6) below [5].

$$Eu = 5,4\left(\frac{d^*}{d_e}\right)^{0.3} Re_{d^*}^{-0.25} C_z \tag{6}$$

After being calculated, the APC studied has a C_2 value of 122.62.

3 Result and Discussion

The qualitative results consist of velocity, pressure, and temperature contour analysis at 172 MW load production and two scenarios (S0 and S1).

In S0 (see Fig. 2), airflow velocity decreased after the inlet because of flow resistance and gradual enlargement of the cross-sectional area when passing the installed modules. According to the flow continuity principle, the decrease in velocity will be proportional to the change in cross-sectional area. The enlargement of the cross-sectional area also causes separation in the airflow to form a low-velocity recirculation area (vortex) close to the wall. Separated flow is also known as a form of secondary flow because the direction of the flow is not the same as the direction of the main flow. This phenomenon can be seen almost in all contours Figs. 2 and 3, except Figs. 2c and 3c. Compared with Figs. 2e–h and 3e–h, it shows that the angle of the horizontal side is sharper than Figs. 2a–d and 3a–d (vertical side), thus it makes secondary flow looks more significant on the left and right side of the duct, close to the wall. Secondary flow nearly not happened in the centre part of the vertical side. That is because the enlargement angle of the cross-sectional area on the vertical side is not as sharp as the horizontal side. It is also because the position is relatively far from the edge of the duct, so secondary flow is relatively huge, whereas

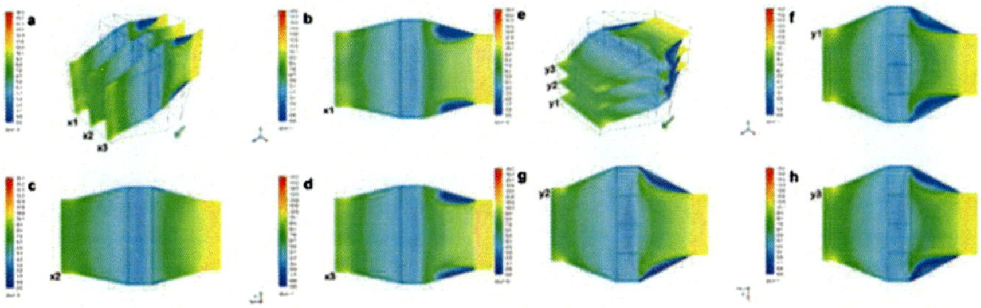

Fig. 2 Velocity contour 172 MW—S0; **a** *x*—coordinate; **b** × 1; **c** × 2; **d** × 3; **e** *y*—coordinate; **f** *y*1; **g** *y*2; **h** *y*3

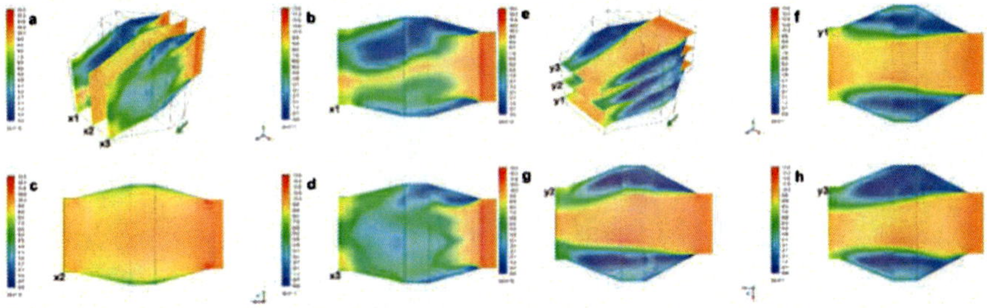

Fig. 3 Velocity contour 172 MW—S1; **a** *x*—coordinate; **b** × 1; **c** × 2; **d** × 3; **e** *y*—coordinate; **f** *y*1; **g** *y*2; **h** *y*3

S1 (see Fig. 3) shows that most of the airflow straight through the centre of the duct to the outlet. The velocity decreases due mainly to the formation of secondary flow, which is huge and uneven, as seen almost in all contours.

In S0 (see Fig. 4), the APC module is modelled as a porous medium, causing flow resistance and pressure drop. The pressure increases due to flow resistance in installed modules. The inlet has the highest pressure and tends to increase until the beginning of the module. It is indicated by the colour of the red contour, which is getting darker. It also happened because the flow starts hitting the front of the module and airflow through the cross-sectional area, which increases gradually so that the pressure increases with the decrease in velocity described in the previous analysis of velocity and streamline contours, whereas S1 (see Fig. 5) shows that the pressure increases due mainly to the formation of secondary flow, which is huge and uneven.

The temperature rises gradually when airflow through the installed APC module (see Fig. 6). It showed a temperature rise along with the flow for S0. The lower side temperature of the APC module increases earlier than the middle and upper sides (see Fig. 6a–d). It is because the inlet of hot water (aux fluid) enters the module from the bottom side, whereas the temperature at the edge of the duct also increases earlier than in the middle side (see Fig. 6e–h). The secondary flow formed at the edge of the duct also increases earlier because it is close to the installed module. Contrary to Fig. 6, Fig. 7 does not show the rising temperature on the APC module. It indicates that the simulation

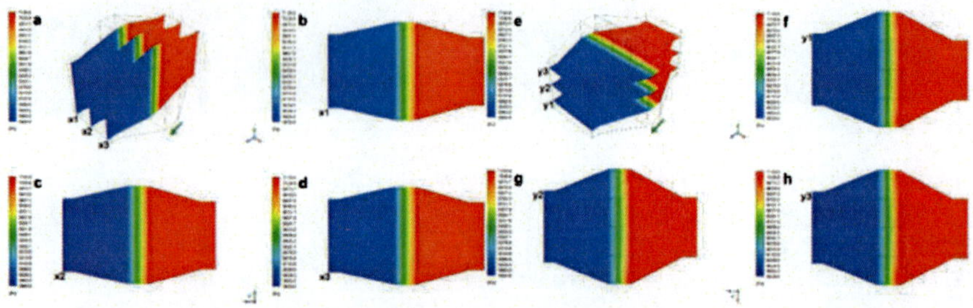

Fig. 4 Pressure contour 172 MW—S0; **a** *x*—coordinate; **b** ×1; **c** ×2; **d** ×3; **e** *y*—coordinate; **f** *y*1; **g** *y*2; **h** *y*3

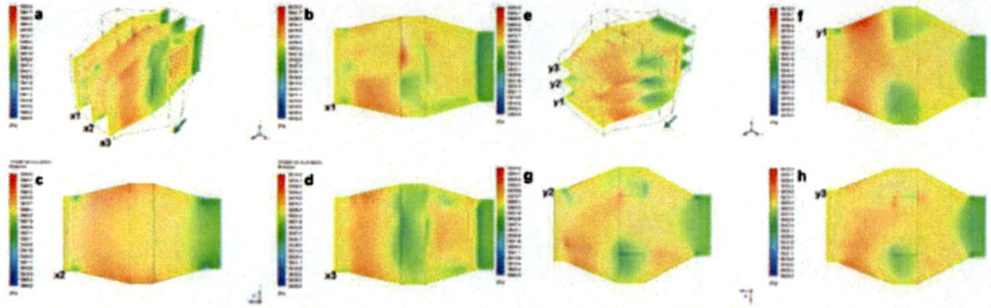

Fig. 5 Pressure contour 172 MW—S1; **a** *x*—coordinate; **b** ×1; **c** ×2; **d** ×3; **e** *y*—coordinate; **f** *y*1; **g** *y*2; **h** *y*3

was successfully applied. It also proves that PM successfully represents APC module as a heat exchanger.

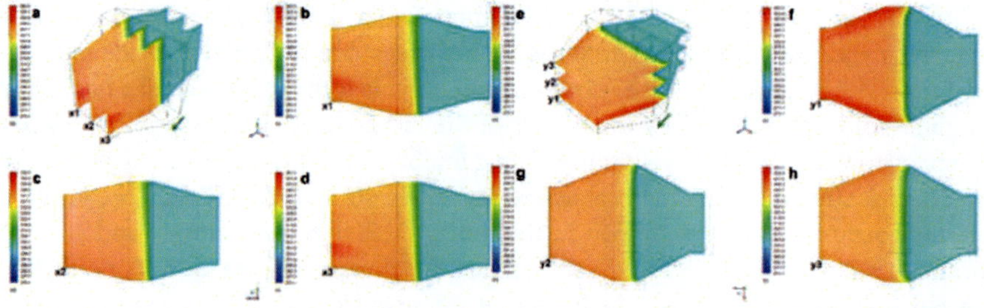

Fig. 6 Temperature contour 172 MW—S0; **a** *x*—coordinate; **b** ×1; **c** ×2; **d** ×3; **e** *y*—coordinate; **f** *y*1; **g** *y*2; **h** *y*3

Comparison data from all scenarios showing the value of pressure drop reduction, temperature rise reduction, and FDF power reduction are shown in Table 3. S1 results in reducing pressure drop compared to its original one (S0). Pressure drop reduction means reduce in power needed by FDF to flow the air pass through APC. The pressure drop value multiplied by its air debit (m^3/s) will result in the value of FDF power reduction (Eq. 7). It will automatically reduce self-energy usage (PS) and improve the productivity

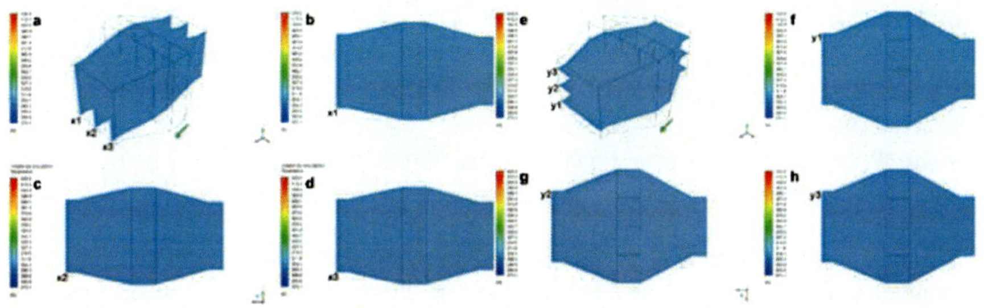

Fig. 7 Temperature contour 172 MW—S1; **a** *x*—coordinate; **b** ×1; **c** ×2; **d** ×3; **e** y—coordinate; **f** *y*1; **g** *y*2; **h** *y*3

of the power plant.

$$\Delta PS_{FDF} = \Delta P_{APC} \times Q \times t \times \eta_{\text{fan}} \times \eta_{\text{motor}} \tag{7}$$

Table 3 Reduction in pressure drop, temperature rise, and FDF power

Parameter	Unit	S0	S1	Reduction (%)
Δ pressure	Pa	1078.4	13.6	99
Δ temperature	°C	43.6	0	100
Δ FDF power	kW	102.4	1.3	99

Table 3 above shows that S1 effectively reduces pressure drop up to 99%, from 1078.4 Pa turns out to 13.6 Pa. However, removing the modules on the S1 caused the APC's ability to raise the combustion air temperature also decrease drastically up to 100%, from 43.6 °C turns out to 0 °C. Reduction of the pressure drop automatically brings down PS up to 99%, from 102.4 kW to 1.3 kW. Reduction of the PS then converted into economic value after being multiplied by the selling price of electrical energy (Rupiah/kWh). So that, it worth total savings up to 1861 million/year.

4 Conclusions

As a porous medium, modelling APC is proven to provide an overview of airflow and heat transfer characteristics in the APC duct. The simulation model is applied from the actual condition and produces parameters that are similar to commissioning data. The scenario of removing the APC module is proven to reduce the power required by the FDF and also reduce self-energy usage (PS). S1 is better than the S0 in terms of reducing pressure drop but lacking in heat transfer capabilities. S1 effectively reduces self-energy usage (PS) and save production costs up to 1,772,124 kWh/year and Rp. 1861 Million/year, respectively.

References

1. Bhutta MMA, Hayat N, Bashir MH, Khan AR, Ahmad KN, Khan S (2012) CFD applications in various heat exchangers design: a review. Appl Therm Eng 32:1–12
2. Wang W, Guo J, Zhang S, Yang J, Ding X, Zhan X (2013) Numerical study on hydrodynamic characteristics of plate-fin heat exchanger using porous media approach. Comput Chem Eng 61:30–37
3. Musto M, Bianco N, Rotondo G, Toscano F, Pezzella G (2016) A simplified methodology to simulate a heat exchanger in an aircraft's oil cooler by means of a porous media model. Appl Therm Eng 94:836–845
4. Ariyanto E, Widodo WA (2019) Steam coil air heater modelling as porous medium to analyze flow characteristic and reduce self energy usage in Gresik unit 1 steam power plant. IPTEK J Technol Sci 30:7–10
5. Yudin VF, Tokhtarova LS, Lokshin V, Tulin SN (1968) Correlation of experimental data on convective heat transfer in cross flow over bundles with transverse spiral and circumferential fins. Trudy TsKTI, No. 82

Numerical Study for the Modified Cooler of SAF Motor at Power Plant

Khoirul Huda[1(✉)], Prabowo[1], Bambang Arip Dwiyantoro[1], and Teguh Widjayanto[2]

[1] Departement of Mechanical Engineering, Institut Teknologi Sepuluh Nopember, Surabaya, Indonesia
huda.pjb@gmail.com
[2] PT PJB Services, Surabaya, Indonesia

1 Introduction

The YKK series motor is a closed cage rotor with an air cooler. The external wind path of the YKK medium-sized high-voltage asynchronous motor comprises an external fan and cooler. The secondary air fan (SAF) that used this motor type had a problem with cooling performance, and the problem was high winding temperature while operated at high load or more than 85% (±93 MW) of the maximum load (110 MW). The temperature of the motor winding has a limit of 130 °C (alarm) and 140 °C (trip) while at that time the temperature was about 120 °C, and to decrease that temperature, the way was installing some portable blower that blown toward to motor body.

The motor cooling mechanism is shown in Fig. 1, the heat generated by the motor winding is on the shell side, and then this heat is facing the cooler above it. The cooler has a shell and tube type, with tube material made of aluminum, inside of tube which contains cold air from ambient and forced by an external fan. While on the shell side, there is hot air which is circulated by the internal fan through the shell and is separated by a baffle in the middle of the cooler.

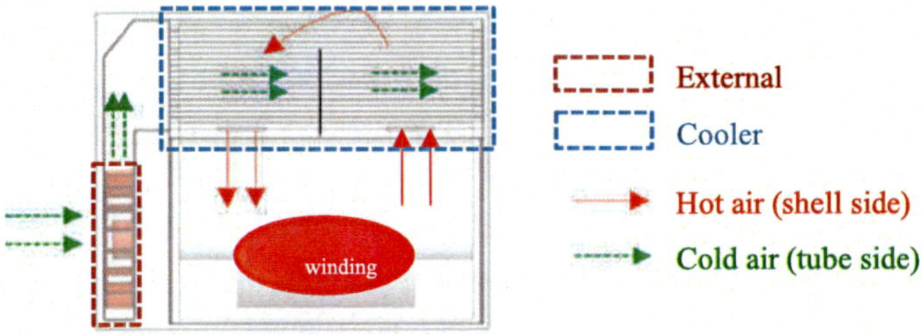

Fig. 1 Cooling diagram for the motor

Cold airflow by an external fan can be optimized by adding a guide plate, modified external fan for tube side, and adding a windshield or baffle in shell side [1], and there are an optimum number of tubes for heat transfer and exit temperatures of shell side [2],

M. Kolhe et al. (eds.), *Recent Advances in Renewable Energy Systems*,
Lecture Notes in Electrical Engineering 876,
https://doi.org/10.1007/978-981-19-1581-9_21

increasing the number of transverse tubes is more effective than increasing the number of longitudinal tubes [5], and by modifying the external fan, adding guide plate in tube side, internal fan settings can improve the cooling process [3].

With the review of previous research, this paper has several points of view, including

- Numerical simulation to combine guide plate designs to find out the best motor cooling performance.
- Numerical simulation to get the optimal number of tubes to get the best motor cooling performance.
- Evaluating the design of the existing cooler and opportunity to improve.

2 Preliminary Experiment

This step is using SOLIDWORKS software to validate the model that will be implemented to cooler. Previously, we learn from textbooks and papers that had related to this theme. Using a short time while the unit was shut down, we decide to decrease the number of tubes by using a guide plate to close the top and step by step, as shown in Fig. 2.

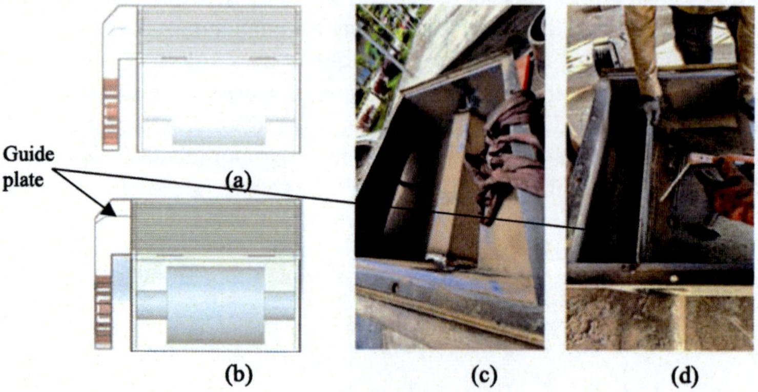

Fig. 2 Preliminary experiment **a** existing model, **b** modified model, **c** welding process, and **d** finished model

From the preliminary model above, as shown in Fig. 2, the winding temperature could be reduced significantly, as shown in Fig. 3, although these conditions were similar to the portable blower, we could conclude that modification was more effective for the cooling process. Therefore, a detailed analysis is required to evaluate it.

3 Numerical Simulation

This study focuses on finding the effect of variation models by changing the guide plate at the duct of tubes side on cooling performance. The model is divided into four

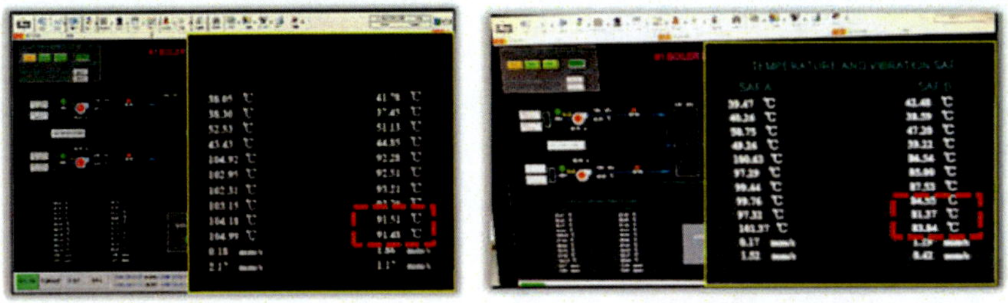

(a) (b)

Fig. 3 Result of preliminary experiments **a** before modification and **b** after modification

models that are using of tubes 100% (17 rows × 2 column tubes), 75% (13 rows × 2 column tubes), 50% (9 rows × 2 column tubes), and 30% (5 rows × 2 column tubes). By numerical analysis with CFD software then the model and equation developed, the model is symmetrical in width and has 70/1000 parts, as shown in Fig. 4.

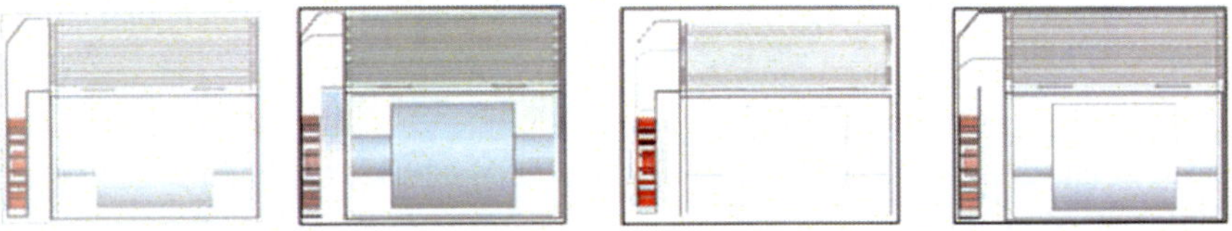

(a) (b) (c) (d)

Fig. 4 Models at **a** 100%, **b** 75%, **c** 50%, and **d** 30%

4 Result and Discussion

4.1 Comparison of Four Models While Using 100%, 75%, 50%, and 30%

Numerical analysis using software CFD and the physical properties of the air are as follows: a density of 1.225 kg/m^3, a specific heat capacity of 1006.43 j/kg.K, a thermal conductivity of 0.0242 W/m.K, and a viscosity of 1.789 e-05 kg/m.s (Fig. 5 and Table 1).

4.2 Analytical Calculation

In this step, we validate the result from numerical simulation, using output from simulation CFD, and then we calculate based on mass flow rate balance and energy balance.

Mass flow rate balance. Both for the tube and shell sides, inlet and outlet should have similar mass flow rates because they are in the steady-state phase.

$$\dot{m}_i = \rho A_i v_i \quad (1)$$

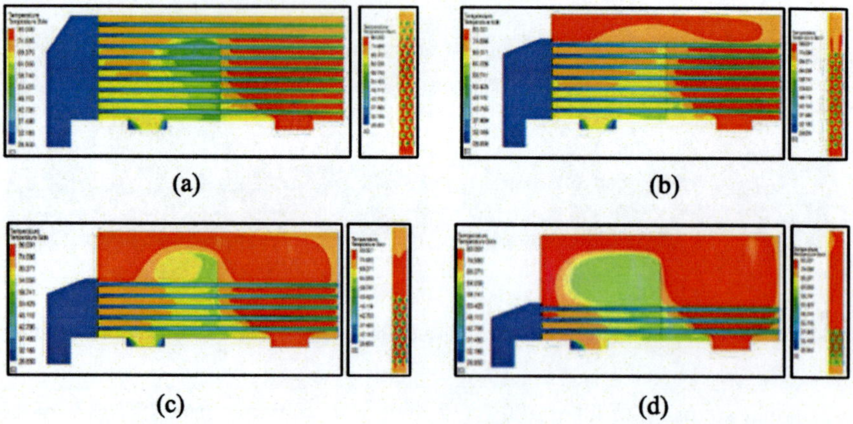

Fig. 5 Temperature contour results for **a** 100%, **b** 70%, **c** 50%, and **d** 30%

$$\dot{m}_o = \rho A_o v_o \quad (2)$$

$$\dot{m}_i = \dot{m}_o = \dot{m} \quad (3)$$

From the table above, the analytical calculation average had almost similar to the result from simulation.

Table 1 Comparison results for all models

Parameter	Model			
	100%	75%	50%	30%
Inlet temperature sisi shell (°C)	80	80	80	80
Outlet temperature sisi shell (°C)	64.503	64.275	65.487	69.757
Inlet temperature sisi tube (°C)	30	30	30	30
Outlet temperature sisi tube (°C)	61.628	62.519	59.408	54.005
Inlet pressure sisi shell (Pa)	532.73	397.67	401.23	380.12
Outlet pressure sisi shell (Pa)	−0.443	−0.499	−0.547	−0.124
Inlet pressure sisi tube (Pa)	51.202	87.381	179.47	540.99
Outlet pressure sisi tube (Pa)	0	0	0	0
Inlet velocity sisi shell (m/s)	6	6	6	6
Outlet velocity sisi shell (m/s)	6.3005	5.7779	5.8983	6.8027
Inlet velocity sisi tube (m/s)	4.7	4.7	4.7	4.7
Outlet velocity sisi tube (m/s)	4.6512	5.8346	8.7654	15.197

Energy balance. Both the tube and shell sides should have an energy balance [4]:

$$\dot{q}_s = \dot{m}_s c_p \left(T_{s,i} - T_{s,o}\right) \tag{4}$$

$$\dot{q}_t = \dot{m}_t c_p \left(T_{t,o} - T_{t,i}\right) \tag{5}$$

$$\dot{q}_s = \dot{q}_t = \dot{q}_1 \tag{6}$$

Equation (3) and Table 2 show that the mass flow rate was balanced and from Eq. (6), and Table 3 can be seen that the calculation energy transfer was similar between shell and tube sides, and for model 75% had higher heat transfer, NTU, and ε (effectiveness).

Table 2 Mass flow rate for all models

Parameter	Outer tube/shell side (Hot)				Inner tube/tube side (Cold)			
	100%	75%	50%	30%	100%	75%	50%	30%
$\dot{m}_{\text{in}}$ (kg/s)	0.1312	0.1312	0.1312	0.1312	0.0645	0.0645	0.0645	0.0645
$\dot{m}_{\text{out}}$ (kg/s)	0.1378	0.1263	0.1290	0.1487	0.0671	0.0644	0.0670	0.0645
$\dot{m}_{\text{average}}$ (kg/s)	0.1345	0.1288	0.1301	0.1400	0.0658	0.0644	0.0657	0.0645
$\dot{m}_{\text{simulation}}$ (kg/s)	0.1311	0.1311	0.1311	0.1311	0.0644	0.0644	0.0644	0.0644

Table 3 Energy balance for all models

Parameter	100%	75%	50%	30%
$\dot{q}$(W)(shell)	2044.76	2074.82	1914.94	1351.54
$\dot{q}$(W)(tube)	2051.23	2108.97	1907.21	1556.85
$\dot{q}$ max (W)	3242.7175	3242.7175	3242.7175	3242.7175
NTU	1.236	1.304	1.075	0.741
ε	0.63	0.65	0.59	0.48
Cmin/Cmax	0.4915	0.4915	0.4915	0.4915

4.3 Discussion

Based on Table 1 and Fig. 6, we see that T_{out} shell side had smaller at model 75% than others. It can be seen in Fig. 8, the vortex at 100% model was appeared much more than the others at duct side before entering the tube. The vortex makes energy loss so that the airflow to the tube was more uniform and effective by decreasing the vortex. Figures 7 and 8 show that the distribution of airflow from the duct to tube at model 100% had random distribution before entering tubes. Therefore, only several tubes at

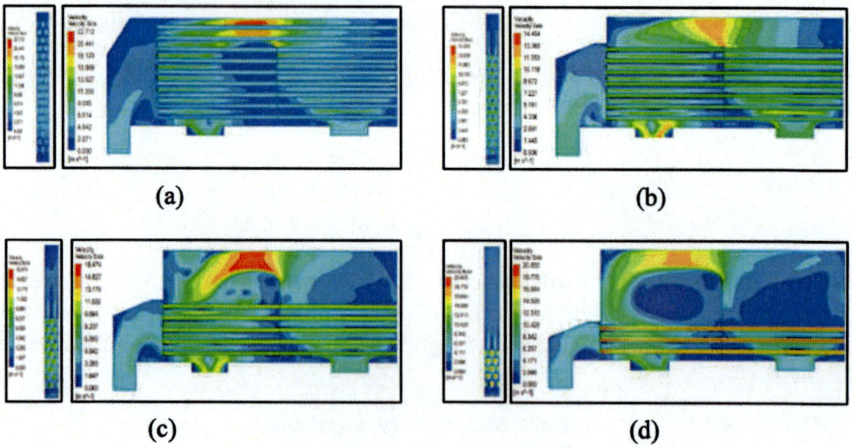

Fig. 6 Velocity contour results for **a** 100%, **b** 70%, **c** 50%, and **d** 30%

the center had a uniform airflow. Even though the number of tubes and area of model 75% is less than 100% model, the distribution of airflow is uniform and has less flow maldistribution. Hence, 50% and 30% models had good airflow; however, it decreased more in the area of tubes and made the performance lower than the performance of 100% and 75% models.

5 Conclusions

From the discussion above, the results show at the 75% model is better than others. This is due to

1. Manifold-induced Flow Maldistribution [6]: Geometry header at tube side makes a flow maldistribution in 100% models, hence the performance decreased.

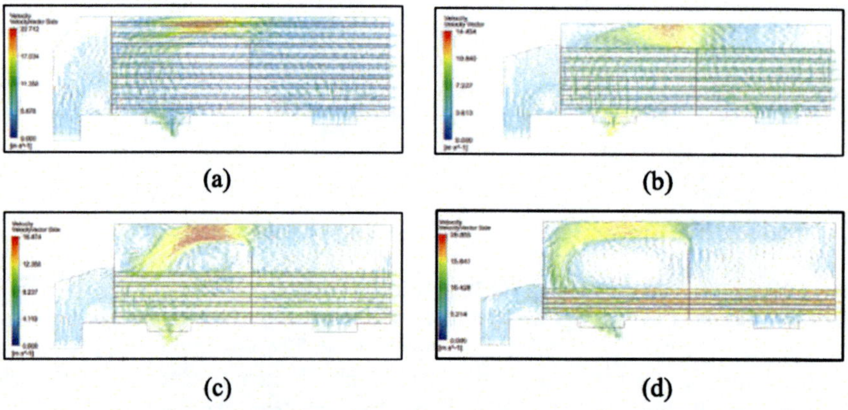

Fig. 7 Velocity vector for **a** 100%, **b** 70%, **c** 50%, and **d** 30%

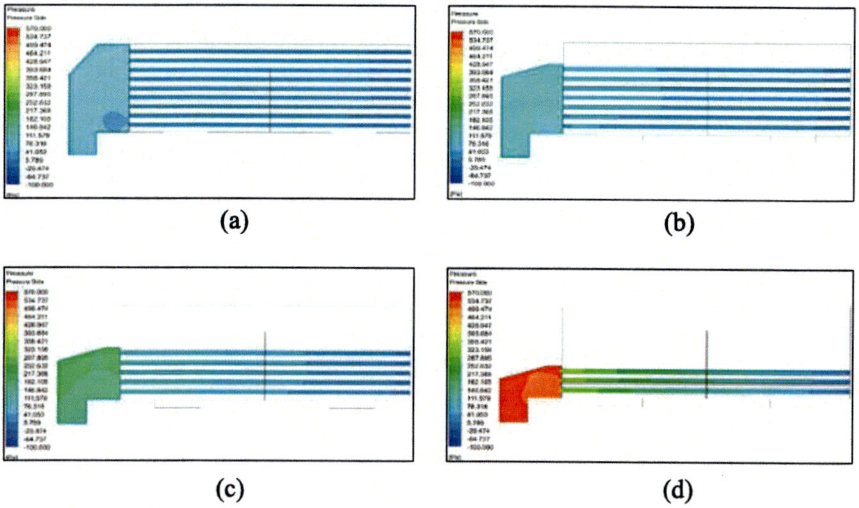

Fig. 8 Pressure contour for **a** 100%, **b** 70%, **c** 50%, and **d** 30%

2. Guide Plate: By adding guide plate makes uniform airflow at all the entrance especially at 75% model. This makes the area more effective for the heat transfer process. Hence, by adding a guide plate makes uniform airflow and reduces the vortex.
3. Number of Tube: 75% model has a certain number of tubes, by adding of several tubes, the area is increasing, and air resistance is increasing too.

Acknowledgements. The author would like to thank Teluk Balikpapan Power Plant, PT Pembangkitan Jawa Bali, for the support and for giving us a chance so this research could be implemented. Moreover thank PT PLN and Department of Mechanical Engineering, Faculty of Industry Technology, Institut Teknologi Sepuluh Nopember that give us a chance to study in this topic.

References

1. Wen J, dan Zheng J (2015) Numerical analysis of the external wind path for medium-size high-voltage asynchronous motors. Appl Therm Eng 869–878
2. Xiwei Y, Dawei M (2018) Design analysis and improvement of cooler in positive-pressure explosion-proof low-speed high-capacity induction motors. Appl Therm Eng 129:1002–1009
3. Chang CC, Kuo Y-F, Wang J-C, dan Chan S-L (2010) Air cooling for a large scale motor. Appl Therm Eng 30:1360–1368
4. Incropera FP, Dewitt DP, Bergman TL, Lavine AS (2007) Fundamentals of heat and mass transfer, 699th edn. Wiley Inc., USA
5. Kumara PDC, Muzathik AM, Jayaweera H (2015) The effect of tube diameter and number of tubes in a cross flow steam condenser on performance. International symposium, 5th. SEUSL
6. Shah RK, Sekulic DP (2003) Fundamental of heat exchanger design. Wiley Inc, USA

Numerical Study of the Effect of Cooling Air on Low-Pressure Air Cooler in a Two-Stage Reciprocating Compressor

Verry Mardiananta Arsana[1(✉)] and Sutardi[2]

[1] The Department of Mechanical Engineering, FTIRS-ITS Collaboration Program With PT Perusahaan Listrik Negara, Surabaya, Indonesia
verryma@gmail.com

[2] Department of Mechanical Engineering, FTIRS-ITS, Surabaya, Indonesia

1 Introduction

The main function of a starting compressor is to produce compressed air to start the gas engine. In Arun Gas Engine Power Plant (GEPP), the starting compressor used is a reciprocating type that consists of two levels of compression: the low-pressure (LP) side and the high-pressure (HP) side. The problem that frequently arises in the starting compressor is the excessive deposit growth rate that clogs the tubing of the HP air cooler, thus reducing the reliability of the equipment. This excessive deposit growth rate is derived from burned lubricants due to the high temperature that occurred in the HP side compression chamber. The cooling capacity of the LP air cooler is enhanced to decrease the inlet air temperature into the HP compressor. Moreover, lower air inlet temperature decreases the temperature in the HP side compression chamber. In Fig. 1, a schematic diagram of starting compressor used in the present study is outlined.

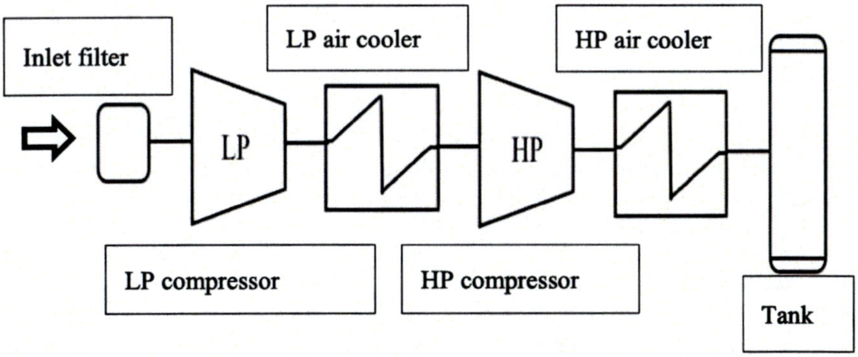

Fig. 1 Starting air compressor diagram

Some research on heat exchangers in general applications, reciprocating compressors, and power generations has been conducted by Ribeiro et al. [1], Wais [2] and

M. Kolhe et al. (eds.), *Recent Advances in Renewable Energy Systems*,
Lecture Notes in Electrical Engineering 876,
https://doi.org/10.1007/978-981-19-1581-9_22

Wibowo et al. [3]. Ribeiro et al. [1] studied the effects of adding louvered fins to compact heat exchangers compared to using only smooth tubes. The study was also conducted by varying the velocity of cooling air and hot air flowing cross-flow on compact heat exchangers. The study obtained the results that using the addition of louvered fins can improve the performance of a cross-flow compact heat exchanger when compared to using a smooth tube.

Wais [2] investigates the effect of fin thickness on tubes in terms of heat transfer rate. Efficiency and dimension are one of the most important aspects to determine the design of a heat exchanger design. The size of the heat exchanger can be more compact by adding fins to increase the rate of heat transfer between the heat exchanger surface and the environment. The heat transfer decreases by increasing fin thickness.

Wibowo et al. [3] investigate the cooling performance improvement in lube oil cooler gas turbine by arranging heat exchangers in series and parallel as well as variations in lube oil flow capacity to determine the best cooling performance. The result of the simulation is known that the arrangement of the lube oil cooler series produces the best cooling capacity with higher pressure drop.

By conducting research on the effect of increasing cooling air velocity on the LP air cooler, it is expected that it will be able to increase cooling capacity and decrease the inlet air temperature into the HP compressor. The limit for operating the starting compressor can be calculated using Eq. (1).

$$T_d = T_s \left(\frac{P_d}{P_s} \right)^{\frac{k-1}{mk}} \tag{1}$$

where

T_d Discharge temperature (K)
T_s Suction temperature (K)
m Number of stages; $m = 1, 2, 3, \ldots$
P_d Discharge pressure (bar)
P_s Suction pressure (bar)
k 1.4

The discharge temperature must be controlled maximum of 270 °C to avoid oil burning.

2 Methodology

This study is based on computational fluid dynamics to obtain contours and distributions of temperature, pressure, and velocity. There are three main steps needed in this numerical method which consist of preprocessing, processing, and postprocessing. The geometry of the LP air cooler, the model of simulation, and the geometry of louvered fins are shown in Fig. 2.

Geometry and boundary conditions are given in Tables 1 and 2. In this study, convection coefficients on the wall tube are calculated based on the velocity of cooling air. The

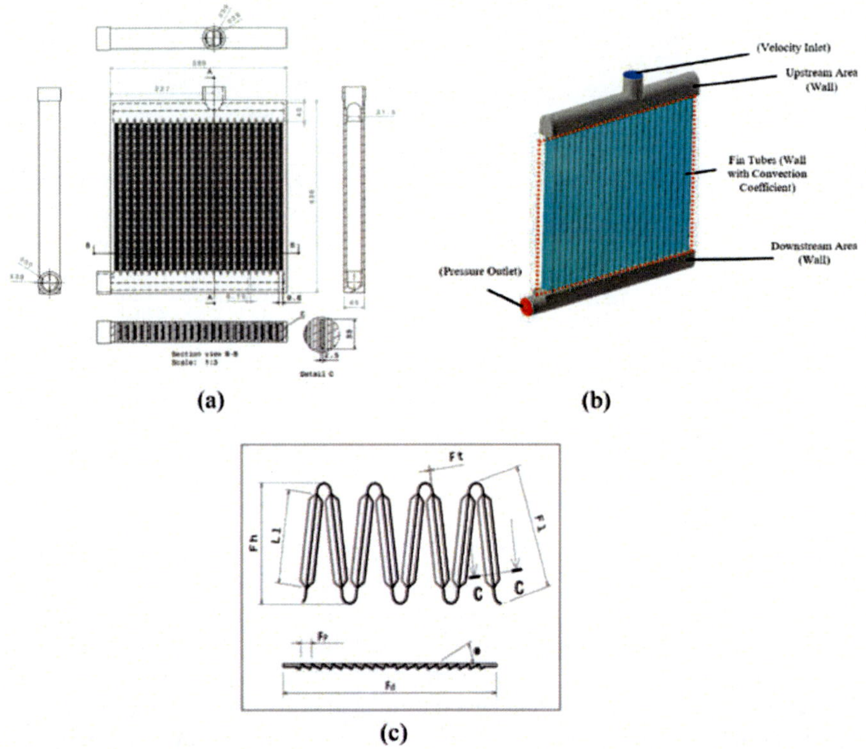

Fig. 2 **a** Geometry of the LP air cooler; **b** LP air cooler model; **c** louvered fins

inlet area is defined as velocity inlet, and the outlet area is defined as pressure outlet. The meshing types used in this simulation are hexahedral mesh with maximum skewness of 0.64 (good) with the number of cells in the fluid domain which is 1.254.802 cells as shown in Fig. 3.

In this study, fin tubes are modeled as a wall with a convection coefficient of 273,9414, 289,2622, 304,4525, and 348,0823 W/m^{2}K as a variation for simulation velocity cooling

Table 1 Geometry of LP air cooler

Fh	10	mm
Fp	3.5	mm
Ft	0.1	mm
Fd	45	mm
Lp	3.5	mm
L1	8.5	mm
θ	27	deg

Table 2 Boundary condition

Free variable			Controlled variable	
Variable	Value	Compressed air temperature	Variable	Value
Cooling air velocity	**4.4 m/s**	**71.62** °C	Velocity inlet Temperature inlet Pressure outlet	9.61 m/s 461.15 K 556,776 Pa
	6.9 m/s	70.74 °C		
	8.5 m/s	69.95 °C		
	14 m/s	80.04 °C		

* Operational data

air 4.4, 6.9, 8.5, and 14 m/s. The free stream temperature is 303 K, obtained from field measurement.

3 Results

The contour temperature obtained from the simulation can be seen in Fig. 4. From the observation of the four variations in the velocity of the cooling air, all models had the same temperature contour characteristics. Wherein the header part of the inlet described as upstream area wall, there is no color degradation (red color), because in the area it is assumed that there is no convection heat transfer and radiation heat transfer is ignored. Heat transfer began to occur on the side of the LP air cooler tube inlet, where the color modeling results began to show a change from red to greenish to the outlet tube and the contour of the color changed to bluish.

Temperature distribution for variation velocity cooling air on isosurface outlet LP air cooler in the direction of the vertical axis and horizontal axis can be seen in Fig. 5.

From Fig. 5, it can be seen that the highest temperature of outlet air through LP air cooler is in the lower-side and right-side area. The distribution temperature at the middle

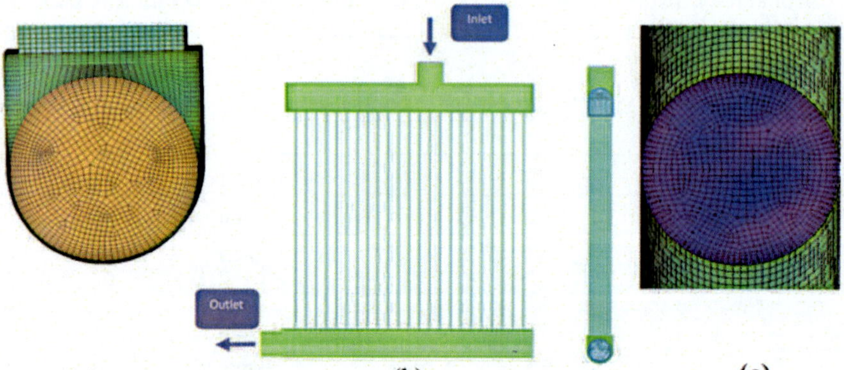

(a) (b) (c)

Fig. 3 Meshing **a** outlet; **b** LP air cooler; **c** inlet

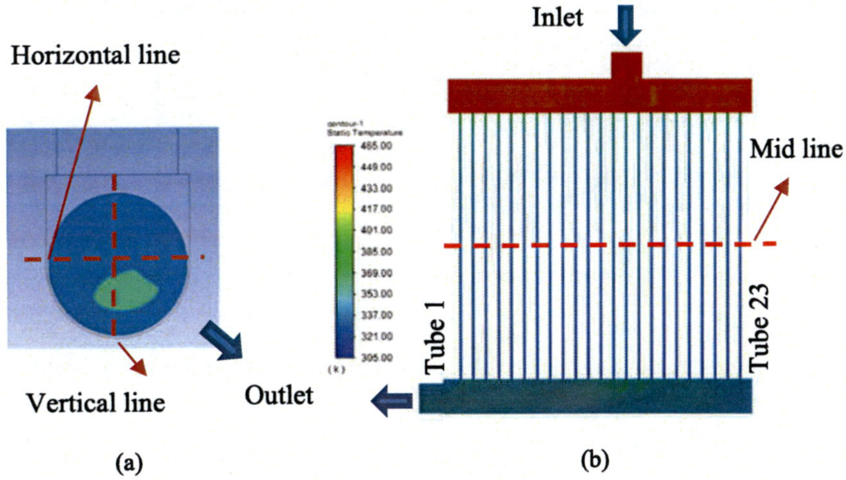

Fig. 4 Temperature contour at cooling air velocity 4.4 m/s **a** outlet; **b** LP air cooler

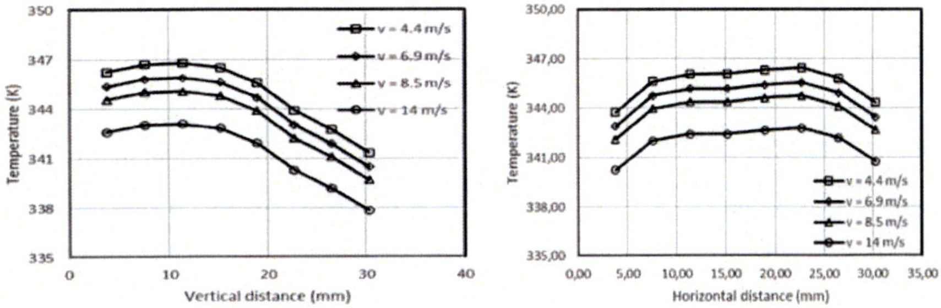

Fig. 5 Temperature distribution for variation velocity cooling air on isosurface outlet of LP air cooler in the direction of the vertical and horizontal axis

line tube number 1 until 23 for all variations velocity cooling air can be seen in Fig. 6. The highest and the lowest temperatures are in tubes number 14 and 8.

The results of the simulation for all variations of cooling air velocity can be seen in Table 3.

Figure 7 shows the pressure contours in the LP air cooler. For all of the variations in cooling air velocity, the pressure contour had the same characteristics. The header part of the inlet is described as an upstream area wall, and the greatest pressure occurs in the area that directly confronts the inlet airflow called stagnation point. This is due to the large mass of air flowing which is not able to directly pass through the tube.

From Fig. 8, the header part of the inlet is described as an upstream wall area; the highest velocity is in the area around the inlet shown by the contours of greenish color. While at the end of the header area inlet both on the left side and the right side, the air flow velocity flowing into the tube is decreased. In the outlet header, the highest air velocity is dominated by the tube on the left side in the same direction of the outlet. The highest velocity that occurs on the LP air cooler outlet is at the bottom position and

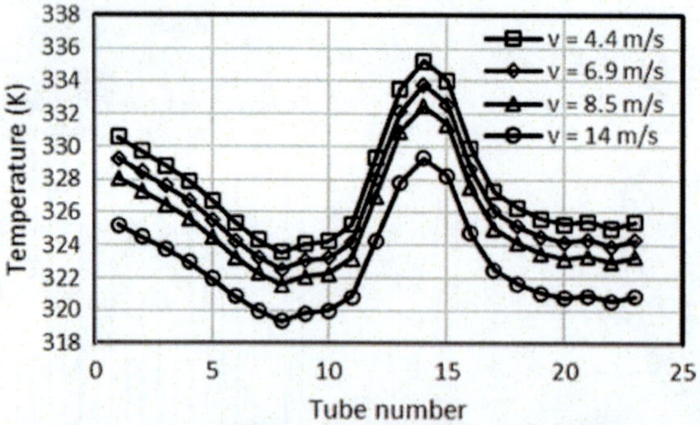

Fig. 6 Temperature at the middle line tube 1 until 23 for all variations velocity of cooling air

gradually decreases toward the top of the vertical axis. The simulation results that the addition of cooling air velocity in the LP air cooler has no effect on the contour of the velocity on the internal flow (compressed air).

Table 3 Simulation result

v (m/s)	h (W/m^2K)	T inlet (K)	T outlet (K)	T outlet (°C)
4.4	273.94	461	344.77	71.62
6.9	289.26		343.89	70.74
8.5	304.45		343.10	69.95
14	348.08		341.19	68.04

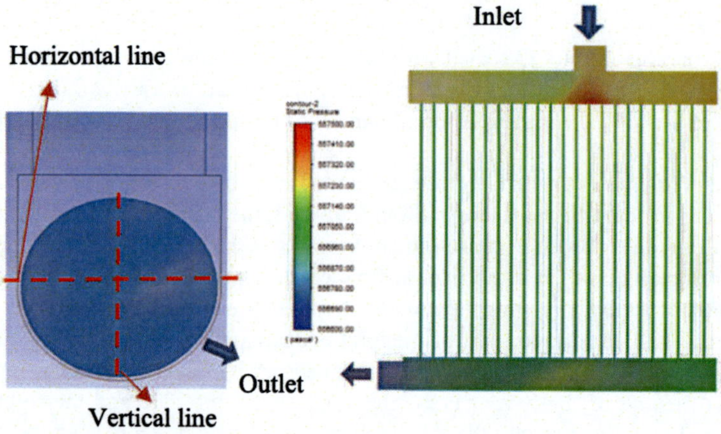

Fig. 7 Pressure contour

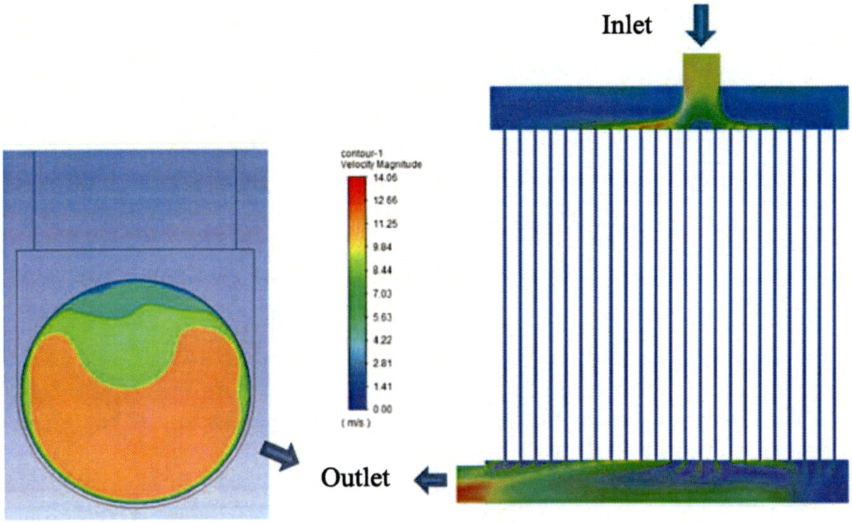

Fig. 8 Velocity contour

The velocity distribution for all variations of velocity cooling air on the outlet of the LP air cooler can be seen in Fig. 9.

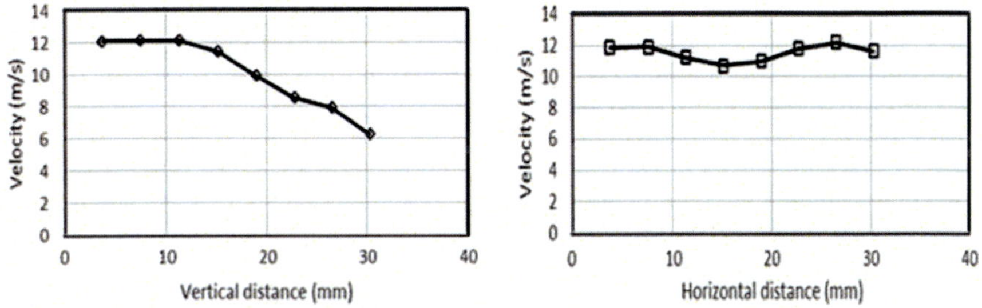

Fig. 9 Velocity distribution at outlet LP air cooler (vertical and horizontal)

From Eq. (1), the maximum operating pressure of starting air compressor that can be operated to avoid burned of lubricant can be calculated as given in Table 4. The maximum temperature resulted from the compression steps in the HP compressor side must be lower than 270 °C.

4 Conclusion

It can be concluded from the simulation results that

1. The cooling air velocity of 4.4 m/s can decrease the compressed air temperature from 188 to 71.62 °C in the LP air cooler outlet. With an outlet temperature of 71.62 °C, the maximum operating limit of the starting air compressor is 23.03 bar gauge.

Table 4 Maximum operating pressure of the starting compressor

V (m/s)	h (W/m^{2}K)	T suction HP side (K)	P discharge HP side (bar)
4.4	273.94	344.77	23.03
6.9	289.26	343.89	23.46
8.5	304.45	343.10	23.86
14	348.08	341.19	24.85

2. The addition of cooling air velocity by 218% (4.4 m/s to 14 m/s) from existing conditions can decrease the compressed air outlet temperature by 3.57 °C (68.04 °C) and increase the operating limit of the starting air compressor to 1.82 bar (24.85 bar gauge).
3. Increasing cooling air velocity by 218% (4.4 m/s to 14 m/s) from existing conditions will result in a temperature in the HP compressor of 284.01 °C if the starting compressor is operated at a pressure of 30 bar gauge. This condition is still higher than the maximum operating temperature limit of lubricating oil (270 °C).

References

1. Ribeiro F, de Conde KE, Garcia EC, Nascimento IP (2020) Heat transfer performance enhancement in compact heat exchangers by the use of turbulators in the inner side. Appl Therm Eng 173(115188):1359–4311
2. Wais P (2016) Correlation and numerical study of heat transfer for single row. Procedia Eng 157:177–184
3. Wibowo KA, Dwiyantoro AB (2014) Studi numerik peningkatan cooling performance pada lube oil cooler gas turbine yang disusun secara seri dan paralel dengan Variasi Kapasitas aliran lube oil. J Teknik Pomits 3(2):2301–9271

Numerical Study of Fin Height Effects with the Staggered Arrangement in Annular-finned Tube Heat Exchangers

Deluxe La(✉), Prabowo, and Tri Vicca Kusumadewi

Department of Mechanical Engineering, Institut Teknologi Sepuluh Nopember, Surabaya, Indonesia
ladeluxe168@gmail.com

1 Introduction

Circular long-distance finned tubes are widely used for heat exchangers connected to various areas of mechanical equipment such as control plants, refrigeration frameworks, hardware cooling, nourishment building, agrarian generation, and livestock upgrades. Due to their straightforward geometry, their generation is expanding with the propels in fabricating advances such as roll shaping and welding. Regarding the material science of finned tube combinations, an expanded surface range of blades contributed to the advancement of heat transfer proficiency, but the higher blade thickness of the contract pitch makes higher stream resistance or weight drop. Subsequently, analysts and engineers are continuously looking forward for the ideal blade setup for this kind of heat exchanger. Past its building importance, investigating normal convection around fundamental geometrical bodies is exceptionally basic to understanding the key material science of convection warm exchange [1].

Kearney [2] depicted that impacts of the bundle design and the height of fin on the local and the heat transfer performance average are conjugated. Low-finned tubes ($df/d = 2$) can perform better in staggered settings, whereas high-finned tubes ($df/d = 4$) do not providing a significant impact on the bundle design. Kearney appeared that the dormant areas covered less of the overall fin range for the inline settings as df/d increments. It should be concluded that the coefficient of heat transfer will be diminished and the pressure drop will be incremented when the height of fin increments. Mon [3] also has been investigated the fin height effect with various dimensions and Reynold number. Recently, Bilirgen et al. [4] the finned tube is determined for a single line with the inline setting in a crossflow so that it can investigate the effect of fin height on overall heat transfer and its pressure drop. Chen et al. [5] also have been conducted numerically about heat transfer by natural convection from a vertical cylinder with circular fins. The simulation was carried out by varying the fin to tube diameter ratio (D/d) in the range of 2–5 and 0.126–5.840, respectively.

The aim of this research is to numerically explore the temperature distributions on the fins surface of four-row staggered bundles. The objective was to determine the fin height effect on the rate of heat transfer and pressure drop. Numerical results validation

M. Kolhe et al. (eds.), *Recent Advances in Renewable Energy Systems*,
Lecture Notes in Electrical Engineering 876,
https://doi.org/10.1007/978-981-19-1581-9_23

is also the secondary aim of the study. It is emphasized that cold fluid is interior the tubes, while the exterior is a hot flue gas. To conduct the numerical simulation, the CFD by ANSYS FLUENT is used for analysis [6]. In this simulation, four lines of the bundle in crossflow were performed.

2 Numerical Simulation

2.1 Computational Domains

The computational domain of two sides view has been modeled for four lines of the bundles as shown in Fig. 1. Within the *y*-coordinate, symmetry boundaries happened in dash lines as shown in Fig. 1a. Within the *z*-coordinate, symmetry planes happened at fin in the middle of the plane as shown in Fig. 1. Within the *x*-coordinate, the computational domain expanded a time of longitudinal tube pitch as upstream and 3 times of longitudinal tube pitch as downstream.

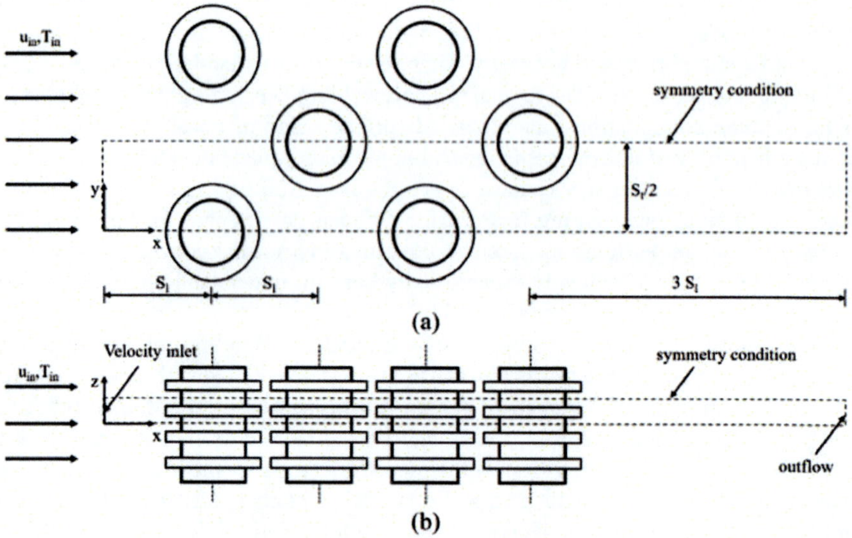

Fig. 1 Computational domains **a** Front view and **b** Top view

2.2 Governing Equations

This stream is considered 3D, steady-state, incompressible, and turbulent flow. The equations governing for the steady, the RNG *k*-ε model are as follows [7]:

Continuity:

$$\frac{\partial \rho}{\partial t} + \frac{\partial}{\partial x_i}(\rho u_i) = 0 \quad (1)$$

Momentum:

$$\rho\frac{Du_i}{Dt} = -\frac{\partial p}{\partial x_i} + \frac{\partial}{\partial x_j}\left[\mu\left(\frac{\partial u_i}{\partial x_j} + \frac{\partial u_j}{\partial x_i} - \frac{2}{3}\delta_{ij}\frac{\partial u_l}{\partial x_l}\right)\right] + \frac{\partial}{\partial x_j}\left(-\rho u_i' u_j'\right) \tag{2}$$

where

$$-\rho\overline{u'_i}\overline{u'_j} = \mu_t\left(\frac{\partial u_i}{\partial x_j} + \frac{\partial u_j}{\partial x_i}\right) - \frac{2}{3}\left(\rho k + \mu_t\frac{\partial u_i}{\partial x_i}\right)\delta_{ij} \tag{3}$$

The RNG k-ε model of FLUENT [7] takes on the following transport equations:

$$\rho\frac{Dk}{Dt} = \frac{\partial}{\partial x_i}\left[\alpha_p\mu_{\text{eff}}\frac{\partial k}{\partial x_i}\right] + \mu_t S^2 - \rho\varepsilon \tag{4}$$

$$\rho\frac{D\varepsilon}{Dt} = \frac{\partial}{\partial x_i}\left[\alpha_p\mu_{\text{eff}}\frac{\partial \varepsilon}{\partial x_i}\right] + C_{1\epsilon}\frac{\varepsilon}{k}\mu_t S^2 - C_{2\epsilon}\rho\frac{\varepsilon^2}{k} - R \tag{5}$$

where σ_k, σ_ε, and C_μ are constants.

$$\mu_t = \rho C_\mu\frac{k^2}{\varepsilon} \tag{6}$$

The theory of RNG presents the values for the constants which are $C_{1\varepsilon} = 1.42$ and $C_{2\varepsilon} = 1.68$.

The energy equation is as follows:

$$\frac{\partial}{\partial t}(\rho E) + \frac{\partial}{\partial x_i}(u_i(\rho E + p)) = \frac{\partial}{\partial x_i}\left(k_{\text{eff}}\frac{\partial T}{\partial x_i}\right) \tag{7}$$

where E is the total energy and $k_{\text{eff}} = k + k_t$ the effective conductivity, including the turbulent thermal conductivity k_t.

2.3 Boundary Conditions

Flue gas with a certain velocity $u_{\text{in}} = 4.7$ m/s flows commonly to the bundle, temperature inlet $T_{\text{in}} = 431.15$ K, the default of turbulent intensity, and turbulent viscosity ratio was set. All velocities at the position of y- and z-coordinate are assumed to zero. The fin and tube materials are considered as aluminum. The thermal conductivity of aluminum is 202.4 W/m.K. A constant temperature, $T_w = 381.15$ K, was set to all tube surfaces including fins bases. The laminar whereas the disparate partitions of the bundle will be served as turbulent zones assumed to the stream between the fins. The symmetry boundary conditions are used at the top and bottom and outlet boundary as an outflow.

2.4 Solution Algorithm

In the current work, there are two fundamental regions of the computational domain that can be considered in terms of appreciating control of the element density. Near the tube wall and fin, it should be finer mesh sizes to determine the second stream (flow separations and horseshoe vortices), where the lofty slopes are to be. The grid generations of the front view and side view are shown in Fig. 2. Indeed, there are a few impediments to the computer resources and CPU time, and 1,800,000 to 2,000,000 cells are used for discrete the computational domains. To decide the degree of an element independence to the results, what is important to be taken for the relative mistakes within the found the means velocity and temperature at a specific point between such grids, which should be less than 5%. The residuals used for velocity components, continuity, turbulent kinetic energy, and turbulent dissipation rates are under 10^{-4} for convergence solution. Numerical investigations are conducted for four various fin heights from 5 to 11 mm increments of 2 mm, which are mentioned in Table 1. All simulations are simulated in certain inlet velocity $V_{\text{in}} = 4.7$ m/s. All the cases are operated in a steady state. In this study, the model of RNG based k-ε turbulence is used to vaticinate the characteristics of fluid flow and heat transfer. To discretize the governing equations, the second-order upwind scheme and SIMPLE algorithm are applied in the computational fluid dynamics (CFD), FLUENT [7].

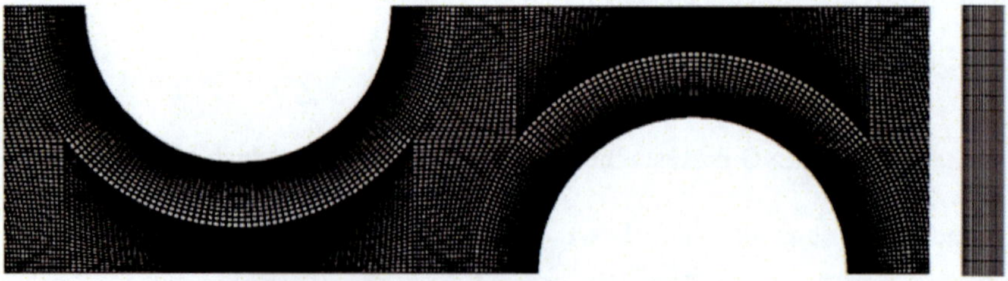

Fig. 2 Grid generation of front view at the middle and side view

Table 1 Dimensions used in simulation

Name	Symbol	Values
Transverse tube pitch	S_t	46.9 mm
Longitudinal tube pitch	S_l	34.3 mm
Tube diameter	D	16.38 mm
Fin spacing	s	2 mm
Fin height	h_f	5, 7, 9, and 11 mm
Fin thickness	t_f	0.25 mm
Number of rows	n	4

2.5 Solution Strategy

The efficiency and the effectiveness of the fin have been calculated as the following equation below:

$$\eta_f = \frac{q_f}{q_{\max}} = \frac{h \times A_f \times \left(\overline{T}_f - \overline{T}_\infty\right)}{h \times A_f \times \left(\overline{T}_b - \overline{T}_\infty\right)} \tag{8}$$

$$\varepsilon_f = \frac{q_f}{q_{wf}} = \frac{h \times A_f \times \left(\overline{T}_f - \overline{T}_\infty\right)}{h \times A_b \times \left(\overline{T}_b - \overline{T}_\infty\right)} \tag{9}$$

where $\overline{T}_f$ is the fin mean temperature, $\overline{T}_b$ is the tube wall mean temperature, the fin face surface area $A_{f,f}$, the fin tip surface area $A_{f,t}$, and A_f represents total fin area:

$$A_f = A_{f,f} + A_{f,t} \tag{10}$$

where $A_{f,f} = 2\pi\left(r_2^2 + r_1^2\right)$ and $A_{f,t} = 2\pi r_2 t_f$.

The fin base area is calculated using the formula:

$$A_b = 2 \times \pi \times r_1 \times t_f \tag{11}$$

where r_1 is tube radius, r_2 is fin radius, and t_f is fin thickness. The total area is obtained as:

$$A_t = A_b + A_f \tag{12}$$

The average air temperature is acquired with the equation:

$$T_\infty = \frac{T_{\text{in}} + T_{\text{out}}}{2} \tag{13}$$

where T_{in} is inlet temperature and T_{out} is outlet temperature. The heat transfer rate equation is:

$$\dot{Q} = \dot{m} \times c_p \times (T_{\text{in}} - T_{\text{out}}) \tag{14}$$

where the specific heat capacity, c_p. The rate of mass flow is given by:

$$\dot{m} = \rho \times v \times A \tag{15}$$

3 Results and Discussion

To prove the fin height effects, numerical simulation was conducted for four-row staggered bundles. For each simulation, simulations were conducted in the conditions of varying four of the fin heights with fixed fin thickness by 0.25 mm, fin spacing by 2 mm, outside diameter of the tube by 16.38 mm, transverse tube pitch by 46.9 mm, and longitudinal tube pitch by 34.3 mm. The results are carried in terms of the contour of fin temperature, heat transfer rate, pressure drop, efficiency, and effectiveness.

3.1 Fin Contour Temperature

Figure 3 shows temperature contour on the fin surface area. Regarding this figure, they noticed that the average temperature on the fin first line is higher than the fin second line. The uniform velocity enters the bundle from the inlet surface to the outlet surface, while velocity flow pass fin first row is lower than velocity flow pass fin second row because of the configuration of the bundle is staggered arrangement. So, air velocity first touching to fin first row before flow passes fin second row. Area after touching to fin first row was narrower, and air velocity will be increased at that position. Based on the correlation of velocity and temperature, heat is the exchange of motor vitality between atoms. In case the speed is more, the motor vitality will be more. So that the heat exchange will be more as well. Since the temperature is the degree of heat exchange, the temperature increments as the speed increments.

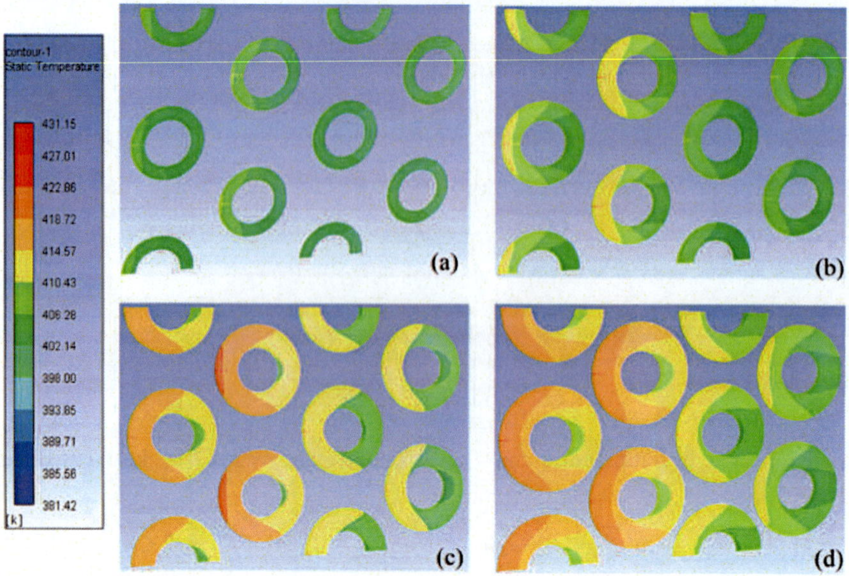

Fig. 3 Fin temperature contour for fin heights **a** 5 mm **b** 7 mm **c** 9 mm and **d** 11 mm

3.2 Heat Transfer Rate and Pressure Drop

Figure 4 represents the effects of fin height on the heat transfer rate and the pressure drop. Figure 4a illustrates certainly that the rate of heat transfer is straight influenced by the height of fin since expanding the fin height will enlarge the heat transfer surface zone of the heat exchanger. The actuality that the utilize of the larger fin provides a larger heat transfer rate is led by delicate decrement air outlet temperature of the bundles. Figure 4b shows that the increasing of fin height delivers an increase of pressure drop about 17.8% for each fin height. The boundary layer becomes thicker, while fin height is increasing.

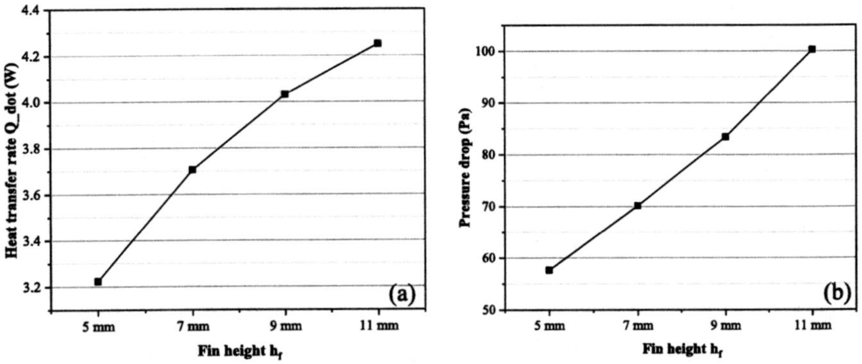

Fig. 4 Effect of the height of fin on **a** Heat transfer rate and **b** Pressure drop

3.3 Efficiency and Effectiveness

The efficiency and effectiveness of the fin for each fin height of the bundles have been calculated as shown in Fig. 5. For fin height (5 mm), efficiency has the highest effectiveness with a percentage of 55%. The gradient decreasing of efficiency from fin height 5 mm to 11 mm of increment of 2 mm is rapidly decreased, respectively, and is presented in Fig. 5a. Nevertheless, the effectiveness of fin height 11 mm was the highest with a value of 35.21. The slope increasing of effectiveness from fin height 5 mm to 11 mm of increment 2 mm is quickly increased in particular and is shown in Fig. 5b.

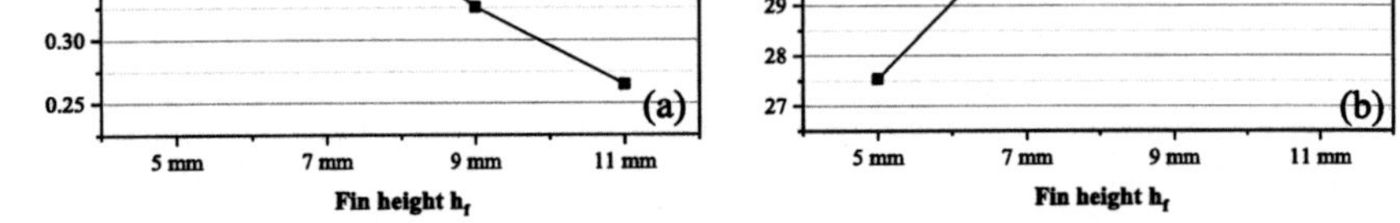

Fig. 5 Effect fin height on **a** Fin efficiency and **b** Fin effectiveness

4 Conclusion

The numerical result on the effects of height of fin indicated that expanding the fin height gives increased rate of heat transfer neither increased the pressure drop of the heat exchangers. Besides that, the result appeared that the fin efficiency will be decreased neither the effectiveness of the annular-finned tube bundle will be increased by expanding the fin height gives. From the first row to fourth row, the temperature distributions in the bundle vary significantly. In addition, the temperature in the mid-plane is higher than the surface plane on the fins. The temperature of the fin tip on the surface of the upstream fin is higher due to the thin boundary layer.

References

1. Kang HC, Chang S-M (2018) The correlation of heat transfer coefficients for the laminar natural convection in a circular finned-tube heat exchanger. J Heat Transfer 140(3)
2. Kearney SP, Jacobi AM (1995) Local and average heat transfer and pressure drop characteristics of annularly finned tube heat exchangers. Air conditioning and refrigeration center. College of Engineering
3. Mon MS (2003) Numerical investigation of air-side heat transfer and pressure drop in circular finned-tube heat exchangers
4. Bilirgen H, Dunbar S, Levy EK (2013) Numerical modeling of finned heat exchangers. Appl Therm Eng 61(2):278–288
5. Senapati JR, Dash SK, Roy S (2017) Numerical investigation of natural convection heat transfer from vertical cylinder with annular fins. Int J Therm Sci 111:146–159
6. Park K, Choi D-H, Lee K-S (2004) Optimum design of plate heat exchanger with staggered pin arrays. Numer Heat Transf Part A Appl 45(4):347–361
7. F. Incorporated (1998) Fluent 5 users guide volume 1 July 1998. Fluent Incorporated

Study of the Effect of Changes in Fill Grid Cooling Tower Unit 2 Salak Geothermal Power Plant

Yuansah[1](✉), Wawan Aries Widodo[1], and Wahyu Somantri[2]

[1] Department of Mechanical Engineering, Institut Teknologi Sepuluh Nopember, Surabaya 60111, Indonesia
yuansah@indonesiapower.co.id
[2] PT. Indonesia Power, Surabaya, Indonesia

1 Introduction

Geothermal power plant is one of the new renewable power plants. One of the uses of geothermal energy is the construction of the Gunung Salak geothermal power plant in 1993 with a capacity of 3 × 55 MW. Gunung Salak geothermal power plant has been operating since 1993, and in 2004, the capacity was upgraded to 3 × 60 MW. The upgrade activities replaced level 1 & 2 diaphragms. Cooling tower Gunung Salak geothermal power plant type's cross-flow mechanical-induced draft. Due to the increasing consumption of steam for generation, the condensate water that is cooled in the cooling tower also increases and causes the cooling tower to suffer damage to the fill grid, drift eliminator, and concrete damage. After operating for 26 years, the cooling tower of Salak geothermal power plant unit 2 was finally rehabilitated. Rehabilitation is carried out to restore performance of cooling tower Unit 2 Gunung Salak geothermal power plant. There are several studies that have been done to solve about review of performance cooling tower. Cooling tower effectiveness is expressed in range and approach. Approach is the difference in temperature of the water coming out of the cooling tower with the wet-bulb temperature entering the cooling tower. Meanwhile, range is the temperature difference between the cooling tower inlet water in the hot basin and the cooling tower water temperature after passing through the fill grid in the cold basin by [1]. Described a detailed methodology for the thermal design of wet, counter-flow and cross-flow types of mechanical and natural draft cooling towers. The fill or packing, natural draft tower, fan design for a mechanical draft cooling by [2]. Developed a mathematical model for predicting the performance of a natural draft cooling tower. The calculated results were validated by the measured data by [3]. Developed heat and mass transfer characteristic of wet counter-flow cooling towers. They are presented by increasing in mass flow ratio; tower effectiveness is increased, but temperature ratio is decreased by [4]. Developed relations for various geometries and configurations and explained that the mass transfer relation could be used to calculate an effective drop diameter by [5]. Investigation involves experimental and two-dimensional computational fluid dynamics analysis of an actual industry operated cooling tower. Inlet water temperature and mass

M. Kolhe et al. (eds.), *Recent Advances in Renewable Energy Systems*,
Lecture Notes in Electrical Engineering 876,
https://doi.org/10.1007/978-981-19-1581-9_24

flow rate of water and air are having main influence on the performance of counter-flow-induced draft cooling tower. Effectiveness of the cooling tower can be increased up to 20% by optimizing the liquid to gas ratio (L/G) of the cooling tower by [6]. Analyses the heat and mass transfer phenomena inside a wet cooling tower with forced air draft and counter-flow arrangement between air and water stream. The scientific contribution of this study is in the application of the porous zone model with appropriate sink terms for the momentum conservation equations in the mathematical model of the cooling tower by [7]. Design and numerical study of variations in the air inlet position on heat and mass transfer in induced draft cooling towers for organic Rankine cycle systems, both cross flow and counter-flow cooling towers by [8]. One of the activities carried out is to replace the fill grid with the new one shown in Fig. 1b and the old type shown in Fig. 1b. It is hoped that with the replacement of the new fill grid, the effectiveness of the cooling tower can increase. Also, the cooling tower rehabilitation can be the basis for the implementation of rehabilitation for the implementation of rehabilitation in other units at the Salak geothermal power plant.

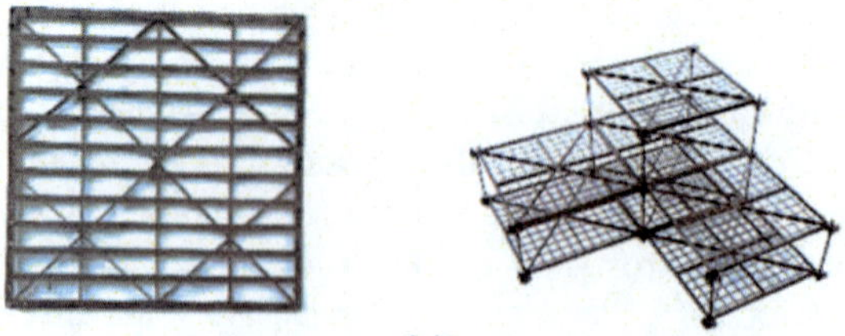

(a)

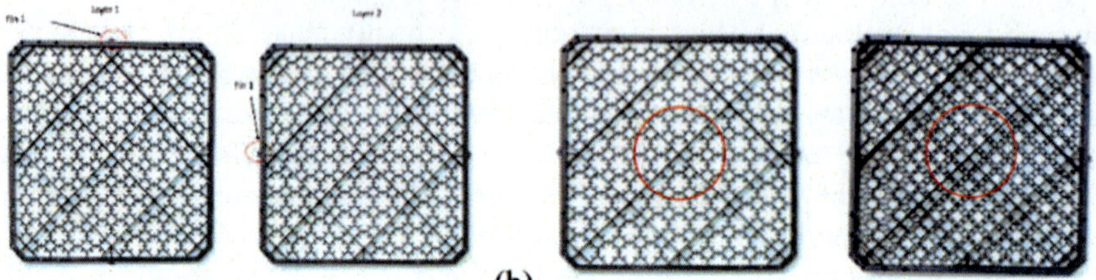

(b)

Fig. 1 **a** Fill grid before rehabilitation **b** Fill grid after rehabilitation

2 Methodology

2.1 Review Cooling Tower After Rehabilitation

Data used manual design, data commissioning before and after rehabilitation for evaluating and calculation of the performance of the cooling tower. Data used are water inlet and outlet cooling tower, mass flow water to cooling tower, temperature wet bulb, flow steam, power generation for calculating the range, approach, effectiveness cooling tower, L/G ratio, and heat capacity.

$$R = T_{W\,\text{in}} - T_{W\,\text{out}} \tag{1}$$

$$A = T_{W\,out} - T_{Wb} \tag{2}$$

$$\eta = \frac{R}{R + A} \tag{3}$$

$$Q = \dot{m} \times (h_2 - h_1) \tag{4}$$

$$EL = 2.47\% \ \times \dot{m} \tag{5}$$

$$DL = 0.05\% \ \times \dot{m} \tag{6}$$

$$\frac{L}{G} = \frac{\dot{m}_{Water}}{\dot{m}_{Air}} \tag{7}$$

$$\text{Tower Capability} = \frac{\text{Capacity Unit Actual}}{\text{Capacity Unit manual}} \times 100\% \tag{8}$$

And then the comparison of design manual, data before and after rehabilitation. And then compare the calculation result. The data of design manual cooling tower are shown in Table 1.

Table 1 Design manual cooling tower

Parameter	Unit	Value
Type of cooling tower		Crossflow
Number of cells		2 × 5
Design waterflow	m^3/h	15,000
Hot water temperature	°C	42.70
Cold water temperature	°C	26.00
Cooling range	°C	16.70
Wet bulb temperature	°C	19.50
Max.evaporation losses	%	2.47
Max.drift losses	%	

2.2 Numerical Method

Numerical simulation of cooling tower fill grid layer arrangement uses Ansys 18.2 software to determine the optimization of cooling tower fill grid layer arrangement. The simulations include making 3D geometry using inventor, meshing, porous zone, preprocessing, and data retrieval/post processing like Fig. 2. Determination of boundary conditions, making modeling domains including water inlet and outlet, ambient air and fan output air and setting numerical porous zone as porosity for layer 1, 2, and 3.

To determine the optimization of cooling tower, fill grid layer arrangement uses five variations. One of them is existing, and four variations are shown in Table 2.

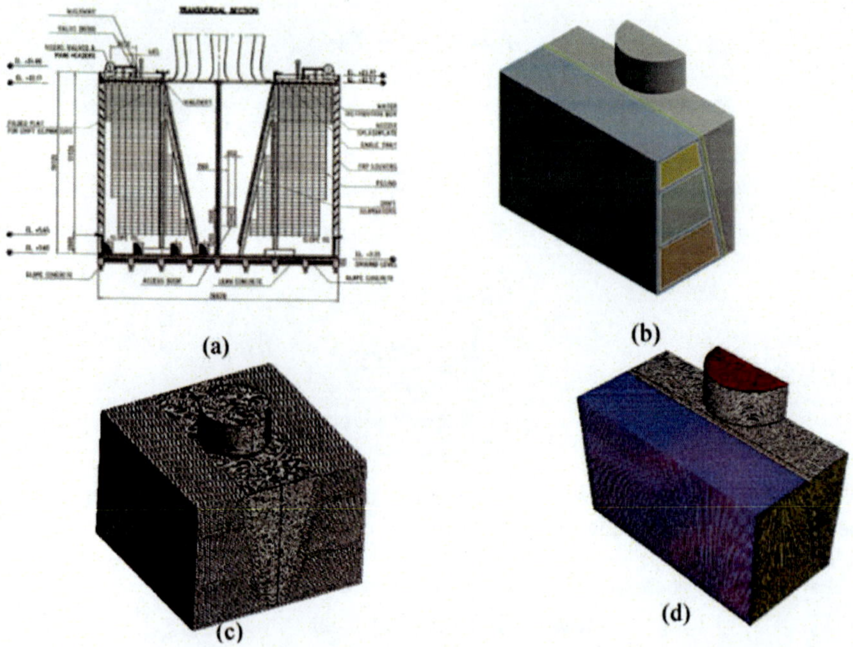

Fig. 2 **a** Geometry cooling tower, **b** Model domain fluida **c** Meshing global, **d** Meshing polyhedra fluent

Table 2 Variation arrangement layer plant for next rehabilitation Unit 1 & 3

Case	Distance layer (mm)		
	Layer I	Layer II	Layer III
Existing	200	300	400
Case I	150	300	400
Case II	300	200	400
Case III	300	300	200
Case IV	200	200	200

3 Results and Discussion

3.1 Performance of Cooling Tower

Before analyzing the fill grid replacement using the numerical scheme method with Ansys 18.2, an evaluation of the cooling tower Unit 2 capacity of the Gunung Salak geothermal power plant before and after rehabilitation was carried out. Data used are water inlet and outlet cooling tower, mass flow water to cooling tower, temperatur wet bulb, flow steam, and power generation for calculating the range, approach, effectiveness cooling tower, L/G ratio, and heat capacity. And then comparison the calculation result. The result showing that the rehabilitation with configuration fill layer grid type new one which indicated increasing effectiveness cooling tower from 74.01% to 83.71% or equivalent 0.37 MW shown in Fig. 3. The use of a new type of fill grid is able to cool the mass flow to 15,329.02 t/h from the manual design of 15,000 t/h so that the cooling tower's cooling capability becomes 102.19%. As a result of the increased mass of water entering the cooling tower and steam entering the turbine, the resulting evaporation loss and drift loss will increase.

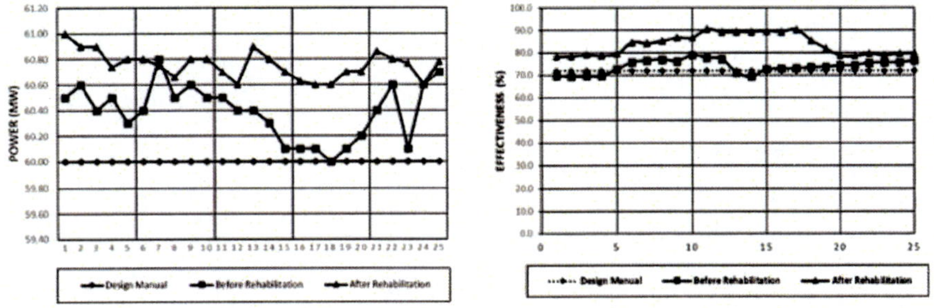

(a) (b)

Fig. 3 **a** Power **b** effectiveness

The water inlet entering the cooling tower has increased from 51.44 °C to 54.64 °C shown in Fig. 4a. The new type of fill grid is able to produce a lower cooling tower outlet temperature when compared to the old type of fill grid from 29.89 °C to 29.04 °C shown in Fig. 4b. The new fill grid is able to produce a better range value from 21.56 to 25.60 shown in Fig. 5b. The higher of the range, the better the cooling tower performance. The smaller approach's the better performance of cooling tower shown in Fig. 5a. For the ratio of water and air mass (L/G) before rehabilitation of 5 for the lowest effectiveness of 69.41% and after rehabilitation, it decreased by 4.68 for the highest effectiveness of 79.63%. To find out the optimization of the arrangement of the fill grid layer, four variations of the distance between layers were carried out to determine the achievable cooling tower output temperature and then potentially non-destructive drift eliminator because temperature distribution pattern.

3.2 Result Simulation

The boundary condition using performance data after rehabilitation is used for simulation arrangement case existing. Determination of boundary conditions, making modeling domains including mass water, water inlet temperature, air inlet temperature and mass air inlet to cooling tower. And finding optimalization setting of porous zone for layer 1, 2, and 3 as value of porosity. The simulation to find temperature water outlet cooling tower with 1000 iteration. And analysis of contour for knowing effect of arrangement fill grid existing for prevent damage other part in cooling tower for a long time is shown in Fig. 6.

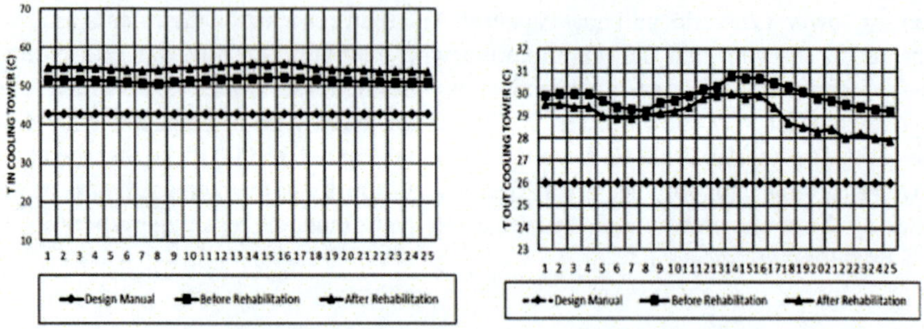

(a) (b)

Fig. 4 **a** Temperature inlet. **b** Temperature outlet

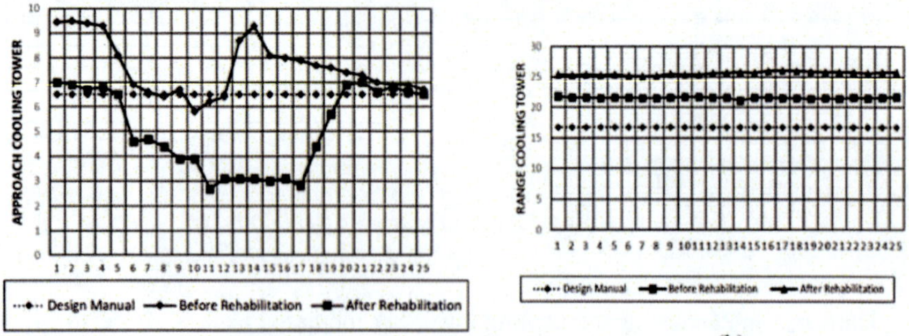

(a) (b)

Fig. 5 **a** Approach, **b** Range

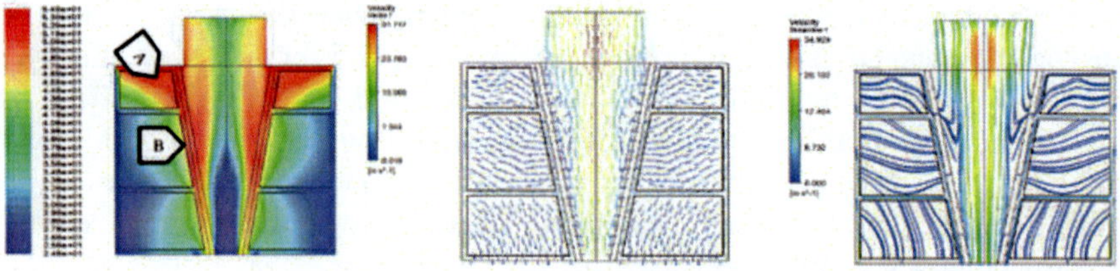

Fig. 6 Contour existing

It can be seen at point A that the highest heat is in the hot basin cooling tower, causing the hot basin cooling tower to experience damage to the concrete. To reduce concrete damage due to hot water temperatures, the surface of the hot basin must be coated to support life time. At point B, a lot of heat from the air and water cooling tower is trapped in the drift eliminator; this causes the drift eliminator to degrade for a long time due to heat. This drift eliminator is made of white PVC as a result of the temperature of the air and water that spreads to the drift area, causing the drift to be damaged by heat, as can be seen from the color of the PVC which changes from white to brownish black like burning. And the surface of the drift undergoes deformation due to heat. In addition, for a long time, there is the potential for drift to escape from the holder due to changes in the length and area of the drift eliminator. And it can cause cooling tower performance to decrease (Fig. 7).

Fig. 7 Hot basin and drift eliminator damage due to water cooling tower temperature

The validation for modeling cooling tower used basic statistic with method one sample t-test for statistic parametrik where the results of the cfd simulation are compared with commissioning data with an accuracy of not more than 5%. And the result is conclusion acceptable when t account is less than t table. The result $t\_account = 0.319 < t\_tabel = 2.10092$. And p-value = 0.0754 more than 5%. In conclusion, modeling is not significantly different from the actual condition. To find out the optimization of the arrangement of the fill grid layer, four variations of the distance between layers were carried out to determine the achievable cooling tower output temperature and then potentially non-destructive drift eliminator because temperature distribution pattern. Simulation is carried out using several variations shown in Table 2. The contour of simulation is shown in Fig. 8 for simulation arrangement fill grid to find best arrangement for outlet temperature cooling tower and temperature result the simulation shown in Fig. 9.

From the simulation and iteration results, it can be seen that the variation of case-1 with a distance of layer I = 150, layer II = 300, and layer III = 400 is able to reduce the cooling tower air to 29.60 °C, and the heat distribution pattern does not spread to the drift eliminator. So, it is hoped that when applied to a cooling tower, the life time of the drift eliminator can be longer.

4 Conclusion

The experimental result was obtained satisfy cooling tower performance with configuration fill layer grid type new one which indicated increasing effectiveness cooling tower from 74.01 to 83.71%. The arrangement fill layer grid as case-1 was utilized numerical simulation by CFD commercial code obtaining good agreement temperature out of cooling tower until 29.60 °C and then has potentially non-destructive drift eliminator.

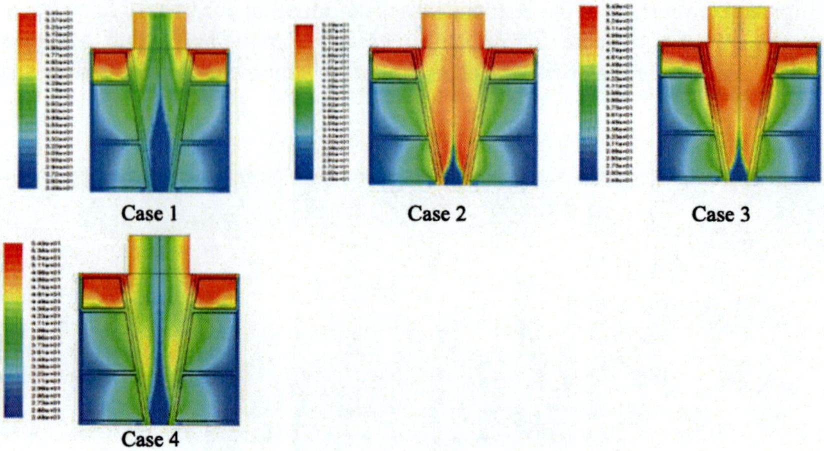

Fig. 8 Comparison analysis arrangement layer fill grid

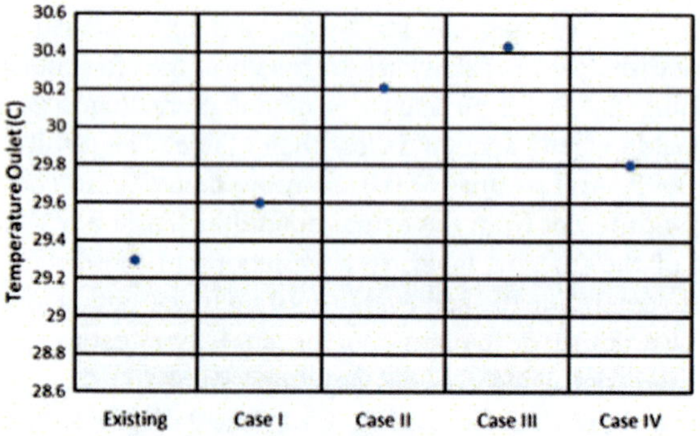

Fig. 9 Result temperature by simulation

References

1. Khan JR, Yaqub M, Zubair SM (2003) Performance characteristics of counter flow wet cooling towers. Energy Convers Manage 44:2073–2091
2. Fisenko SP, Brin AA, Petruchik AI (2004) Evaporative cooling of water in a mechanical draft cooling tower. Int J Heat Mass Transf 47:165–177
3. Fisenko SP, Petruchik AI, Solodukhin AD (2002) Evaporative cooling of water in a natural draft cooling tower. Int J Heat Mass Transf 45:4683–4694
4. Karami M, Heidarinejad G (2008) Investigation of performance characteristics of counter flow wet cooling towers, accepted to oral presentation and publication in 16th international conference of Iranian society of mechanical engineering (ISME), May 14–16. Kerman, Iran
5. De Villiers E, Kroger DG (1999) Analysis of heat, mass and momentum transfer in the rain zone of counter flow cooling towers. J Eng Gas Turbines Power 121(4):751–755
6. Vishnus K, Mathews R (April 2018) Performance analysis and optimization of cooling tower. Int J Sci Res Publ 8(4). ISSN 2250-3153

7. Bleicich P, Sencic T, Wolf I, Bonefacic I (2018) Numerical investigation of heat and mass transfer inside a wt cooling tower. Technical J ISSN 1848-5588
8. Prabowo HF, Prabowo (2013) Perancangan dan studi numerik variasi posisi inlet udara masuk terhadap perpindahan panas dan massa pada induced draft cooling tower untuk sistem organic rangkine cycle. J Sains Dan Seni POMITS 1(1)

Performance Assessment of Drying Machine Using CFD

Mokhammad Fahmi Izdiharrudin(✉), Ridho Hantoro, and Erna Septyaningrum

Department of Engineering Physics, Faculty of Industrial Technology, Intitut Teknologi Sepuluh Nopember, Surabaya, Indonesia
m.f.izdiharrudin@gmail.com

1 Introduction

Dryer is a machine that can be used to get a product of food, vegetables, medicine, and cement, to coal. Some of the methods used to dry these products have a variety that is adapted to the characteristics of the product. Traditional methods using solar heat are used for food and vegetable products, and modern methods using direct fire heating are used for products other than these products. Several studies on the type of dryer were carried out to get the best quality product. In the traditional method, Arman does drying of aloe vera products by utilizing refractance window dryer technology [1]. The drying process of this machine is assisted by hot water to reduce glucose levels up to 19.28% at a temperature of 60. Vigneshkumar drying potato slice products with an indirect solar dryer machine. The method used to obtain the product is phase-change material (PCM). This machine is capable of drying moisture content up to 5.1% [2]. Gupta also dries potato products using a solar greenhouse dryer. In his research, Gupta proved that the machine that had been made was able to dry faster than open drying and got a payback period of 1.9 years [3]. For the modern method, research was carried out by Somchart to determine the increase in the efficiency of the coal-fired utility dryer with the addition of an air heater in the flue gas dryer [4]. This drying aims to reduce the water content of the air. Qi Chen conducts simulations and experiments on fluidized bed dryers. He was able to get an efficiency of up to 92% at a speed of 2.2 m/s [5]. Seung-Hyun optimizes the disk dryer to dry low-grade coal with high moisture content. The high speed of the rotary blade is able to reduce the moisture content at a temperature of 150 °C [6].

Computational fluid dynamics (CFD) method can be used to determine physical parameters such as heat distribution and pressure on the heat transfer medium. Hassan Khodaei has conducted several studies on drying using CFD regarding the drying behavior of biomass in the form of an Arrhenius heat transfer equation model on drying time and decreasing moisture content. A spray-type drying machine is also carried out to determine the prediction of gas flow patterns such as temperature, speed, drying time, and particle position [7]. This design modeling also developed a method of scaling up the drying machine design [8]. Research on optimization of spray-type drying machine on particle size distribution in biochar. The results of this study are contours of temperature, pressure, and velocity [9]. In addition, CFD can also be used to observe

M. Kolhe et al. (eds.), *Recent Advances in Renewable Energy Systems*,
Lecture Notes in Electrical Engineering 876,
https://doi.org/10.1007/978-981-19-1581-9_25

drying in the form of convection heat transfer to temperature distribution, moisture content distribution, moisture diffusivity, moisture transfer coefficient, and drying constant [10]. Research on the effect of humidification on drying has also been carried out with results in the form of physical and chemical parameters [11]. The research variables were the distribution of temperature, drying time, and a decrease in moisture content on the strength of the briquettes after drying [12]. Research on briquette drying modeling has also been carried out to observe the decrease in moisture content and temperature using previous research on heat transfer in agitated contact-convective dryers. It has also been carried out by observing the distribution variables of temperature, humidity, and mass reduction [13]. After that Ekneet also conducted research on dryers with agitated rotary-type dryer designs by observing heat transfer to the concentration of solvents [14]. A review of biomass drying has been carried out by Verma which concludes that the rotary-type dryer design has the largest capacity for drying biomass [15]. Based on this background, this study discusses the performance of the contact-convective type biomass dryer by modeling the agitator design using the CFD method. The choice of contact-convective drying is due to its large drying capacity. Drying performance can be known through two parameters, namely the temperature contour on the heated biochar drying and the exergy quality in the drying process.

2 Mathematical Formulation

To be able to use CFD in the process of calculating momentum, energy, mass equations, and turbulence model, an understanding of the basic properties must be understood. The differential equation that composes the fluid flow profile must be transformed into a simple mathematical equation, which is called the discretization method [16]. The physical aspects of the dynamic flow of fluids are based on the principle of the law of conservation of mass. To calculate the rate of mass increase in a fluid element equal to the net rate of mass flow into a finite element. The left-hand side is expressed as the convection factor and represents the net mass rate of the element across the boundary. The mathematical equation can be written as follows [17].

2.1 Governing Equation

The present analysis is carried out for a steady and incompressible flow of fluid. The governing equations can be written as:

Continuity

$$\frac{\partial}{\partial x_i}(\rho U_i) = 0 \tag{1}$$

Momentum

$$\frac{D(\rho U_i)}{Dt} = -\frac{\partial p}{\partial x_i} + \frac{\partial}{\partial x_j}\left[\mu\left(\frac{\partial U_i}{\partial x_j} + \frac{\partial U_j}{\partial x_i}\right) - \rho u_i u_j\right] \tag{2}$$

$$\frac{p}{\rho} = RT \tag{3}$$

R is the characteristic gas constant = 0.287 kJ/kg-K; ρ is a function of temperature.

Energy

$$\frac{D(\rho T)}{Dt} = \frac{\partial}{\partial x_i}\left[\left(\frac{\mu}{\Pr} + \frac{\mu_t}{\Pr_t}\right)\frac{\partial T}{\partial x_i}\right] \quad (4)$$

k-ε model for turbulent kinetic energy

$$\frac{D}{Dt}(\rho k) = D_k + \rho P - \rho\varepsilon \quad (5)$$

Rate of dissipation of k

$$\frac{D}{Dt}(\rho\varepsilon) = D_\varepsilon + C_{1\varepsilon}\rho P\frac{\varepsilon}{k} - C_{2\varepsilon}\rho\frac{\varepsilon^2}{k} \quad (6)$$

$$u_i u_j = \frac{2}{3}k\delta_{ij} - v_t\left(\frac{\partial U_i}{\partial x_j} + \frac{\partial U_j}{\partial x_i}\right), v_t = 0.09\frac{k^2}{\varepsilon} \quad (7)$$

$$D_\varphi = \frac{\partial}{\partial x_j}\left[\left(\mu + \frac{\mu_t}{\sigma\varphi}\right)\frac{\partial\varphi}{\partial x_j}\right] \quad (8)$$

$$P = -u_i u_j\frac{\partial U_i}{\partial x_j} \quad (9)$$

σ_k and σ_ε are the Prandtl numbers for k and ε.

The constants used in the above k-ε equations are the following: $C_{1\varepsilon} = 1.44$; $C_{2\varepsilon} = 1.92$; $\sigma_k = 1.0$; $\sigma_\varepsilon = 1.3$, $\Pr_t = 1$.

These default values have been determined from experiments with air and water for fundamental turbulent shear flows including homogeneous shear flows and decaying isotropic grid turbulence [18].

2.2 Boundary Conditions

This simulation uses a multiphase Eulerian model with granular viscosity of the Syamlal-O'Brien type. The Gidaspow type was not chosen because of the divergence results. As for the frictional viscosity model, the Johnson type is selected with convergent results. For the turbulence model, k-epsilon was chosen with a turbulent intensity value of 5%. The inlet velocity has a value of 1 m/s with velocity-inlet type. The outlet dryer uses a pressure outlet with a backflow total temperature of 300 K. The base of the wall dryer is imposed no slip and a constant temperature of 353 K.

The modification stage is carried out after validation of the experimental and simulation data obtained. In this process, the temperature distribution can be known. With these data, the disturbance design can be placed in the position of the volume fraction (solid)in the dryer. By using a variation of three angles of 30°, 45°, and 60°. For the 30° of the disturbance design is shown in Fig. 1.

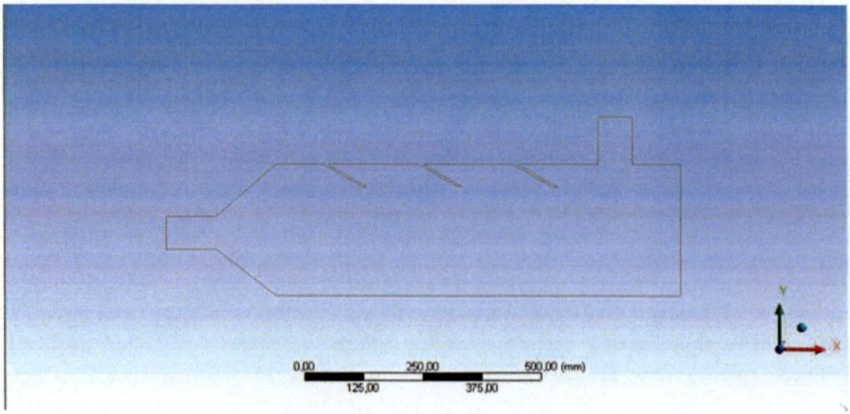

Fig. 1 Disturbance design on drying machine

3 Results and Discussion

3.1 Effect of Grid Density on Dryer Outlet Temperature

To find accurate simulation results, an independence mesh grid is needed using a face sizing meshing type with sizes of 2, 3, 4, 5, and 6 mm. For each meshing size, a uniform Eulerian model simulation is performed. Each meshing size has a different node size from 32,717, 14,383, 8648, 5359, and 3868. The number of nodes does not guarantee the success of a simulation result for the experiment. As shown in Fig. 2, the resulting value difference fluctuates. From taking the grid independence mesh value, it is measured at the outlet dryer temperature with a value of 364.2 K. Sizes 2, 4, and 6 mm have values below the experiment, while the size of 3 and 5 mm has a value above the experiment. It can affect the results of the next simulation.

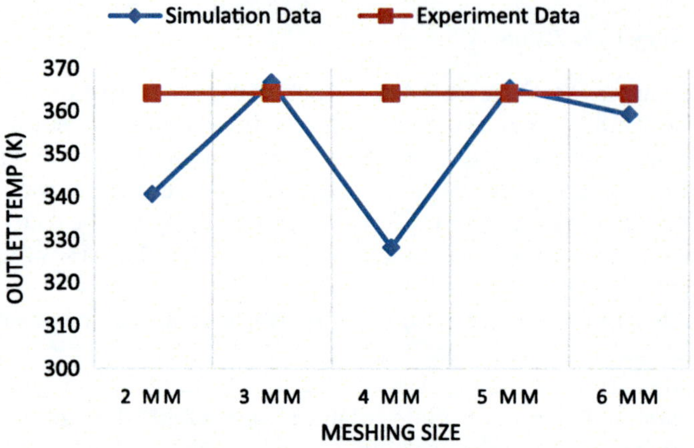

Fig. 2 Grid independence mesh

The largest value difference in the independence mesh grid is at the size of 6 mm at 359.35 K with a difference in value of 1.33%. The value difference that is close to the experimental results is the meshing size of 5 mm. This value is 365.43 K with a difference in value of 0.33%. These results will then be used for simulation of the Eulerian model.

3.2 Effect of Baffle on Drying Performance

In the modification of the disturbance design, the angles are 30°, 45°, and 60°. Figure 3 shows the value of the Nusselts number increases with every increase in angle with values of 5.48, 11.85, and 18.03. The highest value of the Nusselts number is obtained at an angle of 60°. This will certainly affect the temperature contour and the distribution of the fraction.

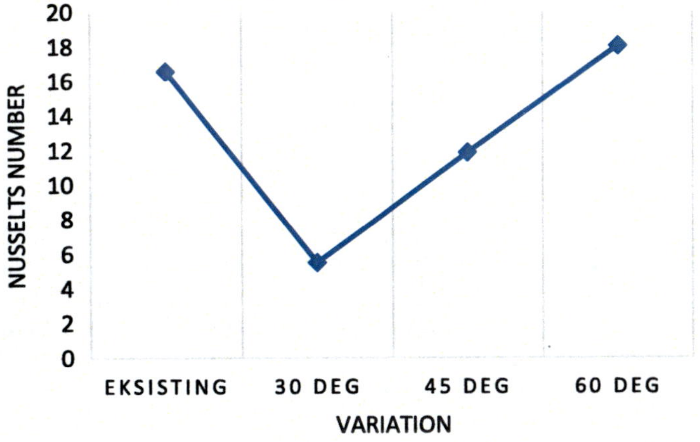

Fig. 3 Nusselts number on variation degrees

3.3 Exergy Analysis

The exergy value of each temperature variation described in graph has a downward trend. This also occurs in the variation of the angle shown in Fig. 4. The decrease occurred at angles of 30°, 45°, and 60° with values of 90.45, 88.35, and 73.34%. The trend is then compared with the existing value to determine the effect of adding a design disturbance. The existing value has an exergy efficiency of 79.56%. This value has a value higher than 60° but smaller than 30° and 45° so that the value of the greatest exergy efficiency is with an angle of 30°.

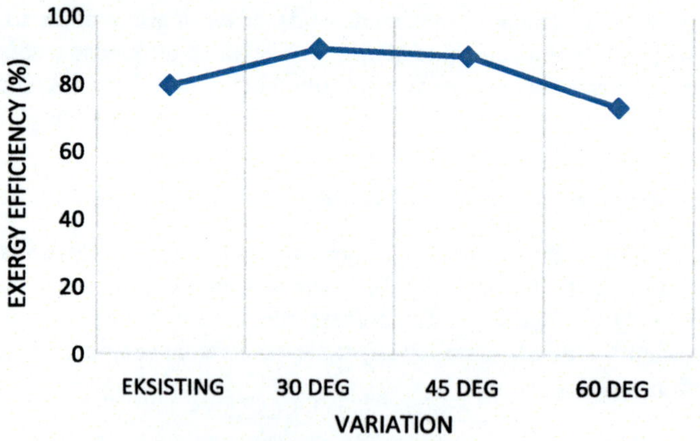

Fig. 4 Exergy efficiency on variation degrees

4 Conclusion

The greatest exergy efficiency value, which is 90.45%, is found at a temperature variation of 364 K with the addition of an angle of 30°. The Eulerian model selected in the CFD simulation with granular viscosity type is Syamlal-O'Brien. Meanwhile, with the turbulence model, the standard *k*-e type is selected with non-viscous heating. The value of the difference in values obtained in this simulation has the smallest value of 1.59%. In the existing condition, it can be seen that the distribution of the contour of the solid fraction is located at coordinate ×0.1 to 0.6. The installation of the disturbance design is carried out on the dryer roof with an angle of 60° to obtain the distribution of the aggregated fraction at ×0.65 to 0.8 coordinates.

References

1. Seyfi A, Rezaei A, Motevali A (2021) Comparison of the energy and pollution parameters in solar refractance window (photovoltaic-thermal), conventional refractance window, and hot air dryer. Sol Energy Dec 2020
2. Vigneshkumar N, et al (2021) Investigation on indirect solar dryer for drying sliced potatoes using phase change materials (PCM). Mater Today Proc (xxxx)
3. Gupta V, Sabharwal Gupta K, Khare R (2021) Experimental analysis for drying of potato slices on detachable solar greenhouse dryer. Mater Today Proc (xxxx)
4. Chantasiriwan S (2021) Optimum installation of flue gas dryer and additional air heater to increase the efficiency of coal-fired utility boiler. Energy 221:119769
5. Chen Q et al (2020) Experiment and simulation of the pneumatic classification and drying of coking coal in a fluidized bed dryer. Chem Eng Sci 214:115364
6. Moon SH, Ryu IS, Lee SJ, Ohm TI (2014) Optimization of drying of low-grade coal with high moisture content using a disc dryer. Fuel Process Technol 124:267–274
7. Khodaei H, Yeoh GH, Guzzomi F, Porteiro J (2018) A CFD-based comparative analysis of drying in various single biomass particles. Appl Therm Eng
8. Kuriakose R, Anandharamakrishnan C (2010) Computational fluid dynamics (CFD) applications in spray drying of food products. Trends Food Sci

9. Jaskulski M, Wawrzyniak P, Zbiciński I (2018) CFD simulations of droplet and particle agglomeration in an industrial counter-current spray dryer. Adv Powder Technol
10. Nguyen MP, Ngo TT, Le TD (2019) Experimental and numerical investigation of transport phenomena and kinetics for convective shrimp drying. Case Stud Therm Eng
11. Zhang M, Zhang S, Wang M, Lu Z, Jia W (2019) Effects of dehumidification drying parameters on physical and chemical properties of biomass brick. Constr Build
12. Paul G, Olivier M, Esther A, Daniel M, Jean CL (2019) Heat and mass transfer local modelling applied to biomass briquette drying. Procedia Manuf
13. Balázs T, Örvös M, Tömösy L (2007) Heat and mass transfer in an agitated contact-convective heated dryer. Food Bioprod Process
14. Sahni EK, Chaudhuri B (2013) Numerical simulations of contact drying in agitated filter-dryer. Chem Eng Sci
15. Verma M, Loha C, Sinha AN, Chatterjee PK (2017) Drying of biomass for utilising in co-firing with coal and its impact on environment—a review. Sustain Energy Rev
16. Versteeg HK, Malalasekera W (1995) An introduction to computational fluid dynamics the finite volume method. Longman Sc & Technical, Malaysia
17. Anderson JD (1995) Computational fluid dynamics the basics with applications. McGraw-Hill
18. Launder BE, Spalding DB (1972) Lectures in mathematical models of turbulence. Academic Press, London, New York

Cold Isostatic Pressing Treatment in the Preparation of Al and Y-Doped LLZO ($Li_{6.15}La_3Zr_{1.75}Al_{0.2}Y_{0.25}O_{12-\delta}$) Solid Electrolyte

Septia Kurniawati Arifah, Fitria Rahmawati(✉), and Yuniawan Hidayat

Research Group of Solid-State Chemistry & Catalysis, Chemistry Department, Sebelas Maret University, Jl. Ir. Sutami 36 A Kentingan, Surakarta 57126, Indonesia
fitria@mipa.uns.ac.id

1 Introduction

A secondary battery is a reversible battery that can be re-charged and discharged many times. The commercial secondary batteries which are dominated by lithium-ion battery used liquid electrolytes such as $LiPF_6$ [1]. Battery with a liquid electrolyte provides some risks such as electrolyte leaking, which is very dangerous because the liquid electrolyte was made from flammable organic solvent [2]. Therefore, some effort to replace the liquid electrolyte with solid electrolyte has been conducted by some researchers [3–10]. A solid electrolyte that has good chemical stability is $Li_7La_3Zr_2O_{12}$ (Lithium Lanthanum Zirconat, LLZO). LLZO with a cubic garnet type shows a high ionic conductivity up to 10^{-3} S cm^{-1} [3]. However, the cubic structure is not stable at room temperature. Therefore, some researchers conducted LLZO modification by doping some elements in order to stabilize the cubic structure at room temperature and to increase the ionic conductivity [11–16]. Alumunium doping was also applied to LLZO aims to obtain a dense cubic LLZO [17]. Meanwhile, another research doped yttrium ion, Y^{3+} into LLZO objected to increase its ionic conductivity [14]. Another element to be doped into LLZO is Ga^{3+} [15, 18] and Ta^{3+} [16] which has 3+ charge, similar to Y^{3+} and Al^{3+}. The element dopant with an oxidation number less than Zr^{4+} has been chosen considering the formation of oxygen vacancies after the dopant replaces the Zr ion in their lattice position. The presence of vacant sites would increase the space for Li-ions to mobile within the LLZO structure. In order to combine the objectives to increase the density and the ionic conductivity of the LLZO, this research conducted double doping of Y-Al into LLZO. The synthesis was conducted by solid-state reaction as a simple method without the requirement of organic solvent.

The density of a solid electrolyte pellet or disk is a very important property related to the ionic conductivity of the material. The required density is usually close to 90% relative to the theoretical density of LLZO [19] in order to reach a high ionic conductivity. In order to gain a high density, the pellet of material is usually treated by heat or named as sintering. The sintering temperature of LLZO is above 1000 °C [20–22] which will be very costly for mass production. Cold isostatic pressing (CIP) is a treatment to increase the density by applying high pressure from all without heat treatment. The method

M. Kolhe et al. (eds.), *Recent Advances in Renewable Energy Systems*,
Lecture Notes in Electrical Engineering 876,
https://doi.org/10.1007/978-981-19-1581-9_26

is much inexpensive for solid electrolyte preparation. In this research, the CIP was applied at various pressure to find the best pressure regarding the density and the ionic conductivity of the prepared-LLZO.

2 Methods

2.1 Synthesis of $Li_{6.15}La_3Zr_{1.75}Al_{0.2}Y_{0.25}O_{12-\delta}$ (LLZAYO)

The synthesis was conducted by solid-state reaction methods refers to the previous research [18]. The precursors LiOH, La_2O_3, ZrO_2, Al_2O_3, and Y_2O_3 were mixed in a ball mill machine with zirconia balls as grinding media. The mass ratio of zirconia balls to the mixed powder was 1:30. Isopropanol was used as the dispersant. The mixture was ball milled for 12 h at 450 rpm speed. The pellet form of mixed powder was then heated to 650 °C for 15 h and followed by solid-state reaction at 1000 °C for 4 h.

2.2 Characterization of LLZAYO

The prepared materials were then analyzed by X-ray diffraction equipped with Le Bail refinement in RIETICA software (a free edition) to understand its phase content, crystal structure, and cell parameters. Meanwhile, the particle size is determined by measureIT software.

2.3 Impedance Analysis and the Ionic Conductivity

Impedance analysis was conducted by the silver blocking method. The prepared powder was pressed with a hydraulic press at 7 metric tons to form pellets before further treatment. There were three different treatments, i.e., the green pellets or pellets without further treatment after hydraulic pressed (LLZAYO-Green), the sintered pellets up to 1200 °C for 6 h (LLZAYO-1200) [17], the sintered pellets at 1050 °C for 1 h (LLZAYO-1050) [18], and the cold isostatic pressing-pellets at 20, 30, and 40 MPa; LLZAYO-CIP 20, LLZAYO-CIP 30, and LLZAYO-CIP 40).

The impedance analysis was conducted by LCR meter (EUCOL U2826, frequency range of 20 Hz–5 MHz). The impedance data were fitted by **ZView** (the software embedded within CS Studio-5, Corrtest Electrochemical Workstation) to find the R–C network model which fit well with the Nyquist plot. The conductivity, σ (S cm^{-1}), was measured from the resistance, R, found by Z view fitting and applying Eq. (1).

$$R_i = \frac{L}{\sigma_i \cdot A} \tag{1}$$

where R is fitted resistance (Ω), L is the pellet thickness (cm), and A is the area of the electrode (cm^2).

3 Result and Discussions

The resulted powder of $Li_{6.15}La_3Zr_{1.75}Al_{0.2}Y_{0.25}O_{12-\delta}$ (LLZAYO) was analyzed by X-ray diffraction instruments. The diffraction pattern is depicted in Fig. 1a. The XRD data were refined by Le Bail method with ***RIETICA*** software plotted in Fig. 1b. The refinement process was conducted until the residual factor (Rp and Rwp) was less than 10%, or the chi-square (x^2) was below 1 [23]. The refinement found that the cubic phase percentage of the sample is 84.89%, with lanthanum compound and yttrium oxide as impurities. The cell parameters of cubic LLZO resulted from the refinement are listed in Table 1.

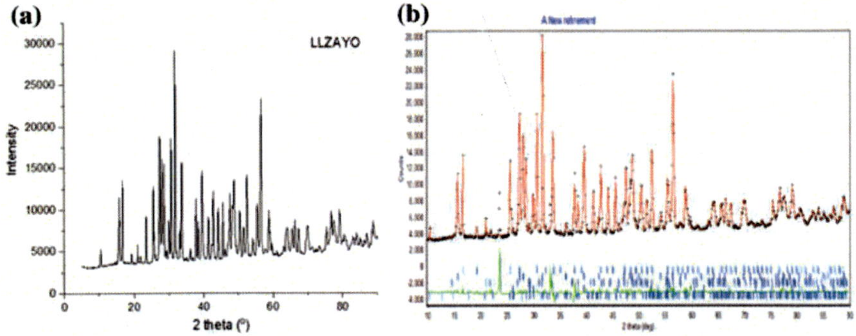

Fig. 1 **a** Diffraction pattern of the prepared-LLZAYO, and **b** Le Bail plot of LLZAYO (red line: calculated peak, black line: experimental peak, green line: difference between calculated and experimental peak)

Table 1 Refinement result of LLZAYO

Standard diffraction	Cell parameters (Å)	Percentage of phases (%)
c-LLZO ICSD#422,259	$a = 13.058627(0)$ $b = 13.058627(0)$ $c = 13.058627(0)$	84.89
Residual factor	Rp (profile factor)	2.88
	Rwp (weight profile factor)	6.85

Based on the refinement results, the percentage of yttrium oxide as impurities is 12.34%, which is sufficiently large. The comparison of the diffraction pattern of LLZAYO with the standard diffraction of cubic LLZO **ICSD#422,259,** and monoclinic yttrium oxide, **ICSD#160,219** is described in Fig. 2. The large percentage of yttrium oxide indicates that the reaction time of the solid-state process needs to be extended in order to optimize the reaction between precursors [24]. Figure 3 shows SEM image of the prepared-LLZAYO powder shows spherical particles which are gathered into aggregates (Fig. 3). The particle size was averagely found to be 0.455 ±0.15 μm.

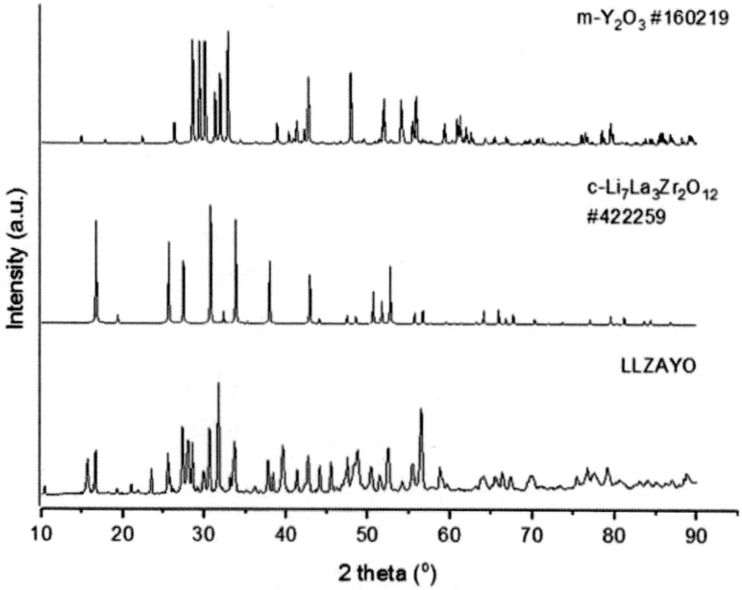

Fig. 2 Diffraction patterns of the sample compared to standard (ICSD)

Fig. 3 Surface image of LLZAYO

The Nyquist plots of impedance data for the LLZAYO-Green, LLZAYO-1000, LLZAYO-1200, and the LLZAYO-CIP are depicted in Fig. 4. The Nyquist plot of LLZAYO-Green is an irregular pattern indicating less contact between particle grains caused by the low compactness. Sintering treatment at 1000 and 1200 °C seems to increase the density of the pellets and allow the decreasing of resistance. Meanwhile, the CIP treatment at 20 MPa increases the material properties shown by a well form semi-circle and the low resistance as described by Fig. 4. The ionic conductivity and the density of the pellets are listed in Table 2. It shows that even though the density of the material is low, however, CIP can improve the particle packing in green bodies [25].

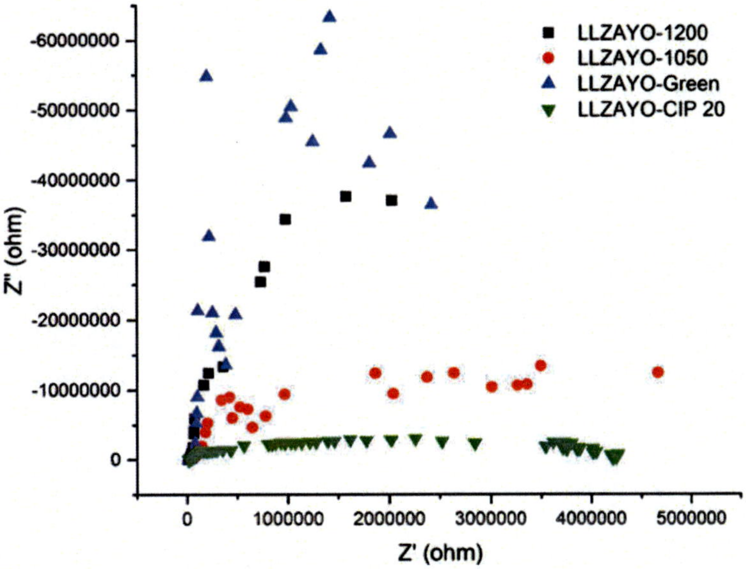

Fig. 4 Nyquist plot of the prepared LLZAYO

Table 2 Ionic conductivity and density of the prepared LLZAYO

Prepared materials	Ionic conductivity (S cm^{-1})	Density (g cm^{-3})
LLZAYO-1200	5.315×10^{-9}	3.9779
LLZAYO-1000	4.190×10^{-8}	5.0838
LLZAYO-Green	3.509×10^{-9}	3.4220
LLZAYO-CIP 20	5.479×10^{-7}	3.3520

The sintering pellet shows high density, but the pellet is less stable which might be caused by the formation of Li_2CO_3 on the pellet surface [26, 27]. The sintering stage has been carried out without an inert atmosphere so that there was a possibility of a reaction between the pellets and CO_2 in the open air to form Li_2CO_3 at high temperature. The reaction undergoes a two-step process that involves protonation of LLZO (i.e., Li^+/H^+ exchange) and formation of LiOH as an intermediate as in reaction 1 and carbonation of LiOH as in reaction 2 [28–30]. The Gibbs-free energy (ΔG) is minus means the reaction may occur spontaneously. Another research found that the decreasing of ionic conductivity of LLZO because of humid air exposure comes from 6.459 × 10^{-4} S cm^{-1} down to 3.619 × 10^{-4} S cm^{-1} [27].

$$Li_7La_3Zr_2O_{12} + xH_2O \rightarrow Li_{7-x}H_xLa_3Zr_2O_{12} + xLiOH\Delta G = -33.0\,kJ/mol \quad (2)$$

$$LiOH + 1/2\,CO_2 \rightarrow 1/2\,Li_2CO_3 + 1/2\,H_2O \quad \Delta G = -33.6\,kJ/mol \quad (3)$$

The condition of the LLZAYO-1200 after impedance analysis is depicted in Fig. 5a, and the LLZAYO-1050 after impedance analysis is depicted in Fig. 5b. The LLZAYO-Green and the LLZAYO-CIP 20 had good stability even after the atmosphere exposure, as it can be seen in Fig. 5c for LLZAYO-Green and Fig. 5d for LLZAYO-CIP 20.

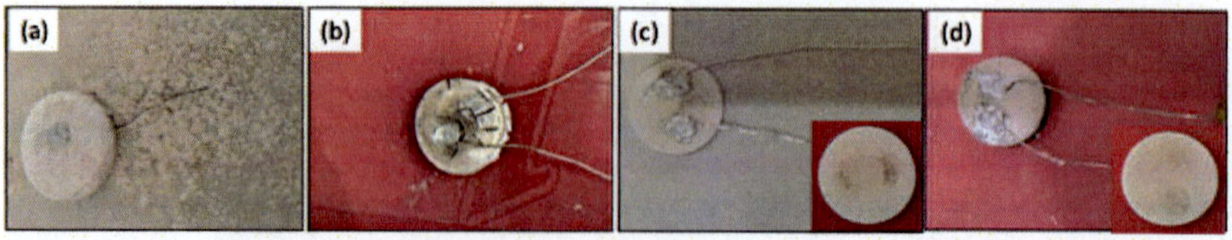

Fig. 5 **a** LLZAYO-1200, **b** LLZAYO-1050, **c** LLZAYO-Green, and **d** LLZAYO-CIP 20 after impedance analysis

The previous research found ionic conductivity of LLZO with a double dopant (Al, Ta) is 3.7 × 10^{-4} S cm^{-1} [18]. In this research, the CIP treatment was conducted, and it is found that after 20 MPa of CIP; the ionic conductivity increases up to 5.479 × 10^{-7} S cm^{-1}. Therefore, various CIP pressures were applied. The Nyquist plots of the LLAZYO-CIPs are described in Fig. 6. The result shows the increase of ionic conductivity, up to 1.06 × 10^{-5} S cm^{-1} after 40 MPa was applied. The density of LLZAYO-CIP 40 MPa is 4.1253, or it is 80.92% relative to the theoretical LLZO density [27]. This percentage is high but not as high as using sintering and CIP methods [19, 27].

The ionic conductivity of all LLZAYO-CIP along with their density is listed in Table 3. Table 3 shows that the ionic conductivity increases with the increase of the applied pressure. It indicates that the CIP treatment successfully increases the ionic conductivity and keep the LLZAYO pellet to be stable even after stored for a long time under room condition. It is challenging for the next research to find the optimum applied pressure for the LLZAYO pellets.

The elemental mapping had been done to observe the elemental distribution in the LLZAYO sample. The result shows that the elements of LLZAYO distributed very well in the whole crystal as shown by Fig. 7.

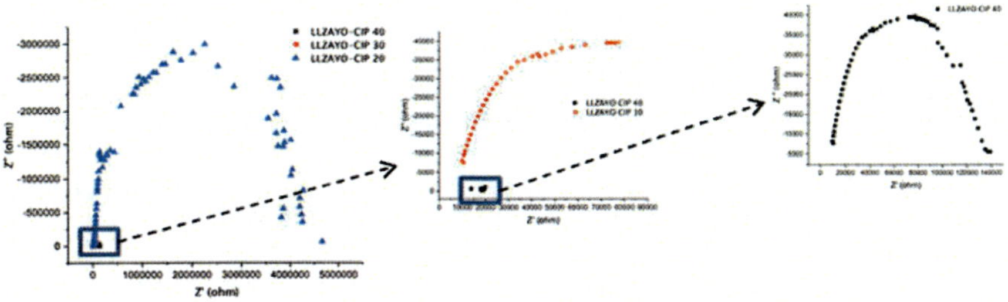

Fig. 6 Nyquist plot of non-sintered LLZAYO pellet with different CIP's pressure

Table 3 Ionic conductivity and density of CIP-treated sample

Pellet's treatment	Ionic conductivity (S cm^{-1})	Density (g cm^{-3})
LLZAYO-CIP 20	5.48×10^{-7}	3.352
LLZAYO-CIP 30	7.08×10^{-6}	3.756
LLZAYO-CIP 40	1.06×10^{-5}	4.125

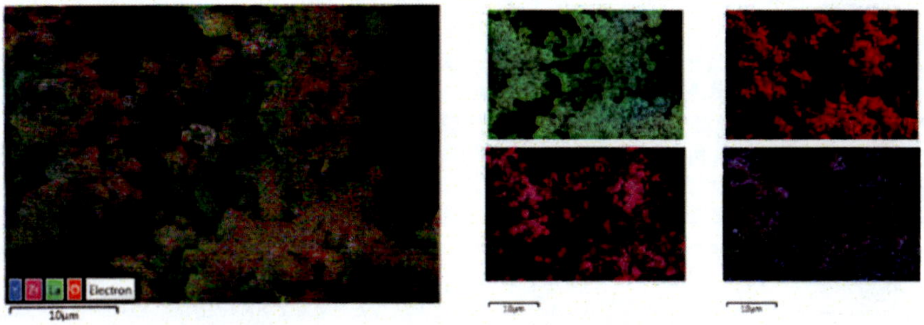

Fig. 7 Elemental mapping of the prepared LLZAYO

4 Conclusion

Another method to obtain dense LLZAYO pellets was by CIP's treatment. The sintering method may give the possibility to the formation of the secondary phase, phase transition, or lithium loss because the evaporation point of lithium is near the sintering temperature (above 1000 °C). Physical treatment like CIP is proven to be an optional method to obtain dense Y-Al-doped LLZO pellet with high ionic conductivity. The highest ionic conductivity is found after 40 MPa treatment, i.e., 1.06×10^{-5} S cm^{-1}. Further, CIP treatment to obtain the optimum ionic conductivity will be interesting.

Acknowledgements. Authors thank The Ministry of Education and Culture, Directorate General of Higher Education, Republic Indonesia for funding this research under the scheme of PDUPT 2021.

References

1. Jiang L, Wang Q, Li K, Ping P (2018) A self cooling and flame retardant electrolyte for safer lithium ion batteries. Sustain Energy Fuels 2018(2):1323–1331. https://doi.org/10.1039/c8se00111a
2. Xu K (2004) Non-aqueous liquid electrolytes for lithium-based rechargeable batteries. Chem Rev 104:4303–4417
3. Awaka J, Kijima N, Hayakawa H, Akimoto J (2009) Synthesis and structure analysis of tetragonal $Li_7La_3Zr_2O_{12}$ with the garnet-related type structure. J Solid State Chem 182:2046–2052
4. Inaguma Y, Chen L, Itoh M, Nakamura T, Uchida T, Ikuta M, Wakihara M (1993) High ionic conductivity in lithium lanthanum titanate. Solid State Commun 86:689
5. Birke P, Scharner S, Huggins RA, Weppner W (1997) Electrolytic stability limit and rapid lithium insertion in the fast-ion-conducting $Li_{0.29}La_{0.57}TiO_3$ perovskite-type compound. J Electrochem Soc 144(6):L167–L169
6. Kanno R, dan Murayama M (2001) Lithium ionic conductor thio-LISICON. J Electrochem Soc 148:A742–A746
7. Stramare S, Thangadurai V, Weppner W (2003) Lithium lanthanum titanates: a review. Chem Mater 15:3974–3990
8. Mizuno F, Hayashi A, Tadanaga K, Tatsumisago M (2005) New, highly ion-conductive crystals precipitated from Li_2S-P_2S_5 glasses. Adv Mat 17:918–921
9. Hayashi A, Minami K, Ujiie S, Tatsumisago M (2010) Preparation and ionic conductivity of $Li_7P_3S_{11-z}$ glass-ceramic electrolytes. J Non-Crystal Solids 356:2670–2673
10. Mo Y, Ong SP, Ceder G (2012) First principles study of the $Li_{10}GeP_2S_{12}$ lithium super ionic conductor material. Chem Mat 24:15–17
11. Rangasamy E, Wolfenstine J, Sakamoto J (2012) The role of Al and Li concentration on the formation of cubic garnet solid electrolyte of nominal composition $Li_7La_3Zr_2O_{12}$. Solid State Ionics 206:28–32
12. Rangasamy E, Wolfenstine J, Allen J, Sakamoto J (2013) The effect of 24c-site (A) cation substitution on the tetragonal-cubic phase transition in $Li_{7-x}La_{3-x}A_xZr_2O_{12}$ garnet-based ceramic electrolyte. J Power Sources 230:261–266
13. El-shinawi H, Paterson GW, MacLaren D, Cussen EJ, Corr SA (2016) Low temperature densification of Al-doped Li7La3Zr2O12: a reliable and controllable synthesis of fast-ion conducting garnets. J Mat Chem. https://doi.org/10.1039/C6TA06961D
14. Bitzer M, Gestel TV, Uhlenbruck S, Buchkremer H-P (2016) Sol-gel synthesis of thin solid $Li_7La_3Zr_2O_{12}$ electrolyte films for Li-ion batteries. Thin Solid Films 615:128–134
15. Li C, Liu Y, He J, Brinkman KS (2017) Ga-substituted $Li_7La_3Zr_2O_{12}$: an investigation based on grain coarsening in garnet-type lithium ion conductors, 695:3744–3752
16. Zhang Y, Deng J, Hu D, Chen F, Shen Q, Zhang L, dan Dong S (2019) Synergistic regulation of garnet-type Ta-doped $Li_7La_3Zr_2O_{12}$ solid electrolyte by Li^+ concentration and Li^+ transport channel size. Electrochimica Acta 296:823–829
17. Xu B, Duan H, Xia W, Guo Y, Kang H, Li H, Liu H (2016) Multistep sintering to synthesize fast lithium garnets. J Power Sources 302:291–297
18. Allen JL, Wolfenstein J, Rangasamy E, Sakamoto J (2012) Effect of substitution (Ta, Al, Ga) on the conductivity of $Li_7La_3Zr_2O_{12}$. J Power Sources 206:315–319
19. Kumar PJ, Nishimura K, Senna M, Duevel A, Heitjans P, Kawaguchi T, Sakamoto N, Wakiya N, Suzuki H (2016) A novel low-temperature solid-state route for nanostructured cubic garnet $Li_7La_3Zr_2O_{12}$ and its application to Li-ion battery. R Soc Chem Adv. https://doi.org/10.1039/C6RA09695F

20. Rettenwander D, Welzl A, Cheng L, Fleig J, Musso M, Suard E, Doeff MM, Redhammer GJ, dan Amthauer G (2015) Synthesis, crystal chemistry, and electrochemical properties of $Li_{7-2x}La_3Zr_{2-x}Mo_xO_{12}$ (x = 0.1−0.4): stabilization of the cubic garnet polymorph via substitution of Zr^{4+} by Mo^{6+}. Inorg Chem 54:10440−10449
21. Rangasamy E, Wolfenstein J, Sakamoto J (2011) The role of Al and Li concentration on the formation of cubic garnet solid electrolyte of nominal composition of $Li_7La_3Zr_2O_{12}$. Solid State Ionics. https://doi.org/10.1016/j.ssi.2011.10.022
22. Kokal I, Somer M, Notten PHL, Hintzen HT (2011) Sol-gel synthesis and lithium ion conductivity of $Li_7La_3Zr_2O_{12}$ with garnet-related type structure. Solid State Ionics 185:42–46
23. Toby BH (2006) R factors in Rietveld analysis: How good is good enough? Powder Diff 21(1)
24. Werner P, Sieber H, Hillebrand R, Hesse D (1996) Time-dependent interfacial reaction mechanism in a spinel-forming solid state reaction studied by TEM. Mat Res Soc Symp Proc 446:191–196
25. Galusek D, Znášik P, Majling J (1999) The influence of cold isostatic pressing on compaction and properties of Mg-PSZ ceramics. J. Mat. Sci. Lett. 18:1347–1351
26. Hu Z, Liu H, Ruan H, Hu R, Su Y, Zhang L (2016) High Li-ion conductivity of Al-doped $Li_7La_3Zr_2O_{12}$ synthesized by solid-state reaction. Ceram Int. https://doi.org/10.1016/j.ceramint.2016.04.149
27. Xia W, Xu B, Duan H, Tang X, Guo Y, Kang H, Li H, Liu H (2017) Reaction mechanisms of lithium garnet pellets in ambient air: the effect of humidity and CO_2. J Am Ceram Soc 1–8
28. Larraz G, Orera A, Sanjuan M (2013) Cubic phase of garnet-type $Li_7La_3Zr_2O_{12}$: the role of hydration. J Mat Chem A 1:11419–11428
29. Boulant A, Bardeau JF, Jouanneaux A, Emery J, Buzare JY, Bohnke O (2010) Reaction mechanisms of $Li_{0.30}La_{0.57}TiO_3$ powder with ambient air: H^+/Li^+ exchange with water and Li_2CO_3 formation 39:3968–3975
30. Sharafi A, Yu S, Naguib M, Lee M, Ma C, Meyer HM, Nanda J, Chi M, Siegel DJ, Sakamoto J (2017) Impact of air exposure and surface chemistry on Li–$Li_7La_3Zr_2O_{12}$ interfacial resistance. J Mat Chem A. https://doi.org/10.1039/c7ta03162a

The Effect of Inlet Air Cooling to Power Output Enhancement of Gas Turbine

Iman Firmansyah(✉) and Prabowo

Department of Mechanical Engineering, Institut Teknologi Sepuluh Nopember, Surabaya 60111, Indonesia

Iman.firmansyah@ptpjb.com, prabowo@me.its.ac.id

1 Introduction

The performance of gas turbines is greatly affected by ambient temperature. Several studies on the effect of compressor inlet air temperature on gas turbine performance have been conducted. De sa and Zubaidy [1] conducted research on the influence of ambient conditions, especially ambient temperature on the output power and the efficiency of gas turbines. The study compared the ambient temperature data against the output power and efficiency of gas turbines. The study was conducted in the Dubai region where the temperature difference in winter and summer can reach approximately 44 K. The study showed that the output power decreased by 1.47 MW and efficiency decreased by 0.1%, for every 1 K increase in ambient temperature.

In addition to research on the effect of ambient air temperature on gas turbine performance, research on the use of technology to lower compressor inlet temperature has also been done. Kamal et al. [2] conducted a feasibility study of the use of a chiller to lower the air temperature of turbine inlets. The feasibility study was conducted in Malaysia. The study was conducted using a simulation of GT Pro software with variations in the chiller capacity used, namely 1700 RT and 1950 RT where the air temperature was cooled to 12 °C. From the study, it was concluded that the increase in the gas turbine output power was 27.5–32.11%, and the heat rate decreased by 2.8–3.74%. Al Hamdan et al. [3] conducted a study on the effects of evaporative cooling utilization on gas turbine performance at the Shuaiba North Electric Generator Power Plant. This study was conducted by comparing the performance of gas turbines when using evaporative cooling with those without it. The gas turbines used as the object in this study were the same unit. The results of the study concluded that the increase in output power was 11.07%, and the heat rate decreased by 4% when the inlet air temperature decreased from 50 to 26 °C.

From the gas turbine operation data, there was an increase in output power when the air temperature dropped. The output power increased by 0.61 MW, and the heat rate decreased by 10.86 kcal/kWh, for every 1 °C decrease in ambient temperature. Based on previous research, this case study aimed to determine the effect of decreasing

M. Kolhe et al. (eds.), *Recent Advances in Renewable Energy Systems*,
Lecture Notes in Electrical Engineering 876,
https://doi.org/10.1007/978-981-19-1581-9_27

the air temperature of compressor inlets using eva-cooler system, fog-spray system, and mechanical chiller system technology on the performance of gas turbines. The performance of the gas turbines observed in this case study was the output power and heat rate. In addition, the most optimal method that can be applied was expected to be discovered (Fig. 1).

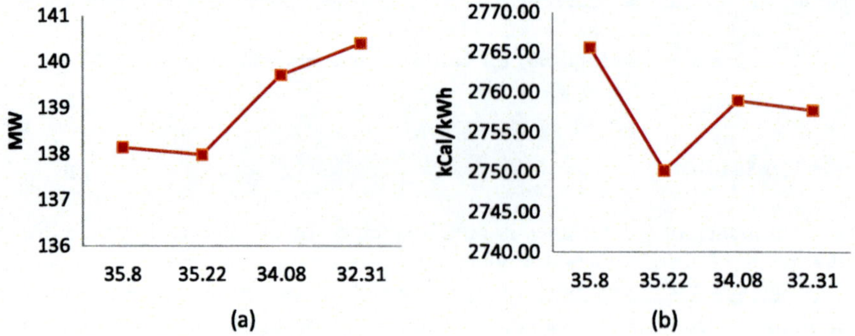

Fig. 1 GT performance versus inlet air temperature variations **a** output power, **b** heat rate

2 Methodology

This research was conducted using a simulation on Cycle-Tempo 5.0 software. Simulations were carried out for several terms, namely existing conditions, modifications using eva-cooler systems, fog-spray systems, and mechanical chillers. Variations of ambient air temperature used in the simulations were according to operating data (see Table 1).

Figure 2 shows the modeling performed on the Cycle-Tempo simulation for each condition. The modeling for the eva-cooler and the fog-spray systems used was the same, which distinguishes only by the effectiveness used in the eva-cooler system which was 85%, while the fog-spray system was 100%.

The effectiveness of the eva-cooler system used in the study was 0.85. As stated by al-Hamdan and Saker [3], the effectiveness of the evaporative cooler can be calculated using formulas.

$$\eta = \frac{T_{DBE} - T_{DBL}}{T_{DBE} - T_{WBE}} \tag{1}$$

where

η: Efficiency in percent
T_{DBE}: Entering dry-bulb temperature
T_{WBE}: Entering wet-bulb temperature
T_{DBL}: Leaving dry-bulb temperature.

Table 1 Operating data

Time	Parameter	Value	Unit
14:00	P air atmosphere	1011,06	mbar
	T air atmosphere	35,8	cel
	Humidity	46,63	%
	Active power	138,14	MW
	Heat rate	2765,52	kCal/kWh
15:00	P air atmosphere	1012,3	mbar
	T air atmosphere	35,22	cel
	Humidity	49,53	%
	Active power	137,98	MW
	Heat rate	2750,14	kCal/kWh
16:00	P air atmosphere	1013,27	mbar
	T air atmosphere	34,08	cel
	Humidity	53,47	%
	Active power	139,71	MW
	Heat rate	2758,91	kCal/kWh
17:00	P air atmosphere	1014,98	mbar
	T air atmosphere	32,31	cel
	Humidity	60,73	%
	Active power	140,39	MW
	Heat rate	2757,72	MW

The amount of water that is evaporated when air passes through the evaporative cooler can be calculated according to a psychometric graph using the formula [4].

$$\dot{m}_w = \dot{m}_a(\omega_2 - \omega_1) \tag{2}$$

where

$\dot{m}_w$: Amount of water evaporated (kg/s)
$\dot{m}_a$: Air flow (kg/s)
ω_2: Humid ratio at leaving evaporative cooling system (kg/kg d a)
ω_1: Humid ratio at entering evaporative cooling system (kg/kg d a).

In the fog-spray modification, the effectiveness used in the study was 100%, where the temperature of the dry ball leaving the fog-spray system was the same as the temperature of the wet bulb.

The mechanical chiller used in this study was a type of water-cooled chiller with ammonia refrigerant (R717). Simulations were used to evaluate the performance of gas

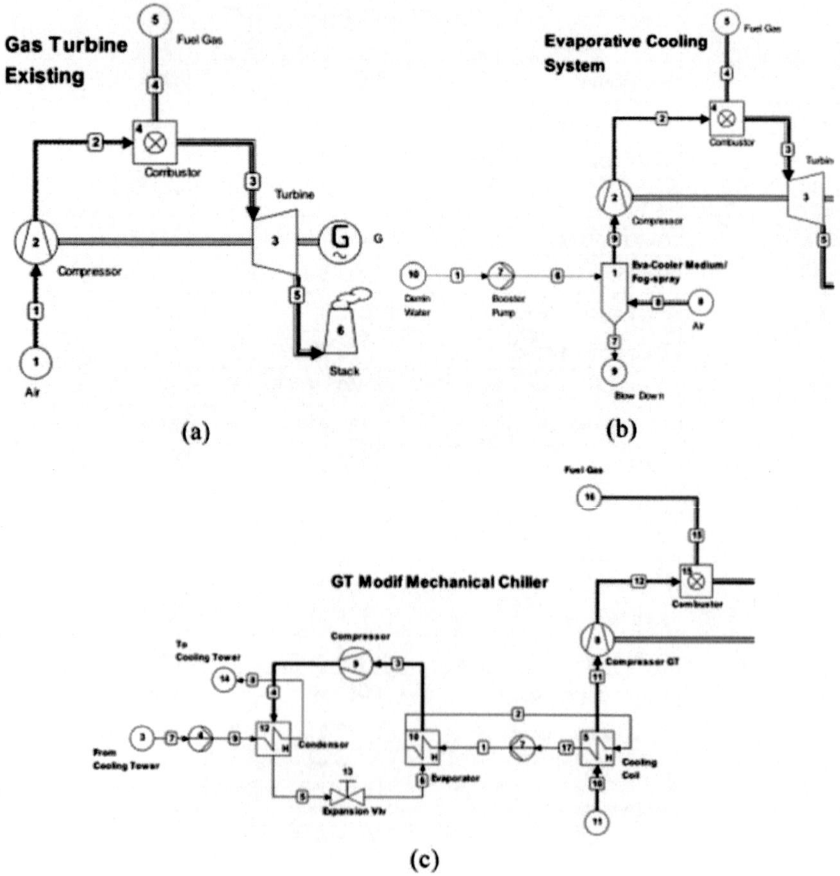

Fig. 2 Cycle-Tempo models. **a** Existing, **b** evaporative cooling (eva-cooler and fog-spray), **c** mechanical chiller

turbines with ambient air temperatures cooled to 24 °C, where the chiller cooling capacity can be calculated by a formula [2].

$$CCL = AF_m(H_a - H_c) \tag{3}$$

where

CCL: Chiller cooling load (Kw)
AFm: Mass flow rate of cooled air (Kg/s)
Ha: Enthalpy of ambient air
Hc: Enthalpy of cooled air.

3 Result and Discussion

The changing of air properties after the cooling process with eva-cooler, fog-spray, and mechanical chiller systems can be explained on psychrometric charts as shown in Fig. 3. The cooling process in evaporative cooling (eva-cooler and fog-spray system) moved at the wet-bulb temperature line. The initial relative humidity affected the collected final temperature. For mechanical chiller systems, the planned outlet temperature was above the dew-point temperature; in this study, it was designed to be 24 °C. The air properties after passing through each of these technologies can be seen on Table 2.

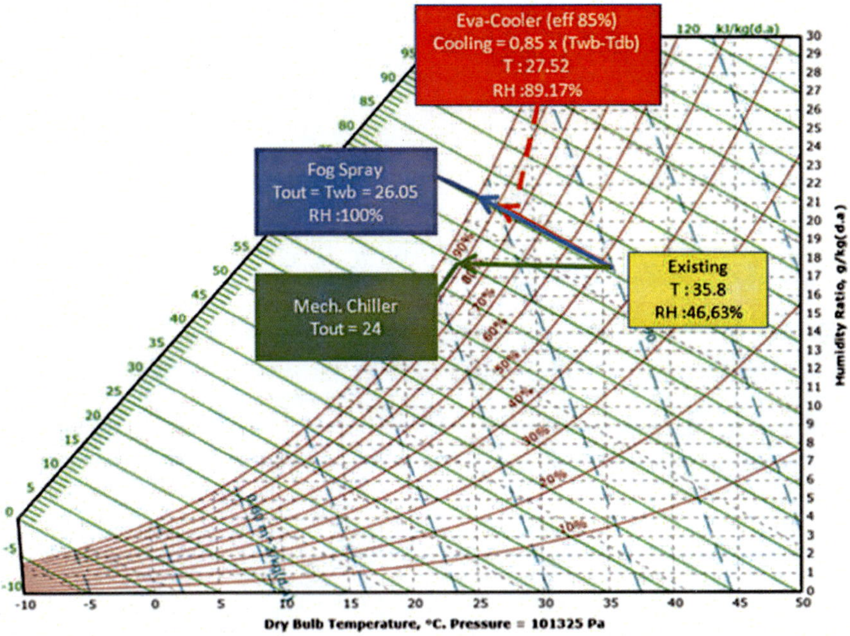

Fig. 3 Comparison of each cooling process at psychrometric chart

From the simulations results on Cycle-Tempo, for each modification, an increase in the performance of gas turbines was obtained. Figure 4 shows the comparison of gas turbine power output between existing and modified conditions. From these results, the maximum increase was obtained by using a mechanical chiller. The net power output of gas turbines for eva-cooler, fog-spray, and mechanical chiller systems increased by 5.01%, 5.37%, and 8.46%, respectively.

In addition to increasing the output power, modifications to the eva-cooler, fog-spray, and mechanical chiller systems required power consumption. Table 3 shows a comparison of the needs for each aux-power consumption. The aux-power consumption of a mechanical chiller is much higher compared to an eva-cooler and fog-spray system. Aux-power consumption was obtained from the compressor and booster pump on the chiller. The aux-power consumption of the eva-cooler systems and fog-spray systems was

Table 2 Properties of air at entering and leaving cooler system

Condition	Parameter	Existing	Eva-cooler	Fog-spray	Mech. chiller
1	T air (°C)	35,8	27.52	26.06	24
	Humidity (%)	46,63	89.19	100	91.89
2	T air (°C)	35,22	27.56	26.21	24
	Humidity (%)	49,53	89.99	100	94.53
3	T air (°C)	34,08	27.28	26.08	24
	Humidity (%)	53,47	91.03	100	95.80
4	T air (°C)	32,31	26.90	25.95	24
	Humidity (%)	60,73	92.79	100	98.53

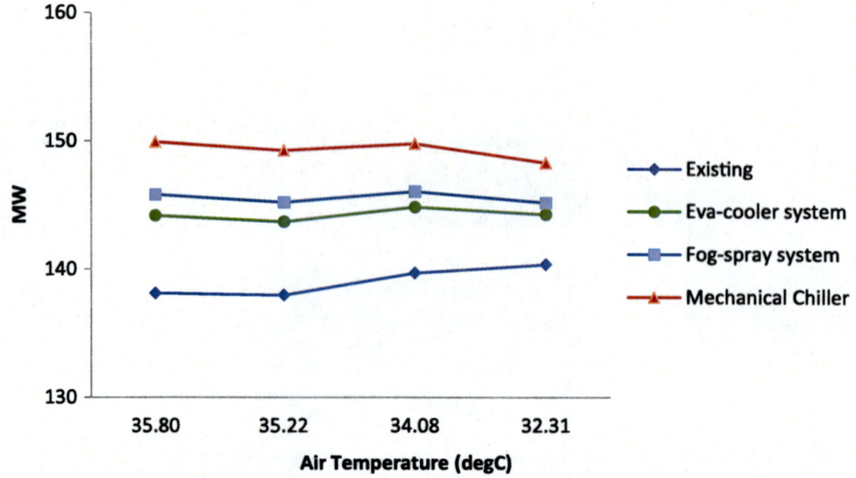

Fig. 4 Comparison between GT powers output versus inlet air temperature

obtained from the use of make-up water supply pump and the make-up water production itself.

Figure 5 shows the need for make-up water in the eva-cooler and fog-spray systems. The need for make-up water in a fog-spray system is greater than the eva-cooler system; the lower the ambient air temperature, the lower the need for make-up water. This is due to a decrease in temperature after the evaporative cooling which is getting smaller. Therefore, it can be seen that the decrease in air temperature in the evaporative cooling systems (eva-cooler and fog-spray system) is strongly influenced by ambient air conditions.

Figure 6 shows the comparison of the gas turbine heat rate between the existing and modified conditions. From these results, the maximum reduction is obtained by using a fog-spray system. The heat-rate reduction of the mechanical chiller system is lower than

Table 3 Auxiliary power consumption of modification system

Air inlet temperature (°C)	Aux-power consumption (kW)		
	Eva-cooler	Fog-spray	Mech. chiller
35,8	1,29	1,53	2081
35,22	1,19	1,41	2163
34,08	1,06	1,26	2103
32,31	0,85	1,00	2059

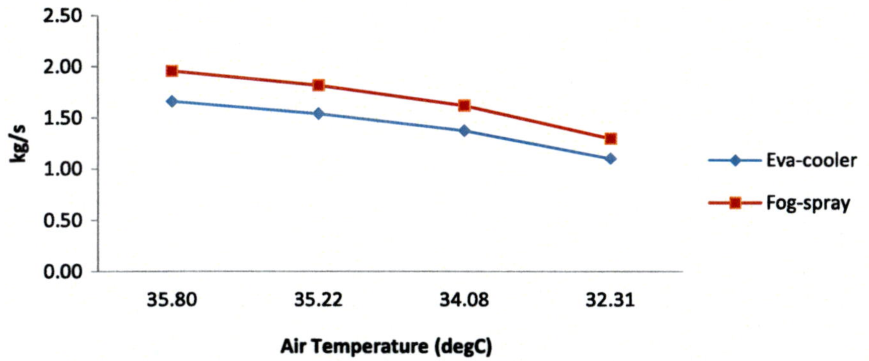

Fig. 5 Comparison between make-up water consumption versus inlet air temperature

the fog-spray system because the aux-power consumption is much higher. The heat rate of eva-cooler, fog-spray, and mechanical chiller systems can decrease by 3.64%, 4.65%, and 3.81%, respectively.

4 Conclusion

Based on the case study conducted, the utilization of an eva-cooler system, fog-spray system, and mechanical chiller system technology could lower the compressor inlet air temperature. Therefore, the output power could increase and the heat rate decreased. The decrease in air temperature in the evaporative cooling systems (eva-cooler and fog-spray system) was strongly influenced by ambient air conditions. The maximum net output power obtained from the utilization of mechanical chiller technology was 8.46%.

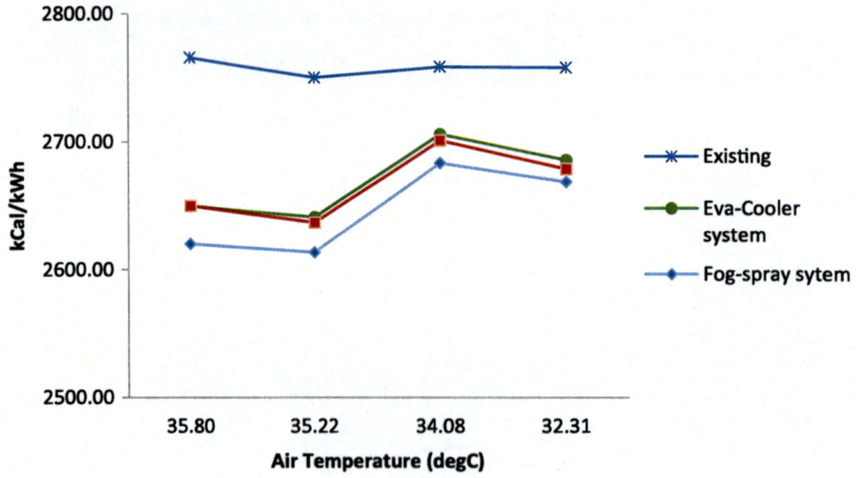

Fig. 6 Comparison between GT heat rates versus inlet air temperature

References

1. De Sa A, Zubaidy SA (2010) Gas turbine performance at varying ambient temperature. Appl Thermal Eng 31:hal 2735–2739
2. Kamal SNO, Salim DA, Fouzi MSM, Yusof DTHKMY (2017) Feasibility study of turbine inlet air cooling using mechanical chillers in Malaysia climate. Energy Procedia 138:hal558–663
3. AL-Hamdan OR, Saker AA (2013) Studying the role played by evaporative cooler on the performance of GE gas turbine existed in Shuaiba North electric generator power plant. Energy Power Eng 5:hal 391–400
4. Moran MJ, Shapiro HN (2006) Fundamentals of engineering thermodynamics, 5th edn. Wiley Inc

Numerical Study for Evaluating Effect of Mass Flow Rate Toward Particle Circulation Rate on Seal Pot in Circulating Fluidized Bed Boiler

Nur Ikhwan(✉) and Denny Oktavianto

Mechanical Engineering Department, Institut Teknologi Sepuluh Nopember, Surabaya, Indonesia
ikhwan@me.its.ac.id

1 Introduction

In 2016, the consumption of coal for fueling steam power plan reaches 69 million tons each year. The consumption will keep increasing until 2035 [1]. The most important pieces of equipment in the power plan are boiler. As the second-largest number, circulating fluidized bed (CFB) boiler promises more efficient combustion and more environmentally friendly. Circulation mechanism keeps unburnt coal being sent back by cyclone and seal pot into furnace for re-combustion process [2]. Present simulation work focuses on investigating the effect of flow characteristic change of loop seal due to flue gas flow rate and bed material change. Case study is taken on CFB boiler of PLTU Tenayan Riau. The boiler experienced overheating in the seal pot due to bed material accumulation [3]. Circulation of bed material in seal pot is managed by secondary air flow rate from supply chamber and recycle chamber. By finding the best composition of secondary airflow rate from both chambers, the occurrence of overheating in seal pot can be eliminated. Simulation is performed using computational particle–fluid dynamics (CPFD) software.

2 Literature Review

Seal pot is a non-mechanical valve that controls the circulation of bed material from cyclone into furnace by controlling flow characteristics of bed material [4]. The function of the supply chamber in the seal pot is for creating bed fluidization, and recycle chamber is for sending/recycling bed material back into the furnace. Improper flow composition in the supply chamber and recycle chamber can reduce the performance of seal pot in circulating bed material.

3 Simulation Setup

The present simulation is focused on loop seal with detailed domain and meshing as shown in Fig. 1. The simulation assumes there is no combustion from remaining coal

M. Kolhe et al. (eds.), *Recent Advances in Renewable Energy Systems*,
Lecture Notes in Electrical Engineering 876,
https://doi.org/10.1007/978-981-19-1581-9_28

(isothermal), heat transfer between fluid, and wall (isothermal), and no presence of limestone and coal. Inlet boundary condition for fluid and solid bed material is based on operational data of boiler operation for cyclone and seal pot. Air inlet for supply and recycle chamber is obtained from secondary air fan (see Fig. 2 left). An increase of air supply for both chambers will not change air mass flow balance since the increase of air supply from both chambers is obtained by decreasing the supply of secondary air for a riser. The initial particle in the bottom side of the seal pot (see Fig. 2 right) is the remaining bed material from the previous boiler operation.

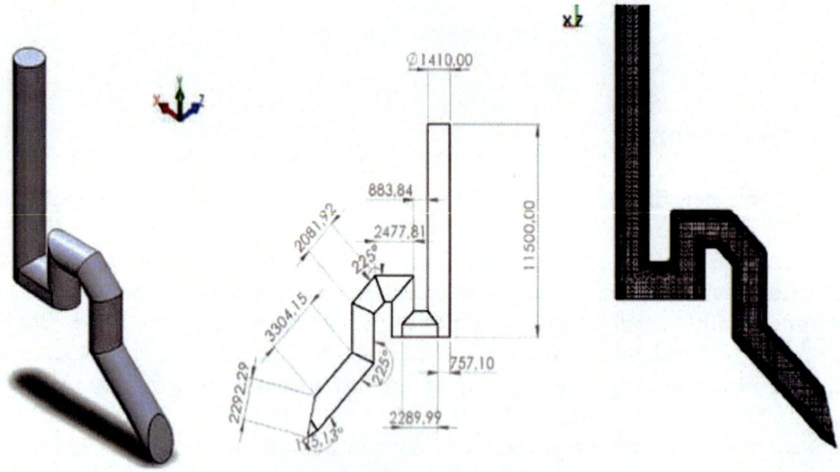

Fig. 1 Simulation domain and meshing on seal pot

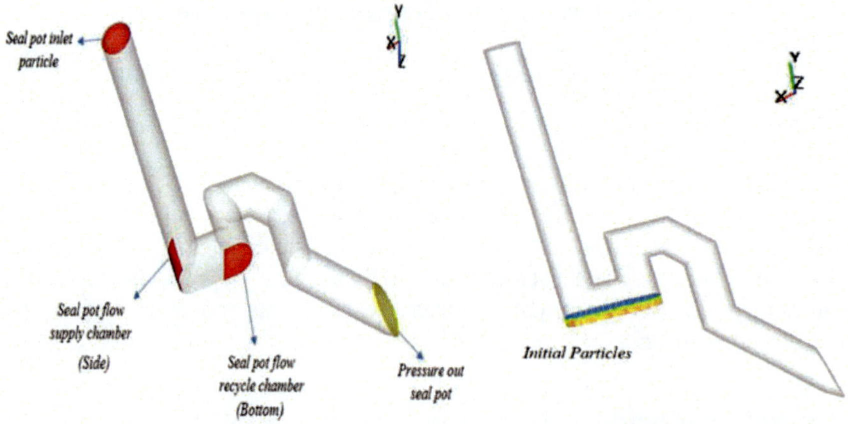

Fig. 2 Boundary and initial condition for present simulation

Simulation is performed to investigate the effect of secondary air supply increase toward circulation rate of seal pot. The variation is 100, 150, and 200% from the existing operating condition (see Table 1). In addition to the flow rate increase, simulation variation is also performed by changing different percentages of flow for supply chamber and recycle chamber, and particle size distribution. Particle size distribution is measured by Sauter mean diameter (SMD) where the size of existing condition is 154 μm and additional size is 308 μm (see Fig. 3).

Table 1 Boundary condition variation for the simulation

Increasing of mass flow rate secondary air seal pot	Total mass flow rate secondary air (kg/s)	Percentage of secondary:primary air seal pot	Locations	Value (kg/s)
Initial conditions ($\dot{m}_1$)	2.3	57%:43% 43%:57%	Supply chamber Recycle chamber Supply chamber	1.3 1 1
Increasing 150% ($\dot{m}_2$)	3.45	57%:43% 43%:57%	Recycle chamber Supply chamber Recycle chamber Supply chamber Recycle chamber	1.3 1.95 1.5 1.5 1.95
Increasing 200% ($\dot{m}_3$)	4.6	57%:43% 43%:57%	Supply chamber Recycle chamber Supply chamber Recycle chamber	2.6 2 2 2.6

Simulation result for inlet and outlet is calculated over the surface area for averaged result and points starting from the center area (point 0) until near the wall area (point 0.75) for transient result (see Fig. 4).

4 Result and Discussion

Fluidization process for the existing condition, at 100% air supply from secondary fan, supply: Recycle chamber flow equals to 57%:43%, and Sauter mean diameter equals to 130 μm, is shown in Fig. 12a. At a time of 50 s, there is a formation of packed bed in the area between supply chamber and recycle chamber with particle volume fraction as high as 0.45. Changing the percentage of supply: Recycle chamber into 43%:57% does not change fluidization condition (see Fig. 5b).

Fluidization for all variations is shown in Fig. 6. Variation of supply chamber: Recycle chamber ratio and particle size on the existing secondary air flow rate (m_1) do not change fluidization process as indicated by the same area of packed bed (volume fraction equals 0.45). On the contrary, an increase in secondary air flow rate (m_2 and m_3) significantly reduces the area of packed bed.

A similar result is also shown for the circulation rate of the solid particles in the seal pot. The cumulative mass of particle remains in seal pot, as shown in Fig. 7, indicates

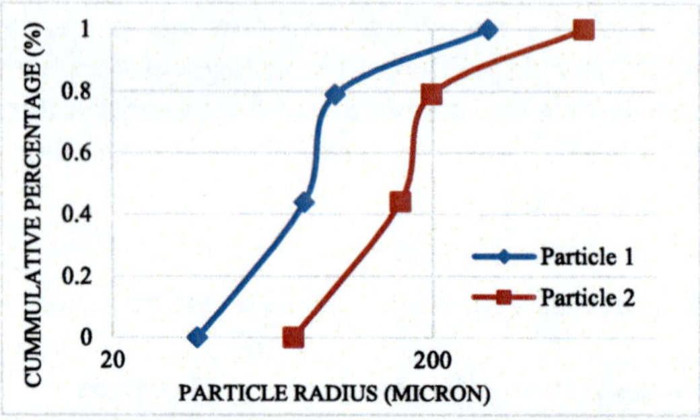


```
# Species 1: pre-defined drag model - Wen-Yu -----
#            with drag multiplier = 1.000000e+00
# Additional parameters are used for this species
# diam       = 1.984800e-04  Particle average diameter (m)
# diamSauter = 1.544524e-04  Particle Sauter mean diameter (m)
# sphericity = 7.500000e-01  Particle sphericity
# density    = 2.200000e+03  Particle density (kg/m^3)
```

SMD_1

```
# Species 1: pre-defined drag model - Wen-Yu -----
#            with drag multiplier = 1.000000e+00
# Additional parameters are used for this species
# diam       = 3.969600e-04  Particle average diameter (m)
# diamSauter = 3.089049e-04  Particle Sauter mean diameter (m)
# sphericity = 7.500000e-01  Particle sphericity
# density    = 2.200000e+03  Particle density (kg/m^3)
```

SMD_2

Fig. 3 Particle size distribution for existing condition (SMD_1) and bigger size (SMD_2)

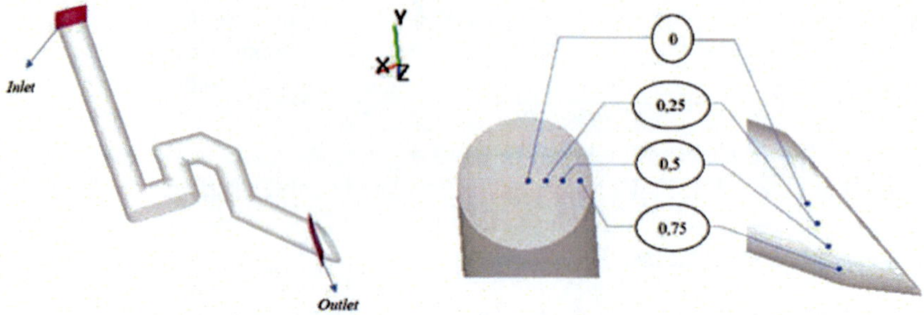

Fig. 4 Data extraction for the averaged result (left) and transient result (right)

circulation rate of particle. Smaller cumulative mass on seal pot indicates that more particles will be sent back to the furnace. It means a better circulation rate of solid particles. On the variation of supply: Recycle chamber equals to 57%:43%; increase of secondary air flow rate will properly reduce cumulative mass particle. For existing particle size (SMD1) and supply: Recycle chamber equals to 43%:57%; increase of secondary air flow rate will reduce cumulative mass particle. However, the reduction tends to be weaker as the secondary air flow rate increases. The worst condition occurs in variation of larger particle size (SMD2) and supply: Recycle chamber equals to 43%:57%. At maximum secondary air flow rate (200%), the cumulative mass particle will increase compared to 150% flow rate. An increase of secondary air flow rate will have a proportional correlation with particle circulation rate if the percentage of flow rate for the supply chamber is higher than recycle chamber.

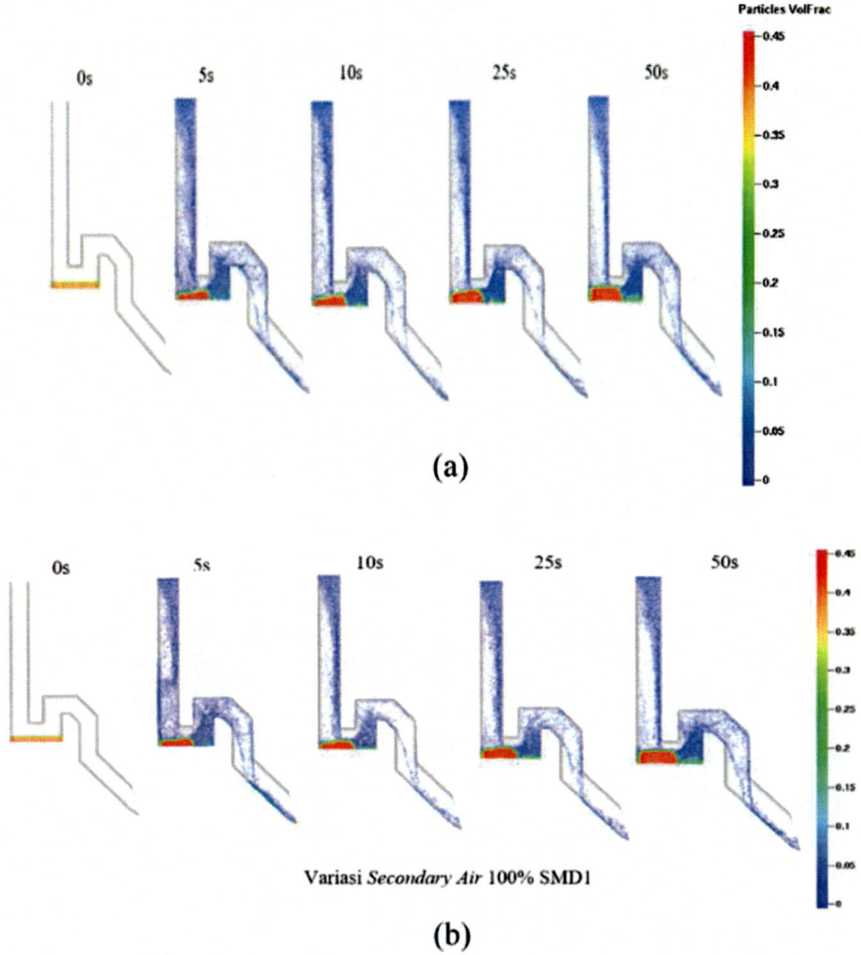

Fig. 5 Fluidization process for existing flow rate (m_1), SMD_1 and supply: recycle chamber ratio = **a** 57%:43%, **b** 43%:57%

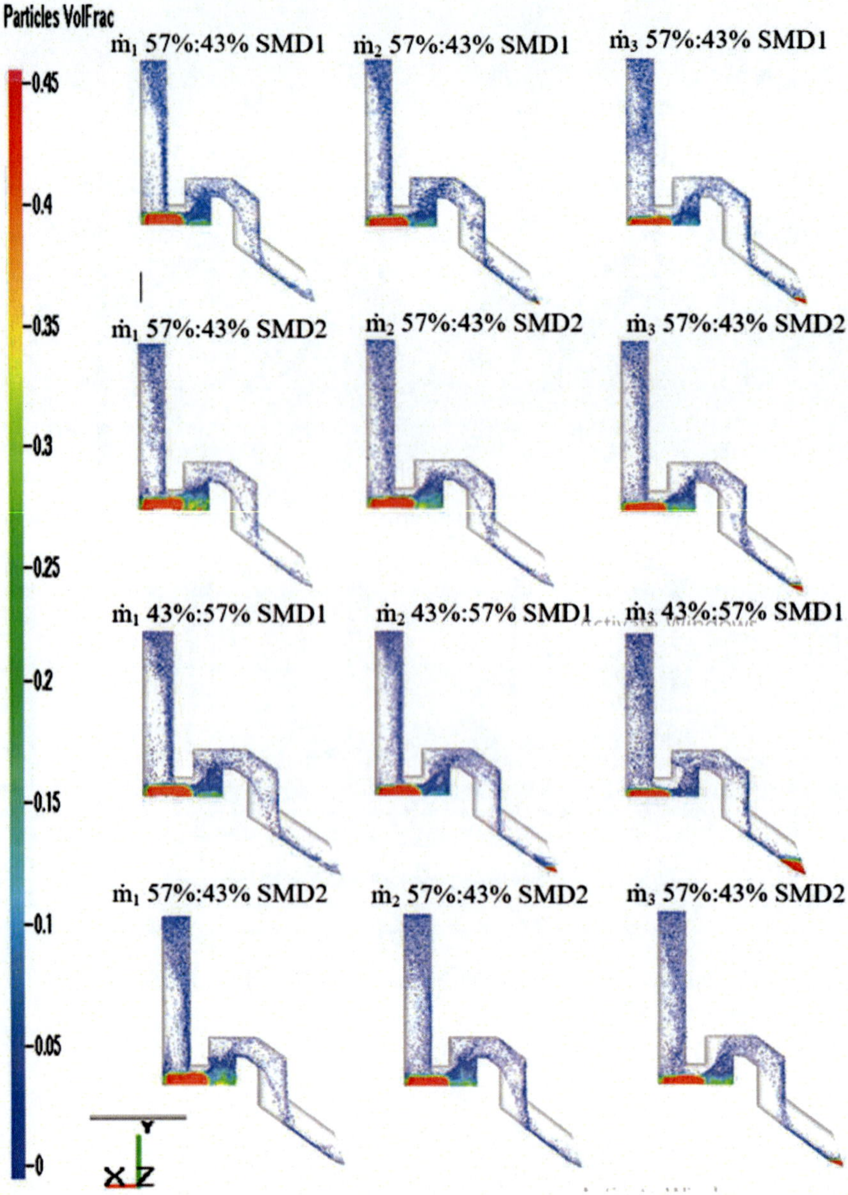

Fig. 6 Particle volume fraction for all variations

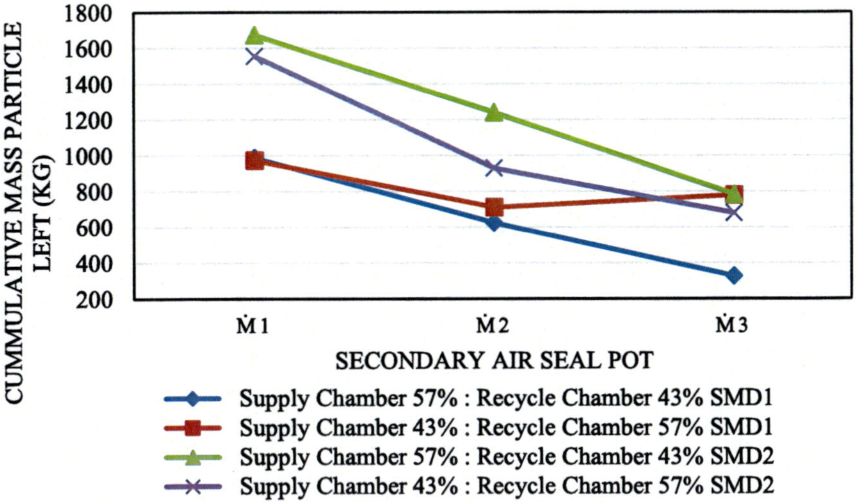

Fig. 7 Cumulative mass of particle remains in seal pot

5 Conclusion

Based on the simulation result and discussion, it can be concluded:

- An increase in secondary air flow rate will reduce the packed bed area in the bottom of the seal pot. It is indicated by narrow area of particle volume fraction equals to 0.45.
- An increase in secondary air flow rate will create a higher fluidization and creates higher fluctuation in the mass flow rate outlet.
- Percentage of supply chamber: Recycle chamber equal to 43%:57% tends to block fluidization and reduce circulation rate of solid particle.
- An increase in secondary air flow rate has proportional correlation with particle circulation rate if the percentage of flow rate for supply chamber is higher than recycle chamber.
- The maximum circulation rate is obtained at a variation of supply chamber: Recycle chamber equals to 57%:43%, secondary air flow rate 200%, and larger particle size (SMD2).
- The worst circulation rate is obtained at a variation of supply chamber: Recycle chamber equals to 43%:57%, secondary air flow rate 200%, and smaller particle size (SMD1).

Acknowledgements. Authors would say thanks to UBJO&M PLTU Tenayan for providing operational data for the CFB boiler, and thanks to DRPM ITS for providing fund for performing this simulation works.

References

1. Hariyadi H, Suciyanti M (2018) Analisis Perkiraan Kebutuhan Batubara untuk Industri Domestik Tahun 2020–2035 dalam Mendukung Kebijakan Domestic Market Obligation dan Kebijakan Energi Nasional. Jurnal Teknologi dan Batubara 2(4):59–73
2. Rayaprolu K (2009) Boiler for power and process, 2nd edn. Taylor & Francis Group, New York
3. PT PJB: review asset failure and improvement planning for 2 × 100 MW CFB boiler PLTU Tenayan. UBJ O&M PLTU Tenayan, Riau (2017)
4. Basu P (2015) Circulating fluidized bed boiler design, operations, and maintenance. Springer, Halifax

Study of Effect of Intake Air Temperature of Compressor Gas Turbine on Exergy Destruction in Tambak Lorok Combined Cycle Power Plants

Marzuki, Ary Bachtiar Krishna Putra(✉), and Christiono Utomo

Department of Mechanical Engineering, Institut Teknologi Sepuluh Nopember, Surabaya 60111, Indonesia
arybach@me.its.ac.id

1 Introduction

The Tambak Lorok combined cycle power plant has 2 blocks, namely Block 1 and Block 2, each block consisting of 3 Gas Turbine Generators, 3 HRSG (high recovery steam generators), and 1 Steam Turbine Generator. Combined cycle power plant can be operated with 3-3-1, 2-2-1, or 1-1-1 operating patterns. Over time, the performance of the Tbrok combined cycle power plant has decreased. This is due to the age of the equipment which has been in operation for more than 20 years and the hot and corrosive environmental conditions due to its location at the seaside. Currently, the supply of natural gas fuel is only capable of operating 3 continuous units with a minimum load or operating 2 continuous units and 3 continuous operating units at peak load by utilizing compressed natural gas (CNG). Therefore, analysis is needed to determine unit operation priorities and unit maintenance priorities to obtain maximum profit.

The first method for analyzing the performance of a combined cycle power plant is based on the first law of thermodynamics which only focuses on the energy quantity balance (Aichmayer et al. 2015). The second method is exergy analysis based on the second law of thermodynamics, which considers the quantity and quality of energy (Mohammad Ameri et al. 2009). Aliyu et al. [1] and Seyyedia et al. have conducted study of increasing environmental temperatures effect in the rate of second-law destruction for all major components. The energy analysis method has been used for a long time, but only reveals the energy transfer process of the whole system, without an analysis of the energy-saving effect of the system. Exergy analysis methods can accurately indicate the type, size, and location of exergy losses in the system (Table 1).

2 Methodology

2.1 Energy Analysis

Energy analysis on CCPP equipment is divided into 2 cycles, and there are the gas cycle and the steam cycle. Each cycle condition uses the following formula (Table 2 and Fig. 1).

M. Kolhe et al. (eds.), *Recent Advances in Renewable Energy Systems*,
Lecture Notes in Electrical Engineering 876,
https://doi.org/10.1007/978-981-19-1581-9_29

Table 1 Installed capacity CCPP Tbrok

Combined cycle power plant	Installed capacity	Engine manufacture	COD	Fuel
Block 1				
GTG 1.1	109,65 MW	GE MS9001E	1993	Gas/HSD
GTG 1.2	109,65 MW	GE MS9001E	1993	Gas/HSD
GTG 1.3	109,65 MW	GE MS9001E	1993	Gas/HSD
HRSG 1.1		Austrian energy	1997	Flue gas
HRSG 1.2		Austrian energy	1997	Flue gas
HRSG 1.3		Austrian energy	1997	Flue gas
STG 1.0	188 MW	General electric	1997	Steam
Block 2				
GTG 2.1	109,65 MW	GE MS9001E	1996	Gas/HSD
GTG 2.2	109,65 MW	GE MS9001E	1996	Gas/HSD
GTG 2.3	109,65 MW	GE MS9001E	1996	Gas/HSD
HRSG 2.1		Austrian energy	1997	Flue gas
HRSG 2.2		Austrian energy	1997	Flue gas
HRSG 2.3		Austrian energy	1997	Flue gas
STG 2.0	188 MW	General electric	1997	Steam

Table 2 Energy analysis on combined cycle power plant

Compressor	Combustion chamber	Gas turbine	Exhaust
$\dot{W}_C = \dot{m}_a(h_2 - h_1)$	$\dot{Q}_{fuel} = \dot{m}_{fuel} \times LHV$	$\dot{W}_t = \dot{m}_g(h_3 - h_4)$	$\dot{Q}_{out} = \dot{m}_g(h_4 - h_1)$
$\dot{W}_{Cs} = \dot{m}_a(h_{2s} - h_1)$	$\dot{Q}_{incc} = \dot{m}_g(h_3 - h_2)$	$\dot{W}_{ts} = \dot{m}_g(h_3 - h_{4s})$	
$\eta_C = \frac{\dot{W}_{Cs}}{\dot{W}_C}$	$\eta_{CC} = \frac{\dot{Q}_{incc}}{\dot{Q}_{infuel}}$	$\eta_C = \frac{\dot{W}_t}{\dot{W}_{ts}}$	

HRSG	Steam turbine
$\dot{Q}_{hrsg} = \dot{m}_g(h_4 - h_5)$ $\dot{Q}_{\text{outstack}} = \dot{m}_g(h_5 - h_1)$ $\dot{Q}_{\text{steam}hp} = \dot{m}_{vhp}(h_6 - h_{13})$ $\dot{Q}_{\text{steam}lp} = \dot{m}_{vlp}(h_7 - h_{11})$ $\eta_{HRSG} = \frac{\dot{Q}_{sthp}+\dot{Q}_{stlp}}{\dot{Q}_{\text{exhaust}GT}}$ $\frac{\dot{W}_{ps}}{\dot{m}} = v(p_{13} - p_{12})$ $\dot{W}_p = \eta_p \dot{W}_p$	$\dot{W}_{sthp} = \dot{m}_{vhp}(h_6 - h_8)$ $\dot{W}_{stlp} = (\dot{m}_{vlp} + \dot{m}_{vlp})(h_7 - h_9)$ $\eta_{st} = \frac{\dot{W}_{stlp}}{\dot{W}_{stlps}}$ $\dot{Q}_{\text{outcod}} = (\dot{m}_{vhp} + \dot{m}_{vlp})(h_9 - h_{10})$ $\frac{\dot{W}_{ps}}{\dot{m}} = v(p_{11} - p_{10})$ $\dot{W}_p = \eta_p \dot{W}_p$

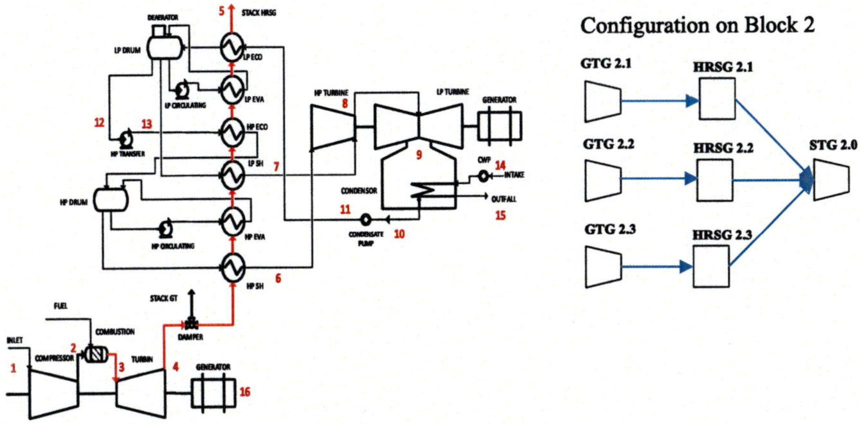

Fig. 1 Piping diagram the combined cycle power plant Tbrok. There are two cycles, namely the Brayton cycle in the gas turbine and the Rankine cycle in the steam turbine. Block 2 unit configuration with 3-3-1, 2-2-1, or 1-1-1 unit operation patterns

2.2 Exergy Analysis

Exergy analysis on CCPP equipment is divided into 2 cycles, and they are the gas cycle and the steam cycle. Each cycle condition uses the following formula (Table 3).

$$\eta_{cc} = \frac{\dot{Q}_{incc}}{\dot{Q}_{\text{infuel}}}$$

2.3 Method of Collecting the Data

The operating parameter data was taken from the combined cycle power plant unit block 2 with a combined cycle operating pattern of 3-3-1 with a net power of 95 MW for each gas turbine. (Fig. 2) Then, the operating data of GTG-HRSG 2.2 with a net power of 75 MW gas turbine was varied with the inlet compressor temperature around 24 to 33 °C. The operating parameter data was obtained from the downloaded DCS (Distributed Control System) and Mark-V in the period of June 2021 with an interval of 15 min. Data validation was carried out using the ranging method (Tables 4, 5, 6, and 7).

3 Result and Discussion

3.1 The Result of Energy Analysis and Exergy Analysis

The results of energy analysis and exergy analysis on Gas Turbine 2.1 show that the total exergy destruction is 177.19 MW (Fig. 3).

The results of energy analysis and exergy analysis on HRSG 2.1 show that the total exergy destruction is 19.29 MW (Fig. 4).

Table 3 Exergy analysis on combined cycle power plant

Compressor:
$\dot{E}_{f2} - \dot{E}_{f1} = \dot{m}_a\left(h_2 - h_1 - T_0\left(s_2^0 - s_1^0 - R\ln\left(\frac{P_2}{P_1}\right)\right)\right)$
$\varepsilon = \frac{\dot{E}_{f2} - \dot{E}_{f1}}{\dot{W}_c}$
$\dot{E}_{dcomp} = \dot{W}_c - (\dot{E}_{f2} - \dot{E}_{f1})$
Combustion Chamber:
$\dot{E}_{f3} - \dot{E}_{f2} = \dot{m}_g\left(h_3 - h_2 - T_0\left(s_3^0 - s_2^0 - R\ln\left(\frac{P_3}{P_2}\right)\right)\right)$
$\dot{E}_{dcomb} = \dot{Q}_{fuelcomb} - (\dot{E}_{f3} - \dot{E}_{f2})$
Gas Turbine:
$\dot{E}_{f3} - \dot{E}_{f4} = \dot{m}_g\left(h_3 - h_4 - T_0\left(s_3^0 - s_4^0 - R\ln\left(\frac{P_3}{P_4}\right)\right)\right)$
$\varepsilon = \frac{\dot{W}_t}{\dot{E}_{f3} - \dot{E}_{f4}}$
$\dot{E}_{dturb} = (\dot{E}_{f3} - \dot{E}_{f4}) - \dot{W}_t$
Exhaust:
$\dot{E}_{f4} - \dot{E}_{f1} = \dot{m}_g\left(h_4 - h_1 - T_0\left(s_4^0 - s_1^0 - R\ln\left(\frac{P_4}{P_1}\right)\right)\right)$
HRSG (High Recovery Steam Generator):
$\dot{E}_{f4} - \dot{E}_{f5} = \dot{m}_g\left(h_4 - h_5 - T_0\left(s_4^0 - s_5^0 - R\ln\left(\frac{P_4}{P_5}\right)\right)\right)$
$\dot{E}_{f6} - \dot{E}_{f13} = \dot{m}_{clhp}(h_6 - h_{13} - T_0(s_6 - s_{13}))$
$\dot{E}_{f7} - \dot{E}_{f11} = \dot{m}_{cllp}(h_7 - h_{11} - T_0(s_7 - s_{11}))$
$\dot{E}_{f5} - \dot{E}_{f1} = \dot{m}_g\left(h_5 - h_1 - T_0\left(s_5^0 - s_1^0 - R\ln\left(\frac{P_5}{P_1}\right)\right)\right)$
$\dot{E}_{dhrsg} = (\dot{E}_{f4} - \dot{E}_{f5}) - ((\dot{E}_{f6} - \dot{E}_{f13}) + (\dot{E}_{f7} - \dot{E}_{f11}))$
Steam Turbine:
$\dot{E}_{f6} - \dot{E}_{f8} = \dot{m}_{sthp}(h_6 - h_8 - T_0(s_6 - s_8))$
$\dot{E}_{dhpturb} = (\dot{E}_{f6} - \dot{E}_{f8}) - \dot{W}_{thp}$
$\dot{E}_{f7} - \dot{E}_{f9} = \dot{m}_{stlp+hp}(h_7 - h_9 - T_0(s_7 - s_9))$
$\dot{E}_{dlpturb} = (\dot{E}_{f7} - \dot{E}_{f9}) - \dot{W}_{tlp}$

(continued)

Table 3 (*continued*)

Condenser:

$$\dot{E}_{f9} - \dot{E}_{f10} = \dot{m}_v(h_9 - h_{10} - T_0(s_9 - s_{10}))$$

$$\dot{E}_{f15} - \dot{E}_{f14} = \dot{m}_{cw}(h_{15} - h_{14} - T_0(s_{15} - s_{14}))$$

$$\dot{E}_{d\text{cond}} = (\dot{E}_{f9} - \dot{E}_{f10}) - (\dot{E}_{f15} - \dot{E}_{f14})$$

$$\dot{E}_{f11} - \dot{E}_{f10} = \dot{m}_{cl}(h_{11} - h_{10} - T_0(s_{11} - s_{10}))$$

$$\dot{E}_{d\text{condpump}} = \dot{W}_p - (\dot{E}_{f3} - \dot{E}_{f4})$$

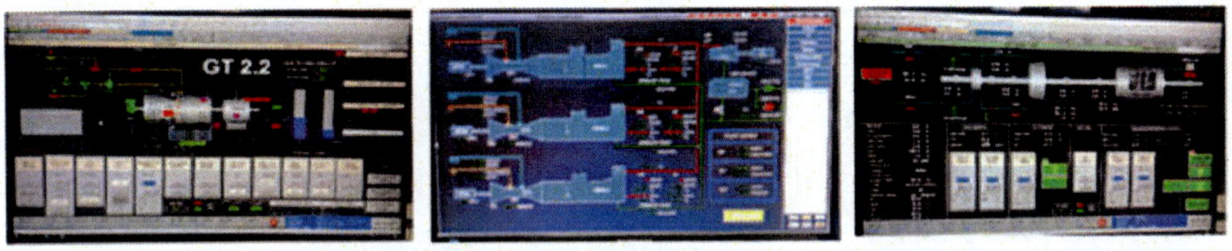

Date	Time	2SGAPI--1006 - FLUE GAS BEF HP SHTR PRESS [mBAR]	2TGOGTD L1010 - GT21 COMPRESSOR	2TGOGTDL1015 - GT21 COMPRESSOR INLET TEMP [DEG C]	2TGOGTDL 1016 - GT21 COMPRESSOR INLET	2TGOGTDL1 035 - GT21 MVAR OUTPUT [MVAR]	2TGOGTDL1 036 - GT21 MW OUTPUT [MW]	2TGOGTD L1079 - GT21 21 FUEL GAS FLOW	2TGOGTDL1 273 - GT21 MED EXHAUST TEMP [DEG	2TGOGTDL 1274 - GT21 COMP DISCH TEMP [DEG	2TGOGTDL 1286 - GT21 TURBINE SPEED RPM []	2TGOGTD L1296 - GT21 BASE/PRE SEL CNTL
6/28/2021	6:00:00 PM	0.105796	7.804523	24.856899	24.335949	-1.156285	69.127113	6.008526	561.644775	319.47205	2996.5916	21.05634
6/28/2021	6:15:00 PM	0.07908	9.325779	24.995823	24.197029	0.625019	93.065338	5.978404	561.297485	341.80347	2996.8416	21.02186
6/28/2021	6:30:00 PM	0.105796	9.399041	25.343121	24.856899	-1.812555	92.127815	5.990807	559.178955	342.94959	2999.5916	20.89257
6/28/2021	6:45:00 PM	0.105796	9.618826	25.030552	25.204201	-0.046876	96.54982	6.175971	559.178955	345.48486	2997.0916	20.89257
6/28/2021	7:00:00 PM	0.132511	9.506779	25.204201	25.343121	0.921903	95.221657	6.127244	559.560974	345.48486	3000.3416	20.6771
6/28/2021	7:15:00 PM	0.105796	9.562802	25.377853	25.030552	-2.031312	94.706017	6.076745	559.873535	345.20703	2996.3416	20.85379
6/28/2021	7:30:00 PM	0.159226	9.480922	25.10001	25.238934	-2.031312	93.846611	6.15648	558.588562	344.51242	3004.0918	20.90119
6/28/2021	7:45:00 PM	0.132511	9.601588	25.238934	25.377853	0.187506	95.487289	6.16977	560.602905	345.24176	3003.5916	20.87534
6/28/2021	8:00:00 PM	0.132511	9.592969	25.586231	25.238934	0.765648	95.487289	6.100665	559.178955	344.51242	2996.5916	20.85379
6/28/2021	8:15:00 PM	0.132511	9.506779	25.273663	25.238934	0.765648	93.049713	6.225585	561.992065	343.05377	3003.8416	21.00462
6/28/2021	8:30:00 PM	0.132511	9.580041	25.447311	25.447311	0.703146	95.768547	6.176857	560.984924	344.26932	2998.3416	20.98739
6/28/2021	8:45:00 PM	0.212659	8.144974	24.995823	25.134743	-1.078158	75.283546	5.205854	560.290344	325.0636	3001.8416	21.63812

Fig. 2 Real-time operating parameter data on combined cycle power plant block 2

The results of energy analysis and exergy analysis on Gas Turbine 2.2 show that the total exergy destruction is 175.92 MW (Fig. 5).

The results of energy analysis and exergy analysis on HRSG 2.2 show that the total exergy destruction is 12.51 MW (Fig. 6).

The results of energy analysis and exergy analysis on Gas Turbine 2.3 show that the total exergy destruction is 182.75 MW (Fig. 7).

The results of energy analysis and exergy analysis on HRSG 2.3 show that the total exergy destruction is 16.40 MW (Fig. 8).

The results of energy analysis and exergy analysis on Steam Turbine 2.0 show that the total exergy destruction is 43.53 MW (Fig. 9).

The results of the exergy analysis on GTG-HRSG 2.2 with variations in environmental temperature of 24–33 °C showed that the exergy destruction rate value is not linear, in which the largest exergy destruction rate occurs at 31 °C of 185.23 MW, and the lowest exergy destruction rate is at 26 °C of 168.32 MW (Fig. 10).

Table 4 Gas turbine operation parameter data 3-3-1 pattern on June 28, 2021, at 8 pm

Operation data gas turbine			GT 2.1	GT 2.2	GT 2.3
State	Symbol	Unit	Value		
	W_{gt}	MW	95.487289	95.3125	94.203125
0	P_0	Bar	1.01325	1.01325	1.01325
	T_0	C	25.238934	25.139977	25.209446
1	P_1	Bar	1.01325	1.01325	1.01325
	T_1	C	25.238934	25.139977	25.209446
2	P_2	Barg	9.592969	9.734971	9.308338
	T_2	C	344.51242	343.458344	337.902802
3	P_3	Barg	9.592969	9.734971	9.308338
	T_3	C	1124	1124	1124
	m_f	kg/s	6.100665	6.082763	6.141234
4	P_4	mbarg	1.045	1.045	1.045
	T_4	C	559.17896	560.854187	560.194458
5	P_5	Bar	1.01325	1.01325	1.01325
	T_5	C	152.46115	160.081741	158.813187

3.2 Analysis of Operating Pattern Modeling

With the availability of natural gas supply of 70 BBTUD, the first operating pattern is to operate 2 units of gas turbine and accommodate the remaining gas using CNG (Compressed Natural Gas) at low load, and then operate 3 units at peak load. The second operating pattern is to operate 3 gas turbine units continuously without using CNG (Fig. 11, Tables 8 and 9).

4 Conclusion

The results of the calculation of the exergy and energy analysis of combined cycle power plant block 2 showed that the efficiency of the open cycle GT 2.1 was 27.76%, GT 2.2 was 27.79%, and GT 2.3 was 27.21%. Meanwhile, the efficiency of HRSG 2.1 was 72.9%, HRSG 2.2 was 82.6%, HRSG 2.3 was 76.3%, and STG 2.0 was 88.25%. The combined cycle efficiency of PLTGU Block 2 pattern 3-3-1 was 42.06%. The results of the research pattern 3-3-1 Block 2 showed that the total exergy destruction in GTG 2.1 was 177.19 MW, GTG 2.2 was 175.92 MW, GTG 2.3 was 182.75 MW, while HRSG 2.1 was 19.29 MW, HRSG 2.2 was 12.51 MW, HRSG 2.3 was 16.40 MW, and STG 2.0 was 43.53 MW. The result of the calculation of the exergy destruction of GTG-HRSG 2.2 at an ambient temperature of 25 °C was 172.72 MW, at an ambient temperature of 28 °C was 180.26 MW, and at an ambient temperature of 31 °C was 185.23 MW. Thus, it could be concluded that the higher the ambient inlet temperature of the compressor, the

Table 5 HRSG operation parameter data 3-3-1 pattern on June 28, 2021, at 8 pm

Operation data on HRSG			HRSG 2.1	HRSG 2.2	HRSG 2.3
State	Symbol	Unit	Value		
6	P_6	barg	69.652313	69.636536	69.659485
	T_6	C	500.87421	511.970642	513.756409
	m_{hp}	kg/h	144,322.53	159,274.5	136,636.609
7	P_7	barg	4.524109	4.30365	4.696503
	T_7	C	315.62863	318.735016	325.52243
	m_{lp}	kg/h	14,424.179	20,490.7598	25,289.3203
12	P_{12}	barg	15.76942	15.87616	15.516357
	T_{12}	C	147.35811	136.32373	136.32373
13	P_{13}	barg	109.30386	109.165039	113.117676
	T_{13}	C	167.34589	158.442352	159.602905
	m_{tr}	kg/h	146,943.11	159,274.5	139,120.266
10	P_{10}	mmHg	66.1791	66.1791	66.1791
	T_{10}	C	41.86729	41.86729	41.86729
11	P_{11}	barg	18.431213	18.431213	18.431213
	T_{11}	C	42.157215	42.157215	42.319336
	m_{cl}	kg/h	173,205.41	187,920.234	174,872.797
	$m_{cl\ st}$	kg/h	558,447		

Table 6 Steam turbine operation parameter data 3-3-1 pattern on June 28, 2021, at 8 pm

Operation data on STG				Unit conversion	
State	Symbol	Unit	Value	Unit	Value
	W_{st}	MW	144.84375	kW	144,843.75
6-g	P_6	Barg	69.145126	kPa	7015.8376
	T_6	C	515.3833	C	515.383301
7-g	P_7	Barg	4.617587	kPa	563.0837
	T_7	C	312.11462	C	312.114624
8	P_8	Barg	4.617587	kPa	563.0837
	T_8	C		C	312.114624
9	P_9	mmHg	66.1791	kPa	8.82315644
	T_9	C		C	43.3798718
10	P_{10}	mmHg	66.1791	kPa	8.82315644
	T_{10}	C	41.86729	C	41.86729
11	P_{11}	Barg	18.431213	kPa	1944.4463
	T_{11}	C	42.157215	C	42.157215
	m_{cl}	kg/h	556,606.56	kg/s	154.612934
14	T_{14}	C	29.248741	C	29.248741
15	T_{15}	C	37.626522	C	37.626522

Table 7 Operation data GTG-HRSG 2.2 with variety of ambient temperature

State	Symbol	Unit	Value									
	T_0	°C	24.03	25.00	26.01	27.19	28.09	29.27	30.07	31.05	32.05	33.13
	Wgt	MW	75.73	75.83	75.83	75.03	75.28	75.73	75.86	75.36	75.86	75.98
0	P_0	Bar	1.01	1.01	1.01	1.01	1.01	1.01	1.01	1.01	1.01	1.01
	T_0	°C	24.03	25.00	26.01	27.19	28.09	29.27	30.07	31.05	32.05	33.13
1	P_1	Bar	1.01	1.01	1.01	1.01	1.01	1.01	1.01	1.01	1.01	1.01
	T_1	°C	24.03	25.00	26.01	27.19	28.09	29.27	30.07	31.05	32.05	33.13
2	P_2	Barg	8.19	8.23	8.19	8.19	8.23	8.24	8.30	8.25	8.24	8.21
	T_2	°C	319.64	320.72	322.73	324.05	325.65	328.56	330.23	331.93	331.72	334.67
3	P_3	Barg	8.19	8.23	8.19	8.19	8.23	8.24	8.30	8.25	8.24	8.21
	T_3	°C	1124	1124	1124	1124	1124	1124	1124	1124	1124	1124
	m_f	kg/s	5.20	5.19	5.11	5.34	5.32	5.34	5.31	5.39	5.21	5.25
4	P_4	Mbarg	1.05	1.05	1.05	1.05	1.05	1.05	1.05	1.05	1.05	1.05
	T_4	°C	558.84	559.01	559.22	559.74	561.51	558.29	561.55	561.44	561.06	561.31
5	P_5	Bar	1.01	1.01	1.01	1.01	1.01	1.01	1.01	1.01	1.01	1.01
	T_5	°C	144.22	144.45	144.22	145.36	143.91	145.14	145.76	145.75	150.04	143.24

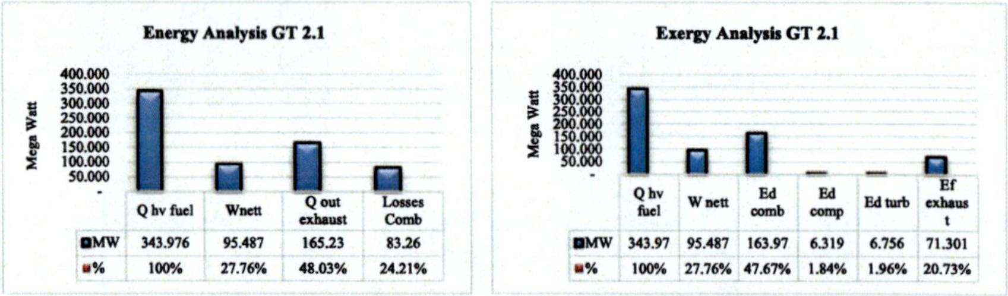

Fig. 3 The results of energy analysis and exergy analysis on gas turbine 2.1

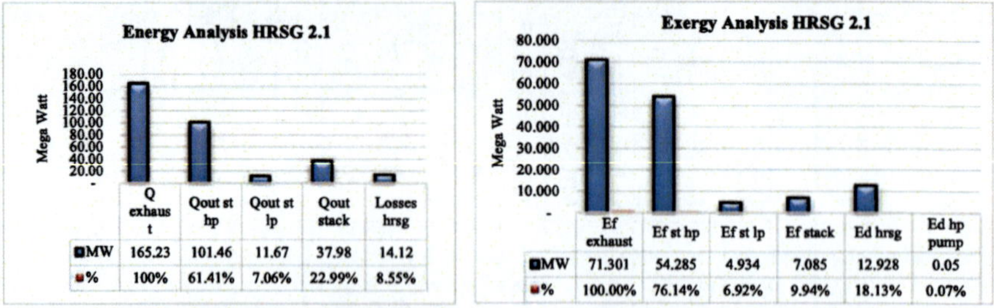

Fig. 4 The results of energy analysis and exergy analysis on HRSG 2.1

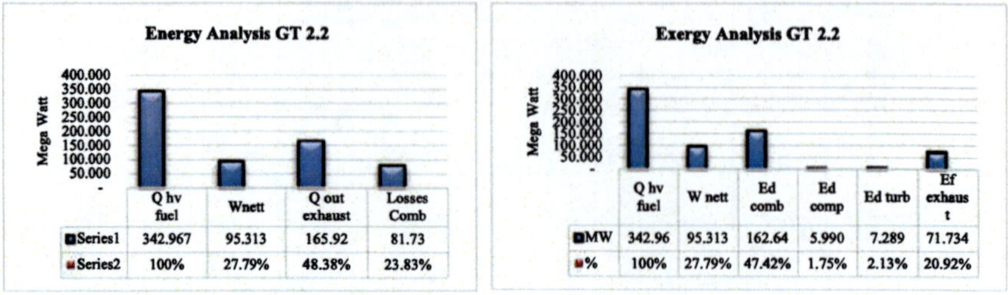

Fig. 5 The results of energy analysis and exergy analysis on gas turbine 2.2

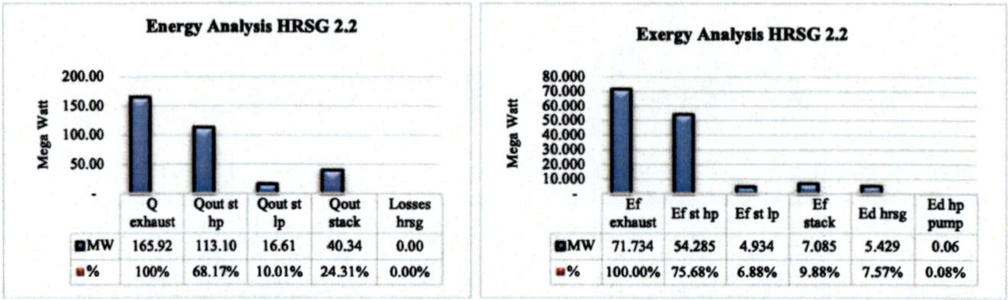

Fig. 6 The results of energy analysis and exergy analysis on HRSG 2.2

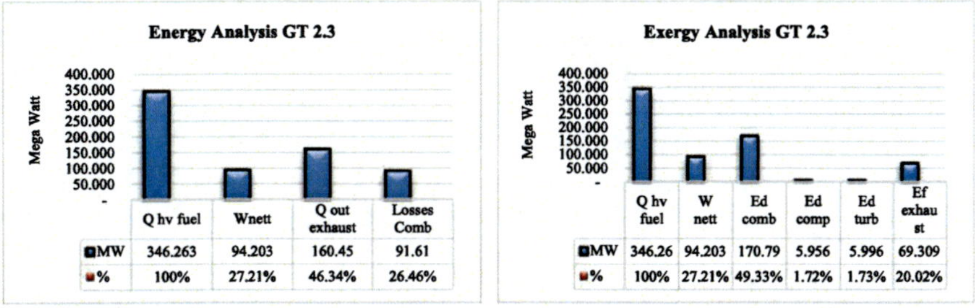

Fig. 7 The results of energy analysis and exergy analysis on Gas Turbine 2.3

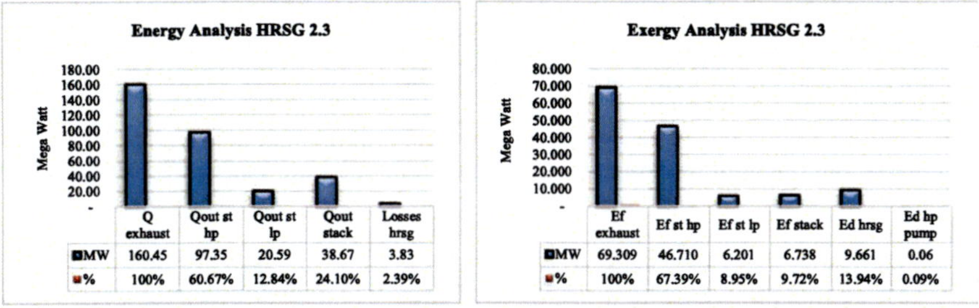

Fig. 8 The results of energy analysis and exergy analysis on HRSG 2.3

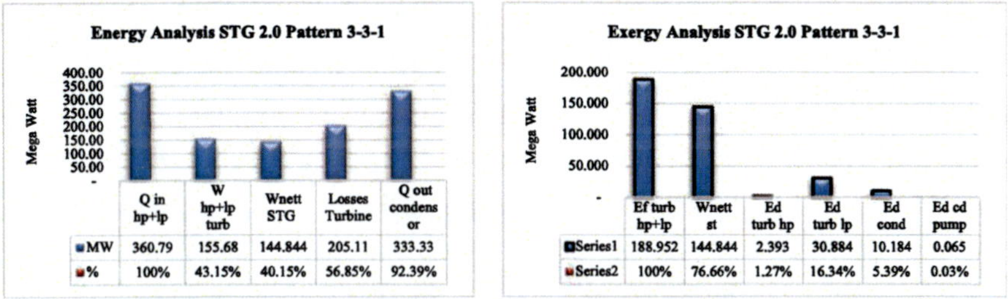

Fig. 9 The results of energy analysis and exergy analysis on steam turbine 2.0

higher the exergy rate destruction. To get the maximum benefit, CCPP Block 2 should operate on 2 gas turbine continuously and 3 gas turbine during peak load (make use of CNG), so that it can result 7520.53 MW per 24 h with cost of goods manufactured 0.064482 $/kWh. Meanwhile, the operation on 3 gas turbine with minimum load result 7403.97 MW per 24 h with cost of goods manufactured 0.065497 $/kWh.

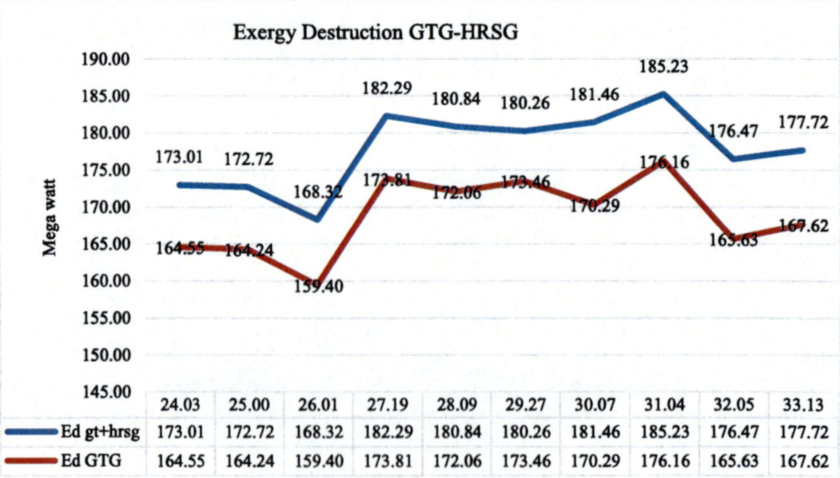

Fig. 10 The results of energy analysis and exergy analysis on steam turbine 2.0

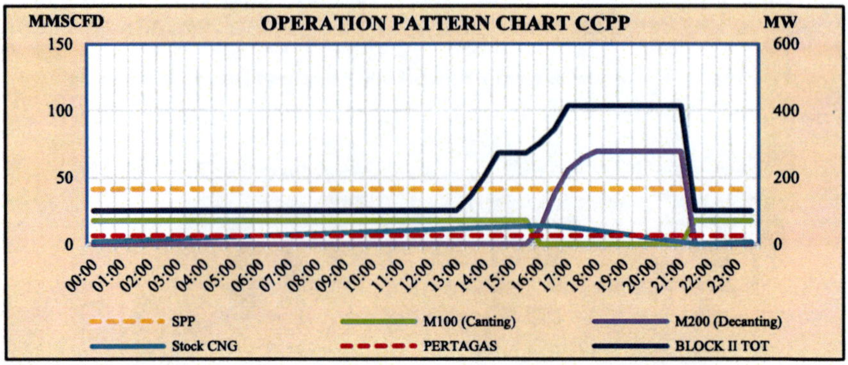

Fig. 11 The operation pattern chart combined cycle power plant

Table 8 The first operating pattern

Early hours	Final hour	Duration hours	Used gas (BBTUD)	Net production (MWh)	Stored in CNG (BBTU)	Use of CNG (BBTUD)
0:00	13:30	13.50	26.05	2,677.28	10.47	0.00
13:30	16:00	2.50	8.34	892.12	1.94	0.00
16:00	21:30	5.50	30.82	3,455.35	0.00	14.40
21:30	0:00	2.50	4.82	495.79	1.94	0.00
Total production				7520.53	MW	

Table 9 The second operating pattern

Early hours	Final hour	Duration hours	Used gas (BBTUD)	Net production (MWh)	Stored in CNG (BBTU)	Use of CNG (BBTUD)
0:00	23:59	23.98	66.30	7403.12	0.00	0.00
Total production				7403.12	MW	

References

1. Aliyu M, AlQudaihi AB, Said SAM, Habib MA (2019) Energy, exergy and parametric analysis of a combined cycle power plant, Mechanical Engineering Department, King Fahd University of Petroleum and Minerals, 31261, Dhahran
2. Aichmayer J, Spelling B, Laumert (2015) Thermoeconomic analysis of a solar dish micro gas-turbine combined-cycle power plant. Energy Procedia 69: 1089–1099. https://doi.org/10.1016/j.egypro.2015.03.217
3. Seyyedia SM, Hashemi-Tilehnoeeb M, Rosenc MA (2018) Exergy and exergoeconomic analyses of a novel integration of a 1000 MW pressurized water reactor power plant and a gas turbine cycle through a superheater. Aliabad Katoul Branch, Islamic Azad University, Aliabad Katoul, Iran, Department of Mechanical Engineering
4. Ameri M, Ahmadi P, Khanmohammadi S (2009) Exergy analysis of a 420 MW combined cycle power plant Int J Energy Res 32 (2):175–183. https://doi.org/10.1002/er.1351
5. Tambak Lorok combined cycle power plant maintenance and system description gas turbine vol III
6. Tambak Lorok combined cycle power plant maintenance and system description high recovery steam generator
7. Tambak Lorok combined cycle power plant maintenance and system description steam turbine vol I

Experimental Study Combustion Air Optimization to Increase Efficiency Due to Changes in Coal Quality with Variation of Load in Labuan Steam Power Plant

Adriska Simon Prayoga(✉)

Department of Mechanical Engineering, Institut Teknologi Sepuluh Nopember, Surabaya 60111, Indonesia
adriska83@gmail.com

Abstract. One of the main components in a steam power plant is a boiler, which has the function to produce steam. Boiler efficiency is defined as the ratio of the amount of heat energy of the fuel to produce steam compared to the amount of fuel input to the boiler. Changes in coal quality influence combustion system settings, boiler efficiency, and impact on the environment. Research with an experimental method is needed to adjust combustion air with a variation of load and O_2 trim values to get a complete combustion and optimum boiler efficiency and to see the effect on temperature and emission of flue gas. This paper has presented the data experiments, analyzing the effect of variations of O_2 trim on the thermal efficiency of the boilers, flue gas temperature, and emission of the flue gas: NO_x, CO, and CO_2. It can be concluded that using the 4100 kcal/kg calorific value of coal, the optimum value of O_2 trim of boiler efficiency at 72, 80, and 88% loads is 2%. Boiler efficiency will increase as the load increases. Boiler loads at 72%, 80%, and 88%, the maximum boiler efficiency is 83.33%, 83.42%, and 83.49%, respectively.

Keywords: Boiler efficiency · O_2 trim · Flue gas · Experimental

1 Introduction

Boiler efficiency measures how much the boiler's ability to convert the value of chemical energy contained in the fuel into heat energy. The thermal efficiency of the boiler is defined as the percentage of heat input that is effectively utilized to generate steam. There are two assessment methods of boiler efficiency that are a direct method and an indirect method. To optimize boiler efficiency, it can adjust the excess air ratio on the burner. By setting excess air at optimum conditions will reduce the total cost of electricity production, fuel consumption, and the impact of environmental pollution [1]. Proper control of the right amount of excess air maintains optimum combustion efficiency. Amounts of CO_2 and O_2 in combustion gases are indexes of excess air. The desirable CO_2 level depends on the fuel and the optimum excess air for the furnace. Desirable O_2 values

M. Kolhe et al. (eds.), *Recent Advances in Renewable Energy Systems*,
Lecture Notes in Electrical Engineering 876,
https://doi.org/10.1007/978-981-19-1581-9_30

depend much less on the type of fuel. This makes O_2 measurement the preferred method for combustion control. If the measured O_2 content is more than that desired, the air supply is to be reduced. If the O_2 measured is less than that desired, the air supply is to be increased [2]. Labuan Steam Power Plant is a coal-fired power plant with a capacity of 300 MW. The coal used has a low calorific value in the range 4200–4600 kcal/kg, which is obtained by blending medium-rank coal (MRC) or sub-bituminous coal with low-rank coal (LRC) or lignite coal. To reduce fuel costs, the dominant mix coal supply is composed of lignite coal with a calorific value of 3500–4200 kcal/kg. These changes in coal composition and calorific value will cause an influence on combustion system settings, boiler efficiency, and impact on the environment. This change also causes power plants to operate dominantly at a lower load than their capacity. Zixiang [3] conducted and studied the effect of changes in coal quality on combustion in boilers. The result is an effect of lignite coal moisture content changed on combustion patterns and NO_x formation characteristics in a 660 MW boiler model. In addition, also, the combustion temperature decreased when the water content in coal increased. Tanetsakunvatana [4] studied experimentally to analyze the effect of operating conditions and fuel quality on boiler thermal efficiency and emissions from boilers with lignite coal fuel. Pachaiyappan [5] researched the optimization of combustion air to increase boiler efficiency. The flue gas temperature will affect the boiler efficiency, where when the flue gas temperature decreases, the boiler efficiency will increase. For every decrease in flue gas temperature 22 °C, the efficiency of the boiler increases by approximately 1%. Ghritlahre [6] investigated the effect of excess air on boiler efficiency by minimizing losses in the dry flue gas. The results showed that maximum boiler efficiency was obtained with excess air in the range of 20–40%. From the research, it was also found that the higher the excess air in the boiler, the flue gas temperature will be higher. Jaiswal [7] conducted a study to improve boiler efficiency based on combustion air. Increasing boiler efficiency can be done by increasing heat recovery from flue gases and controlling excess air. This research was conducted to adjust combustion air with a variation of load and O_2 trim values. The purpose of this investigation is to obtain the optimum value of O_2 trim to get the best boiler efficiency and to see the effect variation of O_2 trim value on the temperature and emission of flue gas.

2 Methodology

The method used in this study is an experimental method. This research needs to identify with actual testing on the boiler. The experimental method data collection was performed on a boiler with low-rank coal which has a calorific value of ±4100 kcal/kg and the load variation of 72, 80, and 88% based on maximum continuous rated (MCR). Based on the ultimate coal analysis, the theoretical air requirement is calculated. Excess air can be calculated from the O_2 trim value so that the actual air requirements can be obtained for each variation of O_2 trim. At each load, the value of O_2 trim was varied from 1.5%, 2%, 2.5%, 3%, to 3.5%, respectively. In this experiment, the variation of O_2 trim set by adjusting the opening percentage of the secondary air damper from the forced draft fan (FDF) is shown in Fig. 1. Secondary airflow settings to achieve the set O_2 trim value are controlled on the distributed control system (DCS) monitor as shown in Fig. 2. In this study, we will obtain the effect of variations in O_2 trim on the value of NO_x, CO, and CO_2 in flue gas, boiler efficiency, and flue gas temperature, respectively.

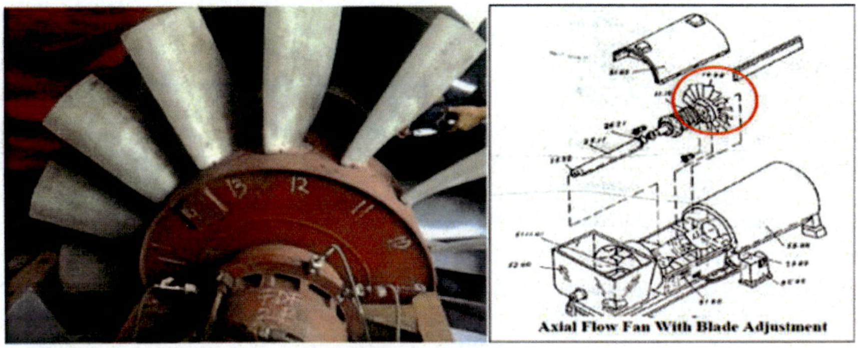

Fig. 1 Forced draft fan (FDF)

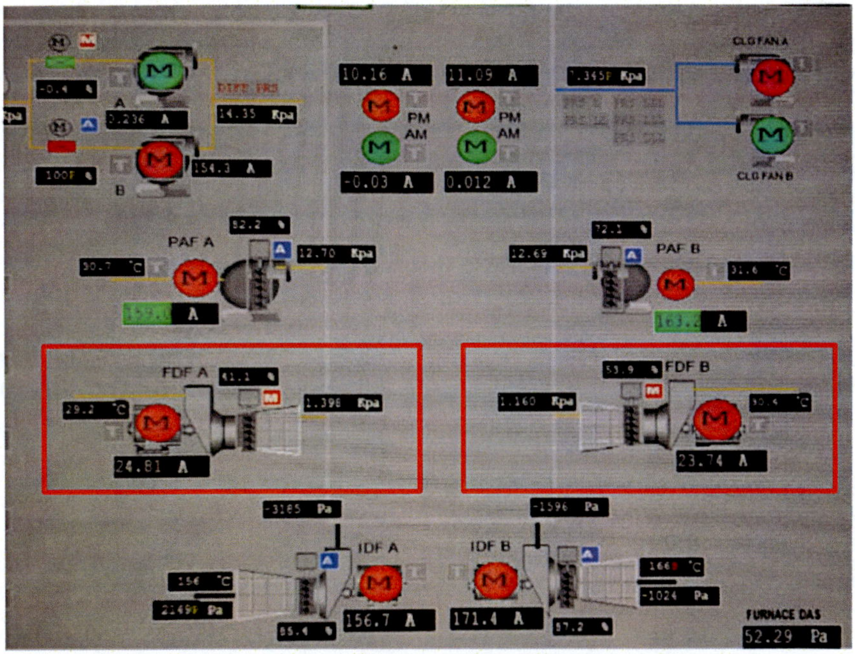

Fig. 2 DCS operation for secondary air controlling

The experimental data sampling points are shown in Fig. 3. The location of flue gas sampling point at the inlet and outlet air heater, measurement of ambient air conditions in the boiler suction fan, coal sampling is taken at the inlet coal feeder, ash sampling is taken at the drain bottom ash silo and electrostatic precipitator, measuring the boiler surface temperature at the point around the furnace, superheater, reheater, and economizer. Experimental measurement devices include the Vario MRU plus industrial portable analyzer to measure temperature, NO_x, CO, and CO_2 level at flue gas, the Fluke 271 to measure temperature and relative humidity ambient air, coal analyzer equipment (LECO AC 600 automatic calorimeter, LECO TGA 701 proximate analyzer, LECO CHN 628 ultimate analyzer, LECO 628 S sulfur analyzer), and GM-9000 infrared thermometer to measure boiler surface temperature. Parameters of coal and ash quality, air conditions for combustion, flue gas conditions, boiler surface temperature, and coal flow, respectively, were obtained in this experiment. Parameter data were obtained that include flue gas temperature, CO and CO_2 levels in flue gases, ambient air temperature and humidity, ultimate analysis results and total moisture in coal, unburned carbon in ash, and boiler surface temperature, respectively. All parameters were taken and recorded as experimental data when varying the pre-determined value of O_2 trim. During the experiment, the coal flow, airflow, and the opening percentage secondary air damper were also recorded at each value of O_2 trim. Boiler efficiency is calculated by the indirect method (heat loss method) [8]. The results of experimental data were then calculated to make chart O_2 trim versus boiler efficiency, NOx, CO, and CO_2 content in flue gas, and also flue gas temperatures. Based on coal calories of ±4100 kcal/kg, the experiment was repeated for several load variations. From each load can be seen the optimal O_2 trim to get the best efficiency of the boiler. The boiler efficiency results have been compared with the varying load as mentioned above.

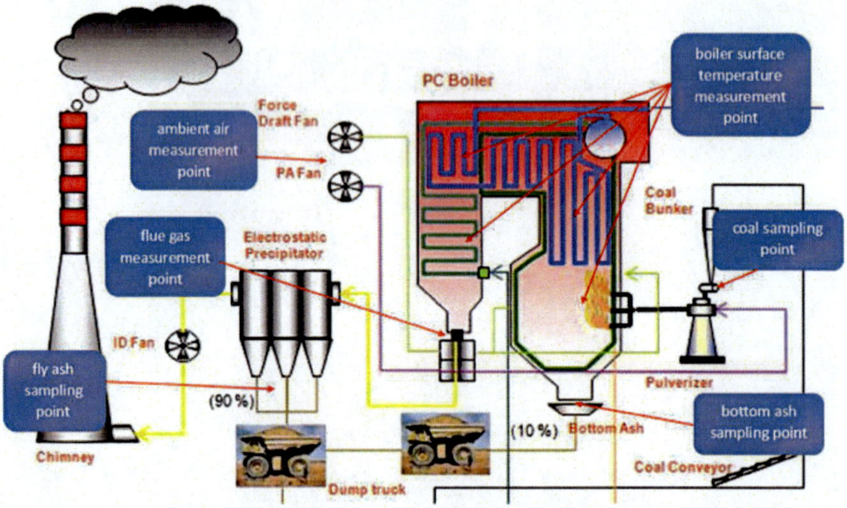

Fig. 3 Location of sampling point experimental study

3 Result and Discussion

The summary data result of boiler efficiency, flue gas temperature, NO_x, CO, and CO_2 level in flue gas for each load with O_2 trim variation is shown in Table 1, Table 2, and

Table 3, respectively. The results of this experimental study indicate that the optimum value of O_2 trim to get the best boiler efficiency at 72, 80, and 88% loads is 2% with a coal calorific value of ±4100 kcal/kg. At the optimum value of O_2 trim can be seen the result of calculation boiler heat loss, and boiler efficiency for each load is shown in Table 4.

Table 1 Experimental data and calculation results at 72% load

Parameter	Run-1	Run-2	Run-3	Run-4	Run-5
Coal calorific value (kcal/kg)	4162	4162	4162	4162	4162
O_2 trim (%)	*1.5*	2	*2.5*	*3*	*3.5*
Excess air (%)	7.69	10.53	13.51	16.67	20
Flue gas temperature (°C)	171.8	172.3	172.7	173.3	173.9
CO level in flue gas (ppm)	3.5	2	1.5	1	0.5
CO_2 level in flue gas (%)	17.37	17.22	16.44	16.17	15.74
NO_x level in flue gas (mg/Nm^3)	382	375	378	400	415
Boiler efficiency (%)	83.31	83.33	83.19	83.07	82.97

Table 2 Experimental data and calculation results at 80% load

Parameter	Run-1	Run-2	Run-3	Run-4	Run-5
Coal calorific value (kcal/kg)	4162	4162	4162	4162	4162
O_2 trim (%)	*1.5*	2	*2.5*	*3*	*3.5*
Excess air (%)	7.69	10.53	13.51	16.67	20
Flue gas temperature (°C)	173.6	174.2	174.6	175.5	175.8
CO level in flue gas (ppm)	2.5	1	1	0.5	0.5
CO_2 level in flue gas (%)	17.80	17.72	17.07	16.75	16.30
NOx level in flue gas (mg/Nm^3)	335	364	370	401	415
Boiler efficiency (%)	83.36	83.42	83.28	83.14	83.02

This study also obtained results related to the effect of O_2 trim on flue gas conditions, where the value of O_2 trim is higher, the temperature of the flue gas will also increase. This can be caused by the increase in secondary airflow, the residence time for the combustion process will be reduced, and a lot of heat energy of combustion will be carried by flue gas so that the flue gas temperature will increase. From previous research, it was also found that the flue gas temperature tends to increase with the increase in excess air supply [6]. With the increase in the value of O_2 trim, the NO_x content in the flue gas will also increase, this is because, with the increase in excess O_2 and high-temperature combustion, the NO_x formed is also increasing. NOx itself is formed at high temperatures, and there is

Table 3 Experimental data and calculation results at 88% load

Parameter	Run-1	Run-2	Run-3	Run-4	Run-5
Coal calorific value (kcal/kg)	4096	4096	4096	4096	4096
O_2 trim (%)	*1.5*	2	*2.5*	*3*	*3.5*
Excess air (%)	7.69	10.53	13.51	16.67	20
Flue gas temperature (°C)	171.8	172.1	172.7	173.8	173.9
CO level in flue gas (ppm)	1	0	0.5	0.5	0.5
CO_2 level in flue gas (%)	17.75	17.42	16.93	16.41	15.92
NOx level in flue gas (mg/Nm^3)	250	306	310	305	367
Boiler efficiency (%)	83.41	83.49	83.33	83.16	83.05

Table 4 Comparison of boiler heat loss for each load at O_2 trim optimum

Losses (%)	Load (%)		
	72	80	88
Dry flue gas loss (L1)	4.98	5.08	5.01
Loss due to hydrogen in fuel (L2)	3.10	3.10	3.26
Loss due to moisture in fuel (L3)	5.52	5.53	5.61
Loss due to moisture in air (L4)	0.19	0.19	0.19
Partial combustion of C to CO (L5)	0.001	0.0003	0.0000
Surface heat loss (L6)	0.13	0.13	0.11
Loss due to unburnt in fly ash (L7)	0.44	0.37	0.29
Loss due to unburnt in bottom ash (L8)	2.31	2.17	2.03
Total losses (%)	16.67	16.58	16.51
Boiler efficiency (%)	83.33	83.42	83.49

still enough excess O2. In previous studies, it was also found that with increasing O_2 trim (excess O_2), the NOx content in the flue gas also increased [5]. NOx is formed maximally at an excess O_2 value about 5–7%. During the experiment, the air supply for Over Fire Air (OFA) was not added when the O_2 trim value was increased or the secondary air was increased, this factor can also cause NO_x in the flue gases to increase. This is in accordance with previous studies where it was found that with increasing OFA, the formation of NO_x can be reduced [9]. Although NO_x increases with the increase in the value of O_2 trim, for all experiments, the value still meets the emission-quality standard in accordance with the Government Regulation Republic of Indonesia. On the other hand, the level both of CO and CO_2 in the flue gas will decrease as the value of O_2 trim increases. The increasing combustion air makes the combustion reaction better so that the CO level in the flue gas decreases. The presence of CO in the flue gas can be caused by the residence time of the fuel in the combustion chamber reduced when excess air is added. For CO_2, the maximum CO_2 level occurs during stoichiometric conditions, so that more excess air is added, and the percentage of CO_2 in the flue gas decreases. For every 0.5% increase in the value of O_2 trim, the flue gas temperature will increase approximately 0.5 °C, NO_x level in flue gas increased 21 mg/Nm^3, CO_2 level in flue gas decreased 0.41%, and CO level in flue gas decreased 0.5 ppm. In terms of air supply to the boiler, for every 0.5% increase in O_2 trim at variations of loads as mentioned above, the average airflow will increase by 27 tons/hour. From the experiment, it was found that the higher the load, the boiler efficiency will also increase and CO levels will decrease. At a higher load, the percentage of CO_2 in the flue gas will increase, but it is also dominantly influenced by the carbon content in fuel. The experimental result on load variations is shown in Figs. 4, 5, 6, and 7.

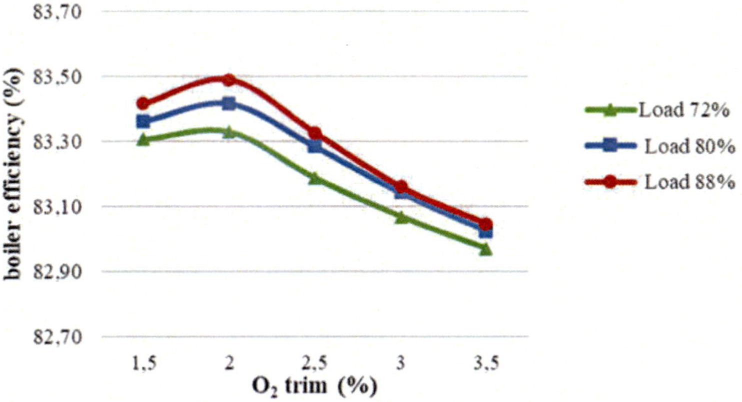

Fig. 4 O_2 trim versus boiler efficiency chart

4 Conclusion

In the case of boiler 300 MW using the 4100 kcal/kg calorific value of coal, the optimum value of O_2 trim is as much as 2% at all loads. At 2% of O_2 trim, the maximum boiler efficiency can reach 83.33%, 83.42%, and 83.49% for the load of 72%, 80%, and 88%, respectively. For every 0.5% increase in the value of O_2 trim, the airflow will increase 27 tons/hour, flue gas temperature will increase about 0.5 °C, NO_x content in flue gas will increase 21 mg/Nm^3, the CO_2 content in the flue gas will decrease 0.41%, and also the CO content in the flue gas will decrease 0.5 ppm.

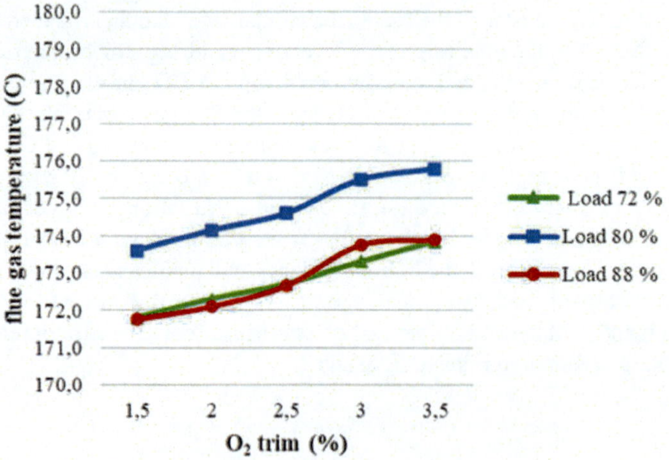

Fig. 5 O_2 trim versus flue gas temperature chart

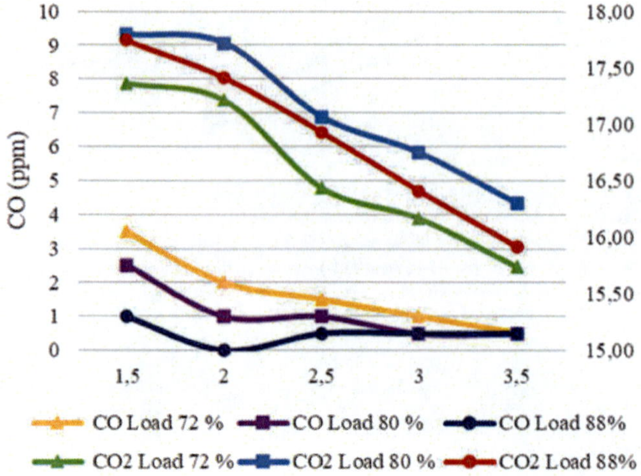

Fig. 6 O_2 trim versus CO and CO_2 content on flue gas chart

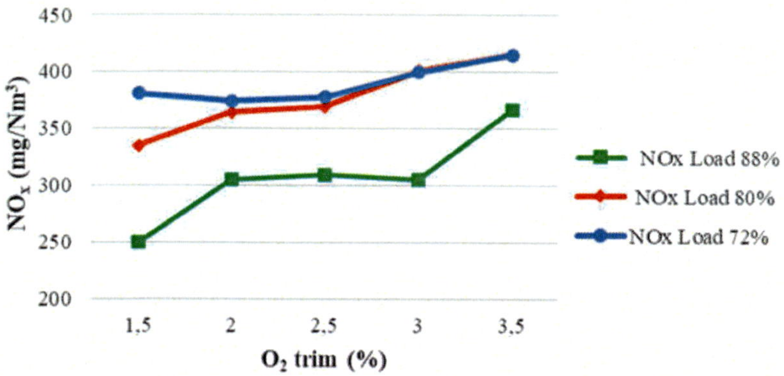

Fig. 7 O_2 trim versus NO_x content on flue gas chart

References

1. Kuprianov VI (2005) Applications of a cost-based method of excess air optimization for the improvement of thermal efficiency and environmental performance of steam boilers. Renew Sustain Energy Rev 9
2. Nag PK (2014) Power plant engineering, 4th edn, McGraw Hill Education (India) Private Limited P
3. Zixiang L, Zhengqing M (2020) Effects of moisture and its input form on coal combustion process and NOx transformation characteristics in lignite boiler. Fuel 266:116970
4. Tanetsakunvatana V, Kouprianov VI (2007) Experimental study on effects of operating conditions and fuel quality on thermal efficiency and emission performance of a 300-MW boiler unit firing Thai lignite. Fuel Process Technol 88
5. Pachaiyappan R, Dasa Prakash J (2015) Improving the boiler efficiency by optimizing the combustion air. Appl Mech Mat 787. ISSN 1662-7482
6. Harish G, Tej Pratap S (2014) Effect of excess air on 30 TPH AFBC boiler on dry flue gas losses and its efficiency. Int J Res Advent Technol 2(6). E-ISSN 2321-9637
7. Jaiswal NK, Tiwari RK (2018) Study the steps of improving the boiler efficency using combution air and its requirements. Int J Future Revol Comput Sci Commun Eng 4(10). ISSN 2454-4248
8. ASME PTC 4.1.: Power test code for steam generating units, New York (1998)
9. Chao L, Ke L, Yongzhen W, Zhitong M, Yilie G (2017) The effect analysis of thermal efficiency and optimal design for boiler system. Energy Procedia 105:3045–3050

Numerical Study Effect Using Low Rank Coal on Flow Characteristics, Combustion, and Furnace Exit Gas Temperature on Tangentially Fired Pulverized Coal Boiler 350 MWe

Arief Laga Putra[1](✉), Wawan Aries Widodo[2], and Ardi Nugroho[3]

[1] Tanjung Awar-Awar CFPP, PT. Pembangkitan Jawa-Bali, Tuban, East Java, Indonesia
arieflaga@ptpjb.com

[2] Department of Mechanical Engineering, Institut Teknologi Sepuluh Nopember, Surabaya, Indonesia

[3] Electricity Technology Development, PT. Pembangkitan Jawa-Bali, Surabaya, Indonesia

1 Introduction

Tanjung Awar-Awar Coal-Fired Steam Turbine Power Plant, which located in Tuban, East Java, Indonesia, is a pulverized coal tangentially fired sub-critical boiler with capacity of 2 × 350 MWe with sub-critical technology which is designed to use 4400–4500 kcal/kg low-calorific coal. To address the need for electricity production coat, the power plant carried out a coal optimization initiative, namely replacing low-calorie coal from 4500 kcal per kilogram to 4150 kcal per kilogram. Financially, replacing coal from 4500 kcal per kilogram to 4150 kcal per kilogram will indeed increase the amount of coal consumption and combustion air (primary–secondary air). However, because the price is cheaper, savings can still be obtained. This initiative has been carried out for more than one year, and there has never been an in-depth study on how the combustion behavior occurs in the furnace when compared to commissioning conditions.

As a type of coal with abundant worldwide reserve and relatively cheaper price, the using of low rank coal in many coal-fired power plant has attracted serious topic from researchers. In general, low rank coal has many benefits over high-medium rank coal, such as comparatively low production cost, high content of volatile substances, and less impurities that produce pollutants [1]. Coal quality rank affected the combustion temperature, especially furnace exit gas temperature, combustion results, temperature distribution of combustion flue gases, flame structure or thermal load and geometry of furnace water walls, higher heating value coal formed high intensity flame, for the same pulverized coal fineness, and almost same residence time of pulverized coal particles [2]. Combined impacts of several parameters, such as the coal specification, combustion of different particle size, and injection of coal through different burner type, were concluded to have significant impact on combustion performance and flame intensity in the furnace [2] as it induces the combustion mixing process and action of coal particles in the main

M. Kolhe et al. (eds.), *Recent Advances in Renewable Energy Systems*,
Lecture Notes in Electrical Engineering 876,
https://doi.org/10.1007/978-981-19-1581-9_31

combustion area in furnace [3]. SOFA angle is quite effective operation indicator to reduce the combustion temperature deviation, and further field tests and CFD simulations on the changes in yaw and tilt angles of over-fire air (SOFA and CCOFA) are necessary to study the optimal operation parameter [4]. The turned OFF coal burners in the opposite position are better than those in the adjacent direction under maximum capacity load, and it may help to form an built ascending flame distribution in the main flame region [5].

The main aim of this study is to predict and analyze combustion temperature and velocity from lowest coal burner elevation until furnace exit gas elevation, which compares the conditions between the combustion of design coal and the combustion of low rank coal with variations in the ratio of primary air-pulverized coal flow. The study provides summary of a numerical study on custom parameter of flue gas-particle stream in the power plant, focused on sub-critical tangentially fired furnace of 350 MW_e boiler.

2 Model

The dimension of the researched 350 MW_e tangentially low rank coal-fired boiler is schematically shown in Figs. 1 and 2. The furnace elevation is 55.89 m, and horizontal cross-section of the burner's region diameter is 0.03 m [6]. Besides the water walls on four wall position, the suspended heat transfer surfaces consist of division panel superheater (DPSH), platen superheater (PSH), finishing re-heater (FRH), platen re-heater (PRH), finishing superheater (FSH), low-temperature superheater (LTSH), and economizer (ECO) in the roof region and back pass flue gas path are conducted in the numerical to analyze the in-furnace heat transfer behavior. However, since only low rank coal combustion behavior and heat transfer inside the furnace are focused in this study, some convective heating surfaces in the flue gas back pass are not concerned, such as air preheater (APH). To simplify the simulation domain and minimize computational cost and time, the physical geometric structure banks of tube of DPSH, PSH, FRH, PRH, FSH, LTSH, and ECO are not performed in this simulation process. Instead, they are simplified into porous medium with the proportional heat absorbing area of them, respectively. The same treatment is also performed to the heat absorbing of water walls tube at the furnace walls. In actual boiler performance, the boiler needs 210.6 t/h low rank coal under TMCR phase to obtain 1153 t/h of superheated steam with the pressure of 15.18 MPa and temperature of 538.8 °C. Detailed physicochemical properties of low rank coal used in both the actual performance and the simulation stage are given in Table 1.

The numerical mesh with 5,089,250 nodes was conducted, providing accuracy and convergence, as well as simulation efficiency. The furnace boiler zones use tetrahedral cells, and all other zones use hexahedral cells. Mesh is tuned in the neighborhood of the coal burners, where the combustion stage was established [7].

3 Result and Discussion

Figure 3 indicates the temperature contour of the flue gas flow in boiler vertical side view, where Fig. 3a—for design coal (medium rank coal) at commissioning data, Fig. 3b—for low rank coal with PA ratio 1.8, Fig. 3c—for low rank coal with PA ratio 2.0, and

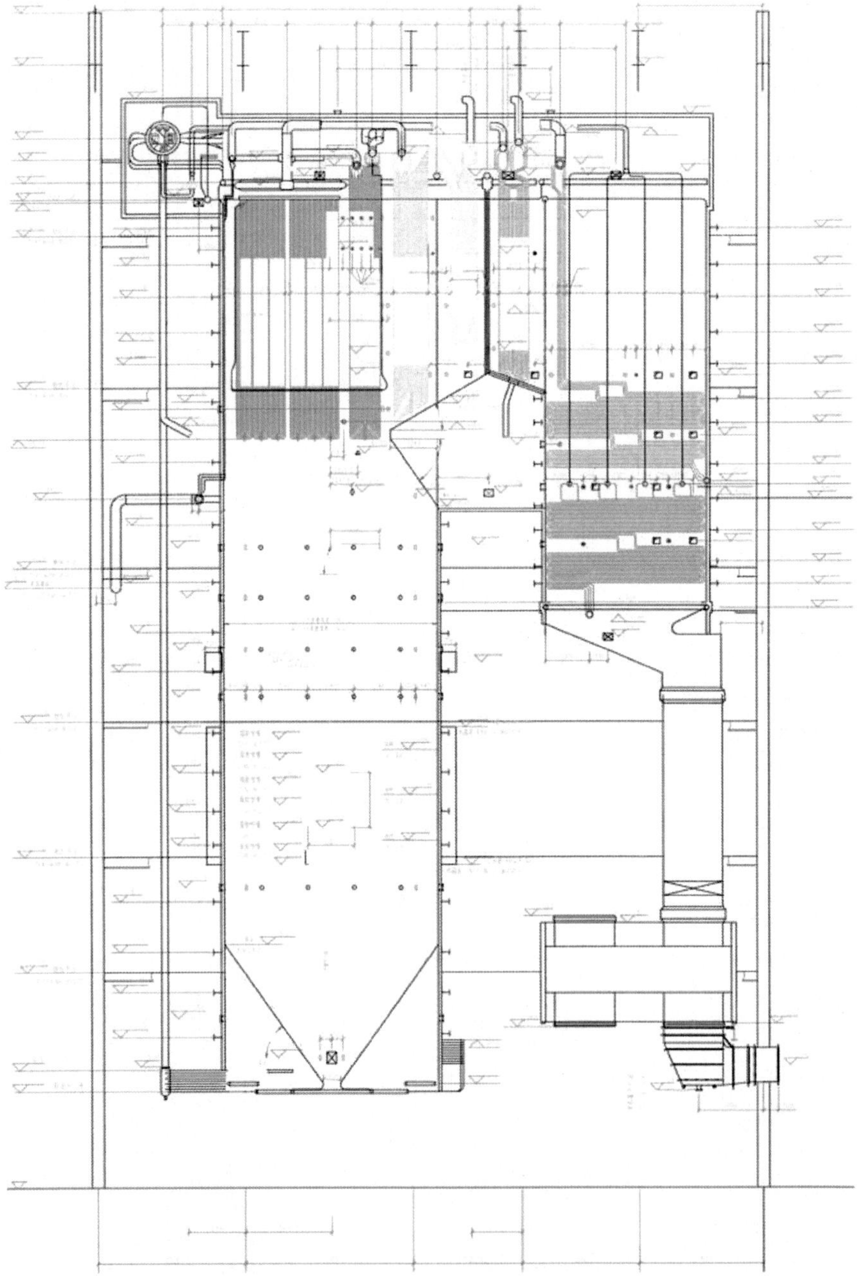

Fig. 1 General arrangement of the Tanjung Awar-Awar's boiler configurations

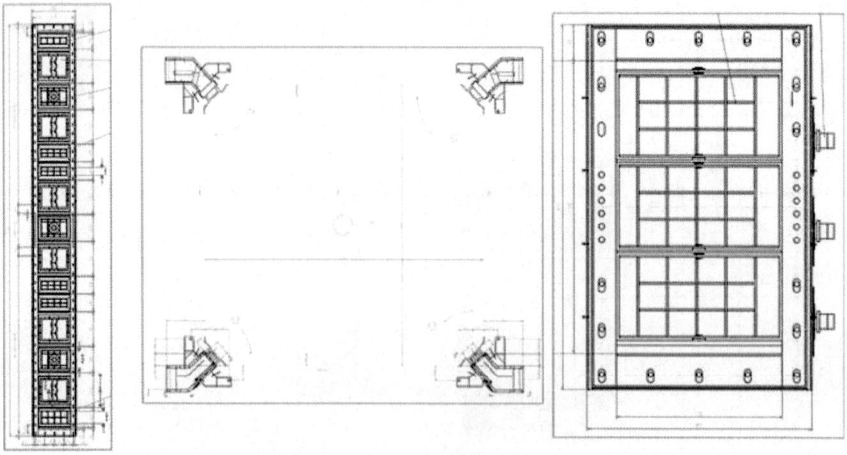

Fig. 2 Coal, secondary air, and SOFA burners configuration

Table 1 Used coal's proximate and ultimate analysis (as received, mass %) and the ash fusion temperatures

Coal parameters	Unit	Used coal
Proximate analysis		
Total moisture	%	35.06
Fixed carbon	%	29.97
Volatile matter	%	31.03
Total moisture	%	35.06
Total sulfur	%	0.20
Ash content	%	3.95
Gross calorific value	kcal/kg	4150
Ultimate analysis		
Carbon	%	44.41
Hydrogen	%	3.12
Total sulfur	%	0.2
Nitrogen	%	0.57
Oxygen	%	12.70
Ash fusion temperature		
Deformation	°C	1370
Spherical	°C	1380
Hemisphere	°C	1390
Flow	°C	1400

Fig. 3d—for low rank coal with PA ratio 2.2. The figure shows that for design coal (MRC), there is no low-temperature regime below A-burner elevation, at all burner elevation center furnace temperature range of 1200–1400 °C, and furnace gas exit temperature range at 1100–1400 °C. For LRC with primary air ratio 1.8, there is a low-temperature regime at burner elevation A, at all burner elevation center furnace temperature range of 1000–1200 °C, and furnace gas exit temperature range of 1100–1200 °C. For LRC with primary air ratio 2.0, there is a little low-temperature regime at bottom furnace, at all burner elevation center furnace temperature range of 1100–1200 °C, and furnace gas exit temperature range of 1100–1200 °C. For LRC with primary air ratio 2.2, there is a little low-temperature regime at bottom furnace, at all burner elevation center furnace temperature range of 1100–1300 °C, and furnace gas exit temperature range of 1200–1300 °C.

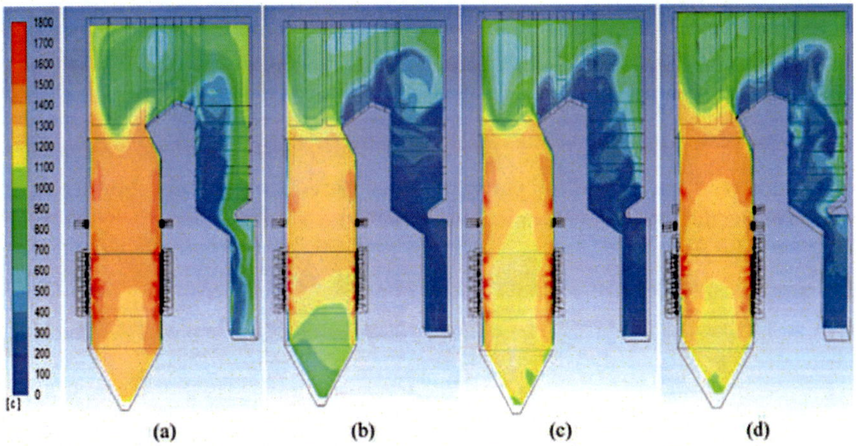

Fig. 3 Temperature contour of the flue gas flow in boiler vertical side view. **a** design coal (MRC) at commissioning data; **b** low rank coal with PA ratio 1.8; **c** low rank coal with PA ratio 2.0; **d** low rank coal with PA ratio 2.2

Figure 4 indicates velocity and trajectory contour of the flue gas stream in boiler, vertical side view, where Fig. 4a—for design coal (medium rank coal) at commissioning data, Fig. 4b—for low rank coal with PA ratio 1.8, Fig. 4c—for low rank coal with PA ratio 2.0, and Fig. 4d—for low rank coal with PA ratio 2.2. The figure shows that for design coal (MRC), there is high velocity particle regime near the furnace wall for all side. For LRC with primary air ratio 1.8, there is medium velocity particle regime near the furnace wall for all side. For LRC with primary air ratio 2.0, there is medium velocity particle regime near the furnace wall for all side. For LRC with primary air ratio 2.2, there is medium velocity particle regime near the furnace wall for all side.

From Fig. 5, LRC coal in each variation of the primary air ratio from the lower elevation burner (A-B) shows an increase in temperature along with the increase in level or elevation to the middle burner elevation (C-D) and then tends to experience a decrease in temperature up to top elevation (E-F) burner. This indicates a more complete

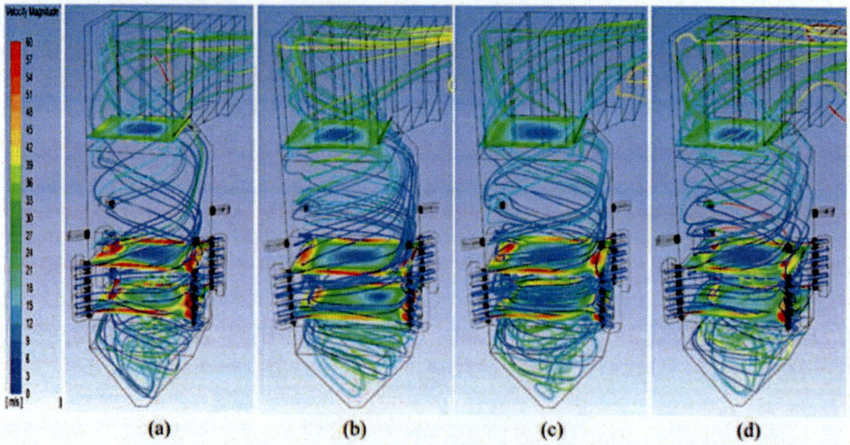

Fig. 4 Velocity and trajectory contour of the flue gas flow in boiler, vertical side view. **a** design coal (MRC) at commissioning data; **b** low rank coal with PA ratio 1.8; **c** low rank coal with PA ratio 2.0; **d** low rank coal with PA ratio 2.2

combustion, and the mixing of pulverized coal particle and combustion air is more homogeneous in the middle elevation burner (C-D).

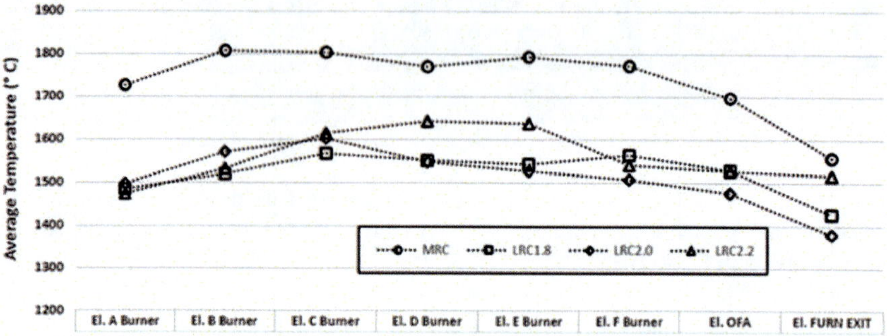

Fig. 5 Average temperature of flue gas along furnace vertical gas path

4 Conclusion

In order to conclude significant factor of combustion flue gas turbulent stream and low rank coal combustion activity in sub-critical tangentially fired furnaces, a combine parameter was conducted by computer code. This research revealed remarkable complexity of two-phase flue gas stream, interphase exchange, and effects between the flame, combustion temperature and velocity. The study conclusion may be summarized as below:

- The addition of primary air by varying the PA/PC has an effect on increasing the average temperature of all burners;
- The addition of primary air by varying the PA/PC has an effect on increasing the average velocity of all burners;
- The increasing of primary air to the pulverized coal (PA/PC) mass flow rate ratio has effect on increasing the velocity of gas–solid particles entering furnace so that it has an impact on flow turbulence.

Acknowledgements. This study was supported by the Indonesian Electricity State Company and its subsidiary company, Pembangkitan Jawa-Bali Ltd. with cooperation of Mechanical Engineering Department of Institut Teknologi Sepuluh Nopember Surabaya, Indonesia.

References

1. Li Z, Miao Z (2019) Primary air ratio affects coal utilization mode and NOx emission in lignite pulverized boiler*. Energy 187:116023. https://doi.org/10.1016/j.energy.2019.116023
2. Belo S, Tomanovi I, Crnomarkovi N, Mili A (2019) Full-scale CFD investigation of gas-particle flow, interactions and combustion in tangentially fired pulverized coal furnace, vol 179, pp 1036–1053. https://doi.org/10.1016/j.energy.2019.05.066
3. Li Z, Miao Z, Shen X, Li J (2018) Prevention of boiler performance degradation under large primary air ratio scenario in a 660 MW brown coal boiler. Energy 155:474–483. https://doi.org/10.1016/j.energy.2018.05.008
4. Young H, Hyun S, Ju Y, Hyung T, Soo D, Woong D (2013) Numerical and experimental investigations on the gas temperature deviation in a large scale, advanced low NOx, tangentially fired pulverized coal boiler. Fuel 104:641–646. https://doi.org/10.1016/j.fuel.2012.06.091
5. Al-abbas AH, Naser J, Kamil E (2012) Numerical simulation of brown coal combustion in a 550 MW tangentially-fired furnace under different operating conditions. FUEL. https://doi.org/10.1016/j.fuel.2012.11.054
6. Harbin, Boiler Maintenance Manual (2012)
7. Ryul C, Nyung C (2009) Numerical investigation on the flow, combustion and NO x emission characteristics in a 500 MW e tangentially fired pulverized-coal boiler. Fuel 88(9):1720–1731. https://doi.org/10.1016/j.fuel.2009.04.001

Numerical Study of Emissions on DDF Engine with 20% CNG with Variation on Compression Ratio

Betty Ariani[1](✉), I. Made Ariana[2], and Aguk Zuhdi M. Fathallah[2]

[1] Naval Architecture, Universitas Muhammadiyah Surabaya, Surabaya, Indonesia
betty.ariani@gmail.com

[2] Department of Marine Engineering, Institut Teknologi Sepuluh Nopember, Surabaya, Indonesia

1 Introduction

The reliability of the marine engineering system was one of the parameters that must be considered in the development process of the maritime and shipping industry. The previous researcher was grouping research on engine performance improvement into three groups, namely optimization of engine design, operating system engineering, and after-treatment conditioning. The three parameters above complement each other to get maximum engine performance and minimum emissions. The exhaust emissions produced by shipping engines include CO_2, NOx, SOx, HC, CO, and PM. The latest data shows that the contribution of the shipping sector to global emissions is 2–3% and is increasing from year to year [1]. Regulations regarding emissions and their application are not only expected to reduce the effect of gas emissions but also to minimize the level of fuel consumption by increasing engine performance and reduce operating costs so that the shipping industry companies are even more competitive [2].

International Maritime Organization (IMO), an organization dealing with pollution from ships, issued its first regulation in 1978. International MARPOL convention in 1973, strengthened again in 1983, was related to emission restriction strategies and prevents and minimizes pollution by shipping activities [3]. As for matters related to air pollution by shipping activities in 1997 regulated in Annex VI MARPOL (Tier 1) with a focus on Sox and NOx, through Tier 2 began to be applied to ships built after January 1, 2011, and continued with Tier 3 for applied to ships built after January 1, 2016. This change involves continuously tightening emission limits [4]. Dual fuel using gas as one of the fuels reduces brake power by more than 30% and increases CO and HC emissions [5], so that good conditioning and treatment are needed to provide optimal benefits. In general, according to [6, 7], the performance of dual fuel is lower than single. So it takes effort to get the desired performance and emissions.

The compression ratio is the ratio of the total volume of the combustion chamber when the cylinder is in the BDC position to the combustion chamber volume at TDC. Theoretically, increasing the compression ratio will result in higher cylinder pressure and heat dissipation that increase the value of overall thermal and engine efficiency [8, 9].

M. Kolhe et al. (eds.), *Recent Advances in Renewable Energy Systems*,
Lecture Notes in Electrical Engineering 876,
https://doi.org/10.1007/978-981-19-1581-9_32

However, increasing the compression ratio will usually increase combustion noise and cause knocking, especially for gases with low ignition temperature. Different things were expressed by [8, 9], and researchers suggested that increasing the compression ratio benefits improved performance and emissions. Experimental trials were on dual fuel with biodiesel fuel—CNG. Bhaskoer [9] has experimental studies on compression ratio, EGR fraction, and temperature on dual-fuel engines. They concluded that increasing the compression ratio will increase fuel substitution and energy efficiency; otherwise, increasing the compression ratio will reduce HC and CO emissions but higher NOx.

In this article, we will discuss how the effect of the compression ratio on exhaust emissions produced on dual fuel with a ratio of 20% CNG to 0% CNG. We will see how the variation of the compression ratio affects the performance, combustion, and emissions when conditions are 20% CNG at a constant speed of 2000 rpm.

2 Material and Method

The simulation test is using single-cylinder engine data with the following specification (Table 1).

Table 1 Engine baseline

Engine (four stroke cycle)	TF 85 MH
Cylinder	1
Combustion system	Direct injection
Bore × stroke	85 × 87 mm
Displacement	493 cc
Compression ratio	18:1
Max engine at full speed	2200
Continuous power output	7.5 kw
Specific fuel consumption	171 gr/hph

The first stage in this process is modeling using solid work. The image made is an existing picture of the condition of the piston and combustion chamber. It is planned to make three piston models with different geometries to produce variations in the compression ratio to 16, 18, and 19. The geometry model created is drawn in two dimensions as many as three models, M1 compression ratio 16, M0 baseline compression ratio 18, and M2 compression ratio 19. The next step is to input the main engine data and import images from solid work to ANSYS while setting the direction of the pilot fuel spray fuel. The next process is meshing or formation into smaller cells, and then, the calculation process with the ANSYS forte solver is started. The calculation begins with determining the fuel to be used, the injection timing and the mass of fuel injected, the boundary conditions and the direction of motion of the piston, the initial conditions, and the gas mixture (Fig. 1).

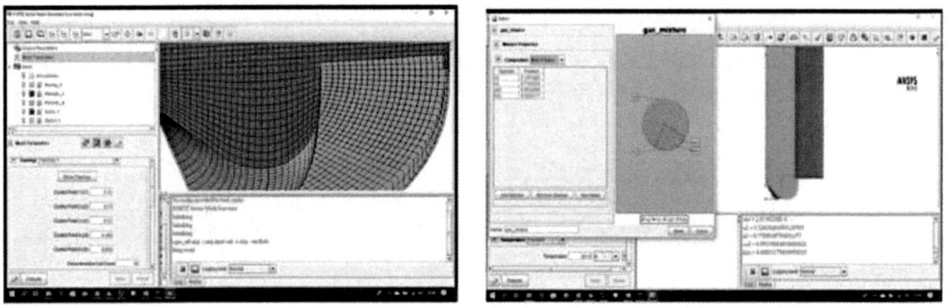

(a) (b)

Fig. 1 Meshing process (**a**) and determination of initial conditions and gas mixture (**b**)

The simulation control process includes determining the crank angle you want to display, and the process running generates a graphic visualization, while the rendering process will generate contour output as we want such as displaying pressure visualization, temperature, and velocity. Variations in the compression ratio are given 16 and 19 at a fixed speed of 2000 rpm. The result of emission data operations in the form of UHC, NOx, and CO.

3 Results and Discussion

The following is the result of reading the unburnt hydrocarbon emissions in three variations of the compression ratio, namely the lower compression ratio of 16 and the higher of 19 against the baseline compression ratio of 18 (Fig. 2).

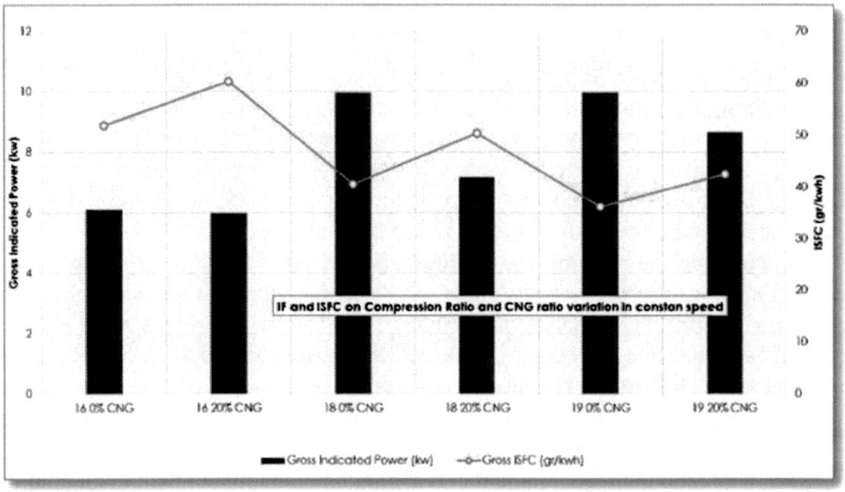

Fig. 2 Engine performance at 0 and 20% CNG on the variation of compression ratio

The more CNG will decrease the indicated power in all compression ratio variations and vice versa. Dual fuels require an increased compression ratio to compensate for the power loss because of CNG due to a single to dual-fuel switch. The level of fuel consumption is getting bigger on dual fuel with a higher percentage of CNG. The increase in consumption occurs at low compression ratios and decreases with increasing compression. The compression ratio 19 has a fuel consumption reduction of up to 15.6% on dual 20% CNG fuel compared to the baseline compression ratio condition (Fig. 3).

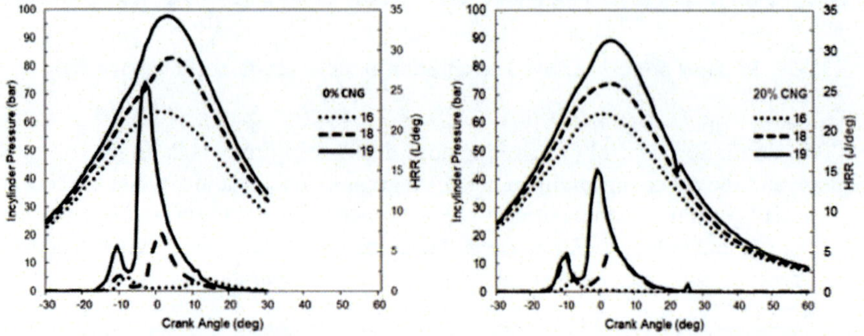

Fig. 3 Cylinder pressure and HRR at 0 and 20% CNG on variation of compression ratio

By entering 20% of the CNG mass in the picture, the cylinder pressure in all combustion chambers decreased compared to when it was 0% CNG. This condition happens because the entry of gas into the combustion chamber causes oxygen intake to be blocked. With reduced oxygen, the process of fuel oxidation is disrupted, causing the cylinder pressure to drop. The fuel mixture becomes too rich so that it reduces the quality of combustion, affects the output, and increases fuel consumption and high emissions. The increase in heat release rate also increased with a peak difference of 24% compared to the HRR at baseline. At a compression ratio of 19, the ignition delay is faster than the baseline with almost the same combustion duration. This ignition delay is faster beneficial for reducing the potential for knocking, especially at low loads. While at the compression ratio of 16, there is a decrease in pressure and temperature in the combustion chamber. In addition, the rate of energy issuance is also the slowest and the shortest (Figs. 4 and 5).

At a compression ratio of 16, an increase in temperature and pressure occurs in the bowl area, a little in the squish area, and a temperature not too high, and uneven distribution makes the potential for methane slip. At low compression pressures, the squish is largest than other models, and the heat range is less than optimal, especially in the tip area. The effect of a lower compression ratio causes a decrease in pressure, temperature, and fluid flow velocity in the combustion chamber so that it does not support the homogenization process fuel mixture (Fig. 6).

Here, at a lower compression ratio, UHC is at a higher value when compared to the baseline. Immediately after injection of diesel fuel at 18 BTDC, the UHC chart peaked at around 11,600 ppm, then sloped toward the top dead center, and continued even though

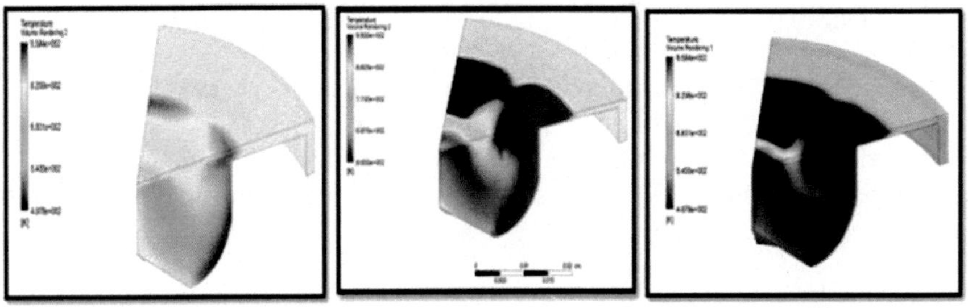

(a) (b) (c)

Fig. 4 Contour temperature at 0% CNG on compression ratio 16 (**a**), 18 (**b**), and 19 (**c**)

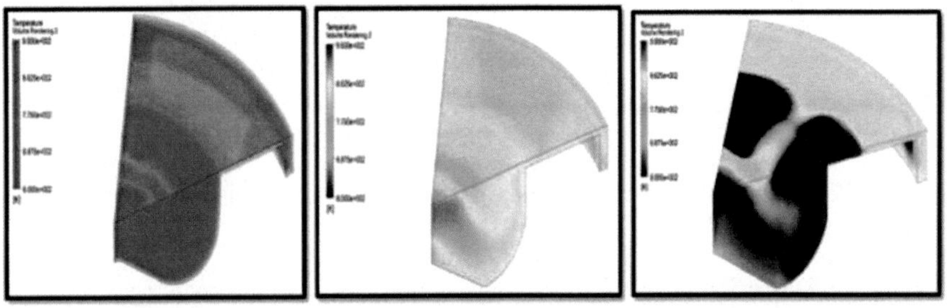

(d) (e) (f)

Fig. 5 Contour temperature at 20% CNG on compression ratio 16 (**d**), 18 (**e**), and 19 (**f**)

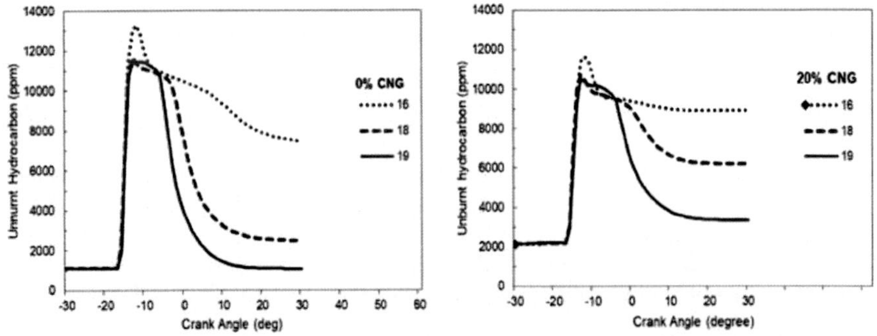

Fig. 6 Unburnt hydrocarbon at 0 and 20% CNG on the variation of compression ratio

it was not as significant as the baseline or when the compression ratio was increased. At a higher compression ratio, there is a decrease in UHC emissions of around 46% compared to the baseline.

4 Conclusion

In general, DF CNG intake in the combustion chamber causes a decrease in performance and combustion due to the reduced amount of oxygen. A high compression ratio promotes accelerated ignition delay resulting in increased performance and minimized UHC. A high compression ratio gives a more significant UHC reduction value in CNG with a low ratio decrease in UHC by 50–56% at 0% CNG and a decrease by 39–46% at 20% CNG.

Acknowledgements. Thank you to the Department of Naval Architecture Universitas Muhammadiyah Surabaya, the Department of Marine Engineering, especially the Laboratory of Marine Power Plant, Institut Teknologi Sepuluh Nopember, Surabaya—Indonesia, and colleagues for their support and thought assistance in this research.

References

1. Bows-Larkin A, Mander S, Gilbert P, Traut M, Walsh C, Anderson K (2014) High seas, high stakes: high seas project final report, tyndall cent. Clim Chang Res. The University of Manchester. England
2. Theotokatos G, Stoumpos S, Lazakis I, Livanos G (2016) Numerical study of a marine dual-fuel four-stroke engine. In: Proceedings of 3rd international conference on maritime technology and engineering, MARTECH, pp 777–783. CRC press
3. Pueschel M (2013) Combination of post-injection and cooled EGR at a medium-speed diesel engine to comply with IMO Tier III emission limits. In: Conseil international des machines a combustion international council on combustion engines, pp 1–9. Shanghai
4. Hu N, Zhou P, Yang J (2017) Reducing emissions by optimising the fuel injector match with the combustion chamber geometry for a marine medium-speed diesel engine. Transp Res Part D Transp Environ (53):1–16
5. Chandra R, Vijay VK, Subbarao PMV, Khura TK (2011) Performance evaluation of a constant speed IC engine on CNG, methane enriched biogas and biogas. Appl Energy 88(11):3969–3977
6. Papagiannakis RG, Hountalas DT (2004) Combustion and exhaust emission characteristics of a dual fuel compression ignition engine operated with pilot diesel fuel and natural gas. Energy Conv Manage 45(18–19):2971–2981
7. Yoon SH, Lee CS (2011) Experimental investigation on the combustion and exhaust emission characteristics of biogas-biodiesel dual-fuel combustion in a CI engine. Fuel Process Technol 92(5):397–492
8. Banapurmath NR, Tewari PG, Hosmath RS (2008) Combustion and emission characteristics of a direct injection, compression-ignition engine when operated on honge oil, HOME and blends of HOME and diesel. Int J Sustain Eng 1(2):80–93
9. Porpatham E, Ramesh A, Nagalingam B (2012) Effect of compression ratio on the performance and combustion of a biogas fuelled spark ignition engine. Fuel (95):247–256
10. Verma S, Das LM, Kaushik SC, Bhatti SS (2019) The effects of compression ratio and EGR on the performance and emission characteristics of diesel-biogas dual fuel engine. Appl Thermal Eng (150):1–8
11. Bora BJ, Saha UK, Chatterjee S, Veer V (2014) Effect of compression ratio on performance, combustion and emission characteristics of a dual fuel diesel engine run on raw biogas. Energy Convers Manage 87:1000–1009

Numerical Study of Gas Mixing Effect on Block 3 and Block 4 Muara Tawar's Gas Turbine Combustion Stability

Danan Tri Yulianto(✉) and Bambang Sudarmanta

Department of Mechanical Engineering, Institut Teknologi Sepuluh Nopember, Surabaya 60111, Indonesia
danantriyulianto@ptpjb.com

1 Introduction

Power plant managed by PT PJB UP Muara Tawar operates as peaker with periodic start–stop operation. Blocks 3 and 4 operate V 94.2 S's type gas turbines. The fuel gas is provided from three different suppliers, Nusantara Regas (NR), Pertamina EP (PEP), and PGN with different gas content. The operation of gas turbine depends on the fixed initial setting during combustion commissioning. Meanwhile, gas turbine must be able to operate using various types of different fuel gas supplier that altered flame characteristics especially on flame stability.

Combustion stability related to fuel and air ratio that maintains stability or in other word the maximum velocity of combustion that still tolerable by combustion system without any flame off [1]. There are several studies that have been done to investigate the effect of gas composition (methane based) on combustion system. Author [2] analyzed the combustion process used H_2, CH_4, C_2H_6 and C_3H_8 by outlet's diameter and fuel jet velocity variation. Measurement showed that the flame lift off will increase linearly with the increase of fuel jet, and the diameter was in fixed diameter. The ratio of air and fuel showed that combustion has a maximum laminar combustion region close to its stoichiometric condition.

The relationship between blow off and lift off is related to the decreasing amount of fuel mass fraction. The less mass fraction will effect on blow-off speed which decreases and makes shorter flame lift distance from the burner. On the other hand, at high flow velocities of mass fraction, the lift-off distance will be further from burner and the flame tends to flame out [3] by lift-off mechanism.

Temperature and NOx emission also become another object of this study on how the fuel composition and excess air effect to combustion. Author [4] presents the simulation on methane combustion and biogas mixture within can-type of gas turbine combustion chamber. The analysis shows that biogas with lower methane content will lead in outlet temperature decrease which also effected the reduction of NOx emission.

Based on these previous studies, the research is carried out by evaluating and optimizing the combustion stability characteristics and temperature distribution by varying fuel gas composition and provides a limitation of the gas mixing composition and ratio of air to fuel (air–fuel ratio) to obtain stability on combustion.

M. Kolhe et al. (eds.), *Recent Advances in Renewable Energy Systems*,
Lecture Notes in Electrical Engineering 876,
https://doi.org/10.1007/978-981-19-1581-9_33

2 Numerical Method

2.1 CFD and Combustion Modeling

This research will be carried out numerically using computational fluid dynamics (CFD) software. There are three main stages needed to be carried out on numerical method preprocessing, processing, and postprocessing. The combustion chamber geometry model consists of eight burners from silo-type combustion chamber. The basic geometry of the gas turbine silo-chamber type is shown in Fig. 1. The fuel gas nozzle diameter is 140 and 799 mm for air inlet. There are eight burners, and each consists of fuel inlet and air inlet and used swirl mechanism.

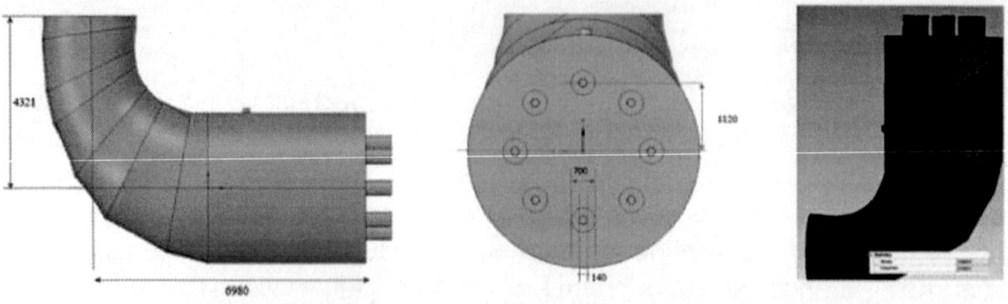

Fig. 1 Combustion chamber model with eight burners

The meshing used the polyhedral-type mesh to accommodate chamber complex models. Meshing on combustion chamber produced 109,004 nodes and 555,857 elements. The optimal number of nodes and element based on the result of outlet combustion temperature reached around 1113 °C. The simulation is carried out inside combustion process where the fuel and air do not mix first but sprayed into the combustion chamber. So non-premixed model is selected as the species model. The non-premixed combustion reaction approach uses the transport equation for one or two conserved quantities (mixed fraction). This approach is specially developed for simulation on flame diffusion and calculated by probability density function (PDF).

2.2 Data Verification

Prior to the combustion simulation, the validation needed to make sure that the simulation and numerical calculation are close to the actual condition based on the parameters in the gas turbine operating parameters.

The result of comparison between operating and simulation condition has shown that the temperature on inlet turbine or outlet chamber from four point of calculation gave measurement around 1 until 5% of error as shown in Table 1, and this means that the simulation has similar measurement compared to operating condition.

Table 1 Validation result

Parameter		Operation	Simulation	Error (%)
Temperature inlet turbine 2	°C	1115	1079	3.20
Temperature inlet turbine 3	°C	1127	1099	2.50
Temperature inlet turbine 4	°C	1142	1213	5.90
Average temperature inlet	°C	1126	1113	1.20

3 Result and Discussion

3.1 Flame Stability

Methane is a fuel gas content that is used as a reference on this analysis. Since the actual composition during operating of the Muara Tawar's gas turbine could not accurately calculate (no gas chromatograph to sense mixing gas), the classification of fuel gas needed to specify the methane-based composition into three categories, high composition (CH_4 94%), medium (CH_4 87%), and low (CH_4 71%). The study of combustion characteristics obtained from the analysis of simulation is carried out from three variations of gas composition and three variations of air ratio as shown in Table 2. The analysis is carried out to analyze the effect of gas composition level on each of the gas suppliers by comparing flame characteristic such as flame lift off and flame length. Lift off occurs due to increase in the rate of heat loss from the flame at the front of burner, so the flame goes out locally and flame appears to jump out until stabilized at a certain distance. The more lift-off distance from the burner tip will effect on combustion stability, so it must be maintained. The flame length is also one of the stability parameters of combustion mechanism which is closely related to the mixing of fuel and air. The increasing of flame length indicated that the combustion happened away from the burner tip and the flame becomes unstable.

Table 2 Fuel gas and excess air composition

Variation	CH_4 (%)	C_2H_6 (%)	C_3H_8 (%)	F. gas flow (kg/s)	Air flow (kg/s)
High	94.53	2.3	1.35	10.24	480
Low	71.41	12.13	7.89	10.24	480
Medium	87.82	2.76	2.06	10.24	480
– Medium 1	87.82	2.76	2.06	10.24	504
– Medium 2	87.82	2.76	2.06	10.24	528
– Medium 3	87.82	2.76	2.06	10.24	576

Combustion with high composition showed that flame lift off has a distance about 0.185 m from the burner tip. The lower composition will increase lift-off distance from burner (0.241 m). However, there is no significant difference on lift-off distance due to less result of laminar burning velocity that is effected by density as can be seen in Fig. 2.

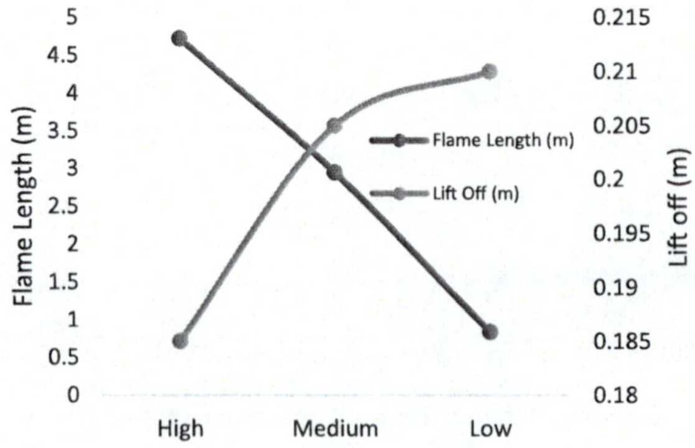

Fig. 2 Lift off by gas composition

The addition of excess air to ensure complete combustion also has an effect on flame lift off as seen in Fig. 3. Simulation resulted from addition of excess air increased lift-off distance with shorter flame length. Flame length tends to decrease due to the process of fuel dilution by more air to aim complete combustion [5].

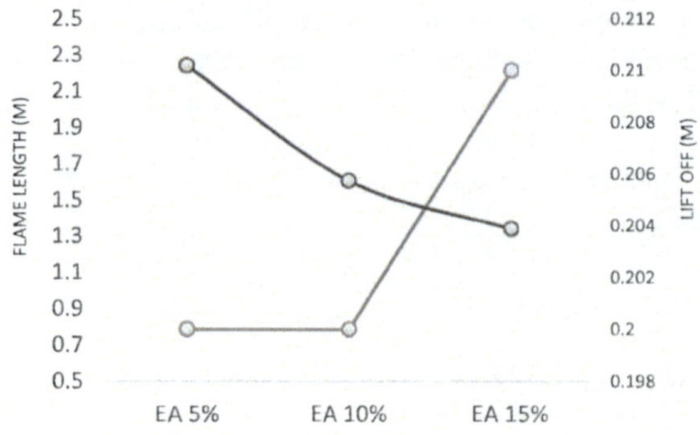

Fig. 3 Lift off by excess air

The combustor design that applied the air swirl helps to increase the combustion intensity and reduce the flame length. Swirl flow has also provided an angular velocity to the axial incoming flow to produce a central recirculation zone (CRZ) which provides the main flame stabilization. The flame length combustion with fuel with high composition produces longest flame length, around 4.7 m from burner tip. Amer and Gad [6] studied the effect of increasing air to fuel ratio on experimental study of LPG combustion. Increasing the air-to-fuel mass ratio (excess air) from 5 to 20% will decrease flame length about 6–16%. The flame length that indicated the lift off will be stabilized by inner recirculation zone. The addition of excess air changes the characteristics of the recirculation zone; the more excess air will create shorter recirculation center distances with a larger recirculation zone area. Hong et al. [7] made some study on recirculation zone as excess air raised. The higher temperature of the products reduces the velocity gradient in the shear layer and thus the reattachment length. The addition of 5% excess air reaching 504 kg/s resulted in a recirculation center distance of 2.38 m from the burner tip.

3.2 Temperature Distribution

Combustion process is a reaction between fuel gas and oxygen in the air. The result of this process was carbon dioxide (CO_2), water (H_2O), and a great deal of energy. The higher the methane content, the higher the maximum temperature reached, and the length of maximum temperature also increases. The temperature distribution from different gas composition and the effect of excess air are shown in Fig. 4.

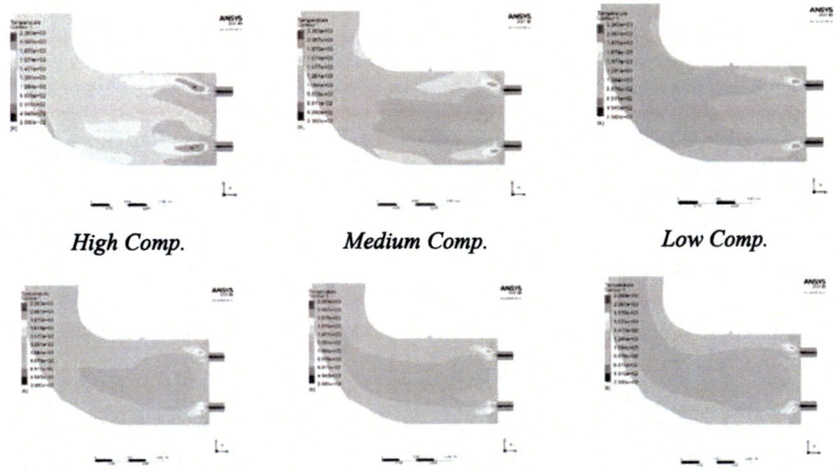

High Comp. *Medium Comp.* *Low Comp.*

Med. *Comp. Excess air* 5% Med. Com. *Excess air* 10% Med. Com. *Excess air* 15%

Fig. 4 Temperature distribution at 137 MW

The addition of excess air in Fig. 5 shows that as the amount of air increases, the temperature to the outlet will decrease. This is due to the combustion that losses a certain amount of energy because too much air enters into the combustion chamber.

The addition of excess air from 15 to 20% does not cause a significant increase in maximum temperature and energy. The combustion efficiency will increase with the addition of excess air until the heat loss in the excess air is larger than the heat from combustion. Munir et al. [8] evaluated the effect of excess air on combustion and concluded that an optimum air–fuel ratio should be maintained to ensure complete combustion as well as effected on excessive heat losses due to surplus air.

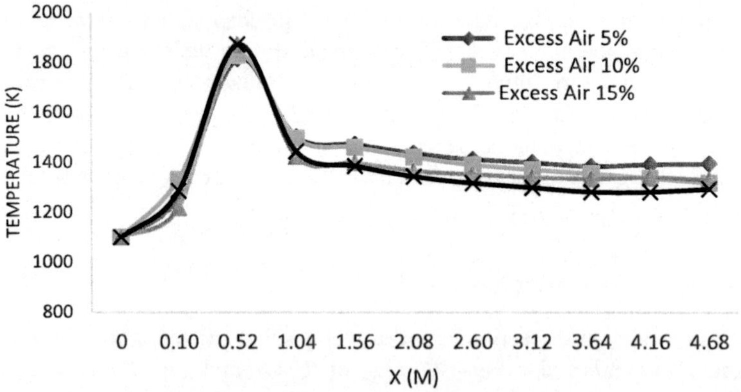

Fig. 5 Temperature distribution on excess air

3.3 NOx Emission Characteristic

NOx emission is produced by the oxidation of atmospheric nitrogen in high-temperature regions of the combustion flame and postflame gases at the outlet. Significant effects on NOx characteristic that the nitric oxide formation rate in postflame gases of hydrocarbon flames ($T > 1800°K$) and follows the Zeldovich chain mechanism. The combustion process will lead to the creation of nitrogen oxides from nitrogen and from air or gas fuel. At higher temperatures, both can react to form NOx in large quantities.

4 Conclusion

The studies and analyses showed that lower gas composition will increase flame lift off and the medium composition increased to 10.8% compared to high composition. The addition of excess air will cause the lift off away from burner tip with shorter flame length. Excess air of 10% will give 4% lift-off distance compared to 5% excess air. Gas composition also effects on gas temperature, and the lower gas composition will give lower maximum temperature. The addition of excess air will give higher maximum temperature, but it will decrease gradually toward chamber outlet. The optimum energy is calculated from outlet temperature when medium gas with 15% of excess air is used as fuel gas. The higher gas composition will tend to higher NOx production, and the addition of excess air will decrease NOx production as a result of lower temperature.

Acknowledgements. Great appreciation to the PT PJB UP Muara Tawar Power Plant including the management, the personnel of maintenance, and operation unit for the arrangements and preparing all of the required information.

References

1. Razak AM (2007) Gas turbine combustion. In: Industrial gas turbines, pp 137–173
2. Kalghatgi GT (1983) Lift-off heights and visible lengths of vertical turbulent jet diffusion flames in still air. Combust Sci Technol 41(1–2):17–29
3. Mahandari CP (2010) Fenomena flame lift-up pada pembakaran premixed gas propane
4. Guessab A, Aris A, Cheikh AM, Baki T (2016) Combustion of methane and biogas fuels in gas turbine can-type combustor model, vol 9, no 5, pp 2229–2238
5. Turns R (2000) An introduction to combustion concept and applications, 2nd edn. McGraw Hill
6. Amer AA, Gad HM, Ibrahim IA, Farag TM (2015) Experimental study of LPG diffusion flame at elevated preheated air temperatures
7. Hong S, Shanbhogue SJ, Ghoniem AF (2015) Science direct impact of fuel composition on the recirculation zone structure and its role in lean premixed flame anchoring. Proc Combust Inst 35(2):1493–1500
8. Likewise O (2019) Thermal reactions, pp 207–218

Numerical Study of the Effects of Burner Tilt and Coal Optimization on Combustion Characteristics of 350 MWe Tangentially Fired Pulverized Coal Boiler

Halim[1,2](✉) and Ary Bachtiar Krishna Putra[2]

[1] PT. Indonesia Power, Jakarta, Indonesia
halim@gmail.com
[2] Department of Mechanical Engineering, Sepuluh Nopember Institute of Technology, Surabaya, Indonesia

1 Introduction

Tangentially fired boilers are most widely used in power generation. Four sets of burners are located at four corners of a furnace. These burners fire tangentially at an imaginary circle at the center of the furnace, forming a whirling fireball. Tangentially burner is usually combined with tilting burner. When tilting burners are used, the gas temperature at the furnace exit can be adjusted by varying their tilting angle. This controls the superheat temperature of the steam. Some studies have been done regarding to the effects of burner tilt [1–7] and coal quality [8, 9] to boiler combustion.

One of the many ways to reduce generation cost is to use lower coal quality than design coal because the lower the quality, the cheaper the price of coal. But using lower-grade coal has some effects [10]. Ash, sulfur, and moisture directly affect the heating value of the coal and limit the capacity of the combustion system.

This study will be focused on investigating the effect of burner tilting on boiler combustion when boiler consumes lower coal quality and then the design coal. The primary objective of this study is to provide deep insight into the effect of the burner tilt angle on the combustion and NO_x emission characteristics of a 350 MW tangentially coal-fired boiler, with the aim of deriving useful guidelines for adjusting the burner tilt angle in such boilers.

2 Boiler Specifications

The 350 MW tangentially fired pulverized coal boiler simulated in this study was designed and built by the Shanghai Boiler Works Ltd. The boiler is 55,550 mm high, 16,100 mm wide, and 14,120 mm deep, with a hopper at its bottom.

There are 24 burners arranged in six layers, A–F, in each corner. However, burner F is not used in this study. Two sets of close-coupled over-fire air (CCOFA), namely CCOFA 1 and CCOFA 2, facilitate the combustion of unburned particles from the lower

M. Kolhe et al. (eds.), *Recent Advances in Renewable Energy Systems*,
Lecture Notes in Electrical Engineering 876,
https://doi.org/10.1007/978-981-19-1581-9_34

part of the furnace. In addition, the tilt angles of the PA, SA, and CCOFA nozzles can be continuously adjusted from −20° to +20°, −30° to +30°, and −5° to +30°, respectively. PA supplies pulverized coal and air at a temperature of 60 °C, whereas SA provides air at a temperature of 319 °C. Total flow rate of combustion air is 1369,53 t/h. The jet angles of primary air and secondary air are 38° and 58°, forming two imaginary firing circles of different diameters in the furnace center. In this work, lignite coal, which has lower quality than that of the design, was selected as fuel. Total coal flow rate is 215 t/h. The coal properties of coal optimization are described in Table 1. For comparison convenience, coal properties of coal design are also involved.

Table 1 Proximate and ultimate analyses of pulverized coal

Parameters	Unit	Coal optimization	Performance coal design	Worst coal design
Total moisture	%Wt	36.85	30.7	34.1
Ash content	%Wt	6.36	3.83	4.43
Volatile matter	%Wt	31.68	33.23	29.08
Fixed carbon	%Wt	25.10	32.97	32.39
GCV	Kcal/Kg	4080	4700	4500
Carbon	%Wt	41.99	49.17	48.69
Hydrogen	%Wt	2.76	2.68	2.15
Oxygen	%Wt	11.08	12.7	9.96
Nitrogen	%Wt	0.06	0.51	0.33
Sulfur	%Wt	0.36	0.23	0.25

3 CFD Model

3.1 Mathematical Models and Numerical Algorithms

The domain of this study is from bottom hopper to the outlet of boiler. The grid number of this study is 3,164,800. Turbulent flow is modeled by the realizable k-ε model because it provides substantial improvements over the standard k-ε model, where the flow features include strong streamline curvature, vortices, and rotation. Coal particle trajectories are tracked at each step of discrete phase model (DPM) calculation. Calculations of the reaction between homogeneous gaseous volatiles were modeled by the finite rate/eddy dissipation (FR/ED) using one reaction that depends on the properties of the coal.

$$C_{1.33}H_{2.59}O_{0.65}N_{0.0405}S_{0.0106} + 1.66O_2 \rightarrow 1.33CO_2 + 1.29C_2O + 0.0202N_2 + 0.0106SO_2 \quad (1)$$

Coal devolatilization is modeled as constant, volatile matter combustion is estimated by species transport model, and char combustion is computed according to the diffusion-limited model. The formation of NO was modeled using thermal mechanisms.

3.2 Simulated Cases and Boundary Condition

There are five burner tilt variations observed in this study. Cases 1–5 (burners tilted up by −30°, −20°, −10°, 0°, and +5°, respectively) were chosen to investigate the effect of the burner tilt angle on the furnace combustion characteristics. The actual range of burner tilt is −20° to +20°. Burner tilt −30° is also investigated in this study to give a comprehensive overview of the boiler combustion and as a reference, if there is any further modification regarding the burner tilt operation range.

Boundary conditions, such as mass flow rate of pulverized coal at burner inlets and volume flow rate of air, are set according to the performance test of the boiler.

The particle size of pulverized coal follows Rosin–Rammler distribution, with a maximum diameter of 0.2 mm, a minimum diameter of 0.07 mm, and an average particle diameter of 0.134 mm. The spread parameter of particle size is 4.52. The SIMPLE algorithm was used to calculate the coupling of the velocity and pressure fields.

4 Result and Discussion

4.1 Validation

The measurement points for this validation are at the point of secondary air layer AB, CD, EF, FF, LTSH inlet flue gas, and boiler outlet. The deviation between simulation and actual data for each point is 4.84%, 5.71%, −2.94%, 0.61%, 4.76%, and 8.17%, respectively.

4.2 Temperature Profile

Temperature distribution of a vertical cross section at the center ($z = 0$ m) is shown in Fig. 1a, and temperature distribution along the furnace height for various burner tilt angles is shown in Fig. 1b.

From Fig. 1a, the fireball moves up as the burner tilt angle increases as well as the temperature at convection area.

From the Fig. 1b, each burner tilt has an identical trend of flue gas temperature distribution. The temperature in the bottom ash area (level 2.5–10 m) has a low temperature because there is no burning coal in that area. The highest temperature in the bottom ash hopper area is at a tilting angle of −30°, while the lowest temperature is at a tilting angle of +5°.

In the burner area, the temperature increases because this area is the center of coal combustion. The highest temperature in the burner area occurs when the tilting angle is −30°, then −20°, −10°, 0°, and +5°, consecutively. The temperature in each burner layer at a tilting angle of −30° and −20° looks more even.

Burner tilt angle also affects FEGT. FEGT for burner tilt angles from −30° to +5° is 1237.01 °C, 1239.99 °C, 1259.21 °C, 1268.13 °C, and 1270 °C, respectively.

To see the effect of coal optimization and burner tilt angle on combustion in the boiler, it is done by comparing the simulation results with the boiler manufacturer's thermal calculation data. Table 2 compares the profile temperature of each burner tilt with worst coal and design coal.

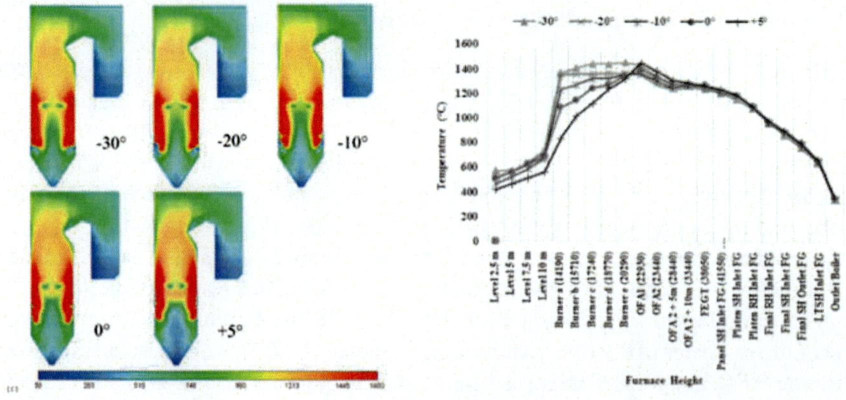

(a) (b)

Fig. 1 **a** Temperature distribution of a vertical cross section at the center ($z = 0$ m), **b** temperature distribution along the furnace height

Table 2 Temperature deviation between simulation results and design in the convection area

Parameters	Worst coal					Design coal				
	−30°	−20°	−10°	0°	+5°	−30°	−20°	−10°	0°	+5°
Platen SH inlet FG	13.51	38.67	46.22	40.27	52.70	28.51	53.67	61.22	55.27	67.70
Platen RH inlet FG	36.21	39.43	46.03	53.52	55.96	53.21	56.43	63.03	70.52	72.96
Final RH inlet FG	23.49	37.78	21.53	27.84	38.89	42.49	56.78	40.53	46.84	57.89
Final SH inlet FG	11.20	29.60	32.82	41.58	47.07	29.20	47.60	50.82	59.58	65.07
Final SH outlet FG	9.31	12.02	38.25	43.22	48.33	26.31	29.02	55.25	60.22	65.33
LTSH inlet FG	42.05	38.59	29.58	26.48	23.57	27.05	23.59	14.58	11.48	8.57
Outlet boiler	55.06	56.06	50.96	47.82	37.79	46.06	47.06	41.96	38.82	28.79
Average	27.26	36.02	37.91	40.11	43.47	36.12	44.88	46.77	48.96	52.33

The smallest deviation to the design data is when the burner tilt is −30°. However, the actual burner tilt angle cannot be lower than −20° because the operating range of burner tilt angle is ±20°. Although it is closest to the design value, the overall value of the tilting angle of −20° is still above the design data, so care must be taken in its operation to avoid overheating in the superheater and reheater tubes.

4.3 Species Profile

O_2 distribution of a vertical cross section at the center ($z = 0$ m) is shown in Fig. 2a, and O_2 distribution along the furnace height for various burner tilt angles is shown in Fig. 2b. The highest O_2 content is in the bottom ash hopper area (2.5–10 m) because in that area, there are no coal particles that react with oxygen. The lower the burner tilt angle, the higher the O_2 content in the hopper area. At burner area (burner A–E), O_2 content begins to decrease because it reacts with coal from the coal burner. At CCOFA level, O_2 increases because of the additional air coming from the CCOFA 1 and CCOFA 2 nozzles. Coming out of CCOFA, O_2 then reacts with coal that has not yet been reacted with coal particles, which causes the O_2 concentration to decrease again.

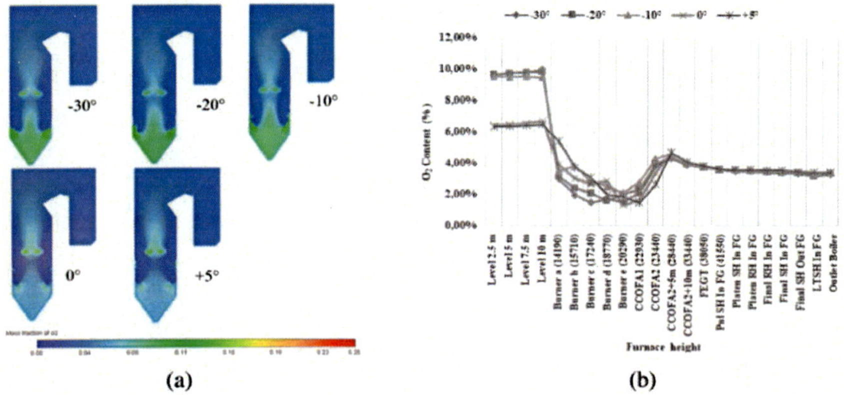

Fig. 2 **a** O_2 distribution of a vertical cross section at the center ($z = 0$ m), **b** O_2 distribution along the furnace height

The smallest O_2 content in the boiler outlet is at a burner tilt of $-30°$, which is 3.27%. O_2 content at burner tilt of $-20°$, $-10°$, $0°$, and $+5°$ is 3.30%, 3.34%, 3.39%, and 3.40%, respectively. It shows that the combustion at burner tilt of -30° is the most complete because most of the O_2 reacted with the coal.

Another method to analyze combustion is to analyze the CO_2 mass fraction. CO_2 distribution of a vertical cross section at the center ($z = 0$ m) is shown in Fig. 3a, and CO_2 distribution along the furnace height for various burner tilt angles is shown in Fig. 3b. The area that has the highest CO_2 content is the burner area (fireball), because complete combustion, which converts C in coal and O_2 to CO_2, occurs in that area.

Mass fraction of CO_2 in the bottom ash area (2.5–10 m) is quite low because only a small amount of coal reacts with oxygen in the area. In the burner area (burner A–E), the CO_2 content increases and reaches its highest mass fraction because in that area a combustion process occurs where CO_2 reacts with coal from the coal burner. This combustion reaction, if it occurs completely, will convert O_2 into CO_2. That is why, in the burner area, O_2 has the lowest mass fraction, while CO_2 has the highest mass fraction.

At OFA level, CO_2 is reduced due to additional air coming from the CCOFA 1 and CCOFA 2 nozzles. Additional O_2 from CCOFA will then react with coal that has not had time to burn in the coal burner area so that the CO_2 mass fraction increases slightly in the CCOFA to FEGT area.

The tilting position that gives the largest CO_2 mass fraction in the boiler outlet area is the tilting position of $-30°$, which is 17.60%. CO_2 content at $-20°$, $-10°$, $0°$, and $+5°$ is 17.56%, 17.51%, 17.42%, and 17.40%, respectively. This shows that the combustion

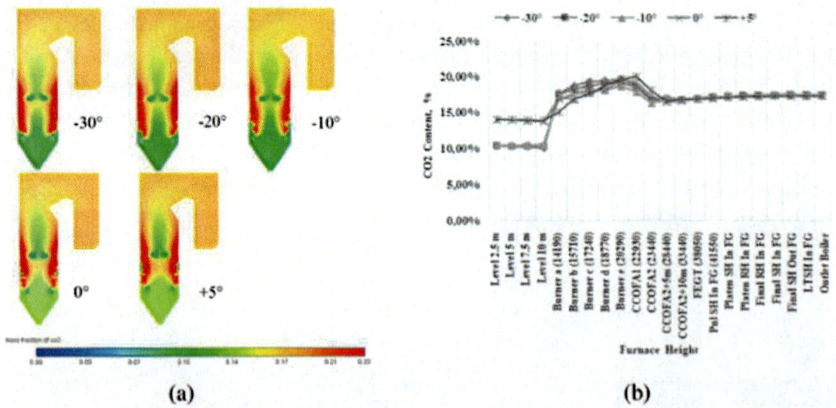

(a) (b)

Fig. 3 **a** CO_2 distribution of a vertical cross section at the center ($z = 0$ m), **b** CO_2 distribution along the furnace height

at a tilting position of −30° is more complete because most of the coal reacts with O_2 and produces CO_2.

4.4 NO_x Emission

NO_x mass fraction contour analysis can be used to predict combustion emissions that occur in the boiler. NO_x distribution of a vertical cross section at the center ($z = 0$ m) is shown in Fig. 4a, and NO_x distribution along the furnace height for various burner tilt angles is shown in Fig. 4b.

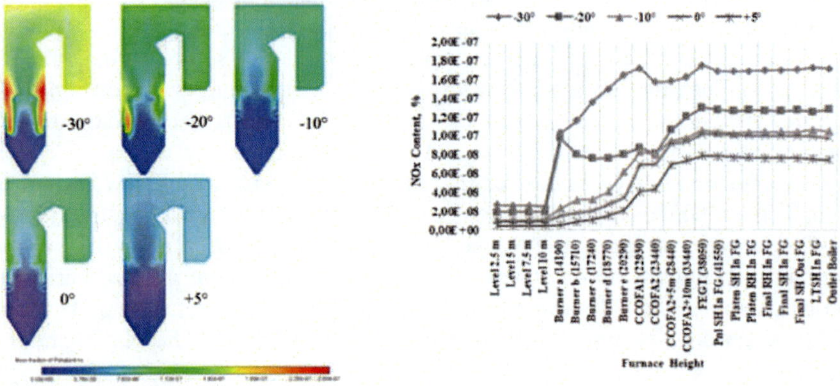

(a) (b)

Fig. 4 **a** NO_x distribution of a vertical cross section at the center ($z = 0$ m) and **b** NO_x distribution along the furnace height

NO_x in the bottom ash hopper area has a minimum mass fraction and is close to 0 because at the bottom of the boiler, although there is an excess of O_2, the temperature in the bottom ash hopper area is not sufficient for NO_x formation. In the burner area, NO_x begins to form because there is an excess of O_2 and the temperature in that area is sufficient for the formation of NO_x. This NO_x will continue to accumulate until the boiler outlet.

The tilting angle that produces the highest NO_x mass fraction is at a tilting angle of −30° which is 1.72e−07% because it has the highest fireball temperature. Meanwhile, NO_x content at −20°, −10°, 0°, and +5° is 1.29e−07%, 1.05e−07%, 9.82e−08%, and 7.56e−08%, respectively.

5 Conclusion

1. As the burner tilt angle increases, the temperature distribution in the furnace changes significantly, and the high-temperature region shifts toward the central and upper parts of the furnace. The heat transfer in the superheater and the reheater improves, which causes the steam temperature to increase.
2. Coal optimization (reducing coal quality below the worst value recommended by boiler manufacturers to reduce fuel cost) affects combustion in the boiler where the flue gas temperature in the convection area exceeds the design. The lowest average temperature deviation between the simulation results with design coal and the worst coal in the convection area is -30° with a deviation of 36.12 °C and 27.26 °C, respectively. While the angles of −20°, −10°, 0°, and +5° are 44.88 °C and 36.02 °C, 46.77 °C and 37.91 °C, 48.96 °C and 40.11 °C, and 52.33 °C and 43.47 °C, respectively.
3. The lowest O_2 content at the boiler outlet is at burner tilt −30°, which is 3.27%. Meanwhile, the angles of −20°, −10°, 0°, and +5° are 3.30%, 3.34%, 3.39%, and 3.40%, respectively.
4. The lowest CO_2 content at the boiler outlet is at burner tilt −30°, which is 17.60%. Meanwhile, the angles of −20°, −10°, 0°, and +5° are 17.56%, 17.51%, 17.42%, and 17.40%, respectively.
5. The lowest NO_x content at the boiler outlet is at burner tilt +5°, which is 1.72e−07%. Meanwhile, the angles of −20°, −10°, 0°, and +5° are 1.29e−07%, 1.05e−07%, 9.82e−08%, and 7.56−e08%, respectively.
6. Based on the analysis of temperature, O_2, and CO_2, the best tilting angle to accommodate coal optimization with a calorific value of 4080 kcal/kg is −30°. But because −30° is not applicable and the range of burner tilt angle is only ±20°, the most optimum tilting angle that can be applied in actual condition is −20°.

References

1. Tian D, Zhong L, Tan P, Ma L, Fang Q, Zhang C, Zhang D, Chen G (2015) Influence of vertical burner tilt angle on the gas temperature deviation in a 700 MW low NO_x tangentially fired pulverised-coal boiler. Fuel Process Technol 138:616–628
2. Tan P, Tian D, Fang Q, Ma L, Zhang C, Chen G, Zhong L, Zhang H (2017) Effects of burner tilt angle on the combustion and NO_x emission characteristics of a 700 MWe deep-air-staged tangentially pulverized-coal-fired boiler. Fuel 196:314–324

3. Thrangaraju SK, Munisamy KM, Baskaran S (2017) Research in varying burner tilt angle to reduce rear pass temperature in coal fired boiler. J Phys Conf Ser (822):012038
4. Chang J, Wang X, Zhou Z, Chen H, Niu Y (2021) CFD modelling of hydrodynamics, combustion and NO_X emission in a tangentially fired pulverized-coal boiler at low load operating conditions. Adv Powder Technol 32:290–303
5. Sankar G, Dhannina CS, Santhosh KD, Balasubramanian KR (2020) Numerical simulation of the heat transfer and NO_X emissions in a 660 MW tangentially fired pulverised coal supercritical boiler. Heat Mass Transf (56):2693–2709
6. Kumar M, Sahu SG (2007) Study on the effect of the operating condition on a pulverized coal-fired furnace using computational fluid dynamics commercial code. Energy Fuels (21):3189–3193
7. Kumar RP, Ramchandra RV, Ravi KN (2013) Effect of parameters in once-through boiler for controlling re-heat steam temperature in supercritical power plants. Res J Eng Sci 2(1):27–34
8. Lidija Joleska Bureska (2017) Influence of coal quality to the boiler efficiency and opportunity for its improvement. Termotehnika 43:59–65
9. Lafanechere L, Basu P, Jestin L (1995) Effects of fuel parameters on the size and configuration of circulating fluidized bed boilers. J Inst Energy 68:184–212
10. Dongke Zhang FTSE (2013) Ultra-supercritical coal power plants: materials, technologies, and optimisation. Woodhead Publishing, UK

Thermogravimetric and Kinetic Analysis on Peat Combustion Through Coats-Redfern Fitting Model

Ardianto Prasetiyo[1] and Sukarni Sukarni[1,2(✉)]

[1] Center for Renewable and Sustainable Energy Engineering (CRSEE), Department of Mechanical Engineering, Universitas Negeri Malang, Jl.Semarang No 5, Malang 65145, Indonesia
sukarni.ft@um.ac.id

[2] Center of Advanced Materials for Renewable Energy (CAMRY), Universitas Negeri Malang, Jl.Semarang No 5, Malang 65145, Indonesia

1 Introduction

Peat is a soil layer of organic matter [1] that forms from the accumulation of vegetations such as *Sphagnum* or other biomass, which has undergone anaerobic decomposition via natural atrophy and incomplete disintegration of the dead vegetations at a limited humidity and air availability [2, 3]. Around 2–3% of the earth's surface is covered in peatlands, most in the northern hemisphere and tropical regions. Peatlands contribute 25% to the world's soil carbon storage [4]. Indonesia possesses 21 Mha of peatlands, which makes up 36% of tropical peatlands [5]. Out of the entirety of the world's peatland, only 0.1% has been utilized for energy, horticulture, medic, and other industries [1].

Utilizing peat as a fuel has several benefits, among which are (a) it cuts off the time needed for coal formation since peat itself is an early form of it, (b) peat could be used in existing power stations without requiring extensive modifications, (c) peat has a higher calorific value compared to wood biomass [6], making it more cost-effective, and (d) it is cheaper than oil and gas. Furthermore, the utilization of peat as a fuel is also environmentally safe due to its lack of mercury and sulfur contents [2]. Peat also produces a small amount of ash, which lowers the risk of fouling and heat exchanger corrosion when used in power plants reactor [2, 7]. However, the characteristics of peat material from Indonesia concerning its potential as an alternative fuel have yet to be documented in the literature.

Some earlier research into the potential of peat as a fuel mixture has been conducted by scholars. Among them is Fagerström et al. [8], which studied the impact of the ash variant from peat on the formation of slag throughout peat combustion with biomass in a boiler in the form of wood pellet. Sommersacher et al. [8] investigated the effect of adding peat into Miscanthus upon the formation of ash throughout the combustion process in a laboratory-scale plant. The ash and gas emission from the combustion of pellets made of a mix of peat with softwood barks and stems inside a 500 kW scale boiler has been examined by Sippula et al. [7]. Finally, Vershinina et al. [9] researched ignition

M. Kolhe et al. (eds.), *Recent Advances in Renewable Energy Systems*,
Lecture Notes in Electrical Engineering 876,
https://doi.org/10.1007/978-981-19-1581-9_35

characteristics using a fuel made of a mixture of peat, crude oil, and water. The research focuses on the major issue related to the formation of slag and gas emission that comes from the simultaneous combustion of peat and other materials. However, they have yet to disclose the characteristics and the reaction kinetics of the peat itself.

Additionally, because the composition of peat depends on its geological, topographical, and meteorological environment [8], different peatlands would produce peats with different makeups. The difference in the compositions would impact the thermal behavior throughout the combustion process. As far as the literature review we conducted goes, there has yet to be any research on the thermal behavior and reaction kinetics of peat combustion that specifically uses peat from Indonesia.

The thermogravimetric (TG) analysis technique is widely used to examine the thermal decomposition and kinetic analyses of various fuel materials [10–12]. The thermogravimetric analysis provides a means to measure the initial and final combustion temperatures, as well as the maximum reactivity temperature and the total combustion period. Moreover, the TG test also allows the dataset to calculate the activation energy required for a material's combustion. Therefore, it is essential to conduct a thermogravimetric test on peat combustion to study its combustion behavior, reaction kinetics, and thermodynamic parameters.

This paper presents the decomposition characteristics of peat combustion by way of the TG method. It proceeds to elaborate on the temperature distribution at every stage of decomposition, as well as the analyses of the kinetic and thermodynamic parameters.

2 Material and Method

2.1 Material

The peat sample was taken from the peatland area of Sebangau, Palangkaraya, Central Kalimantan, located at the coordinate points of GPS2°01'55.6" E 113° 42.06',3". It was acquired at a depth of 80 cm below the soil's surface using a Macaulay peat sampler with a diameter of 40 mm. The wet sample was then sun-dried for 8 h and oven-dried at a temperature of 90 °C for 6 h. Afterward, it was crushed and filtered using a size 60 mesh. Finally, the peat sample is stored in an airtight bottle and tested in a laboratory.

2.2 TG Analysis Method

The combustion characteristics of the peat sample were studied using a thermogravimetric (TG) analyzer (Mettler Toledo TG/DCS1). Air was streamed in throughout the combustion test, and it was kept at a constant flow rate of 100 ml/min, while the heating rate was set at 10 °C/min. The combustion was performed at a temperature range of 25–1000 °C. About 10 mg of peat was put in an Al_2O_3 crucible. The mass-loss rate, relative to the increase in temperature and time, was measured automatically. Throughout the experiment, the sample's mass loss as a function of time or temperature is demonstrated as a thermogravimetric (TG) graph, while the mass-loss rate in relation to time or temperature is defined as a derivative thermogravimetric (DTG) graph.

2.3 Kinetic Parameters Evaluation

The fitting methods of kinetic analysis were used for the single heating rate. The thermal reaction rate is defined as the conversion (α) per unit time and expressed as the following equation [13]:

$$\frac{d\alpha}{dt} = k(T)f(\alpha) \tag{1}$$

in which $k(T)$ represents a rate constant that determines the dependence of reaction rate to the temperature, and $f(\alpha)$ stands for a reaction model. The conversion is calculated with Eq. (2):

$$\alpha = \frac{m_a - m_b}{m_a - m_c} \tag{2}$$

in which α is the fraction of degraded material due to the thermal attach, m_a is the initial mass before thermal degradation, m_b is mass at a time t, and m_c is the final mass of the sample. The experiment is conducted in a non-isothermal condition ($\beta = dT/dt$), so that Eqs. (1) and (2) are rewritten into:

$$\frac{d\alpha}{dT} = \frac{A}{\beta} e^{\frac{-E}{RT}} f(\alpha) \tag{3}$$

The reaction rate constant, showing the dependence of reaction rate on temperature, according to Arrhenius, is as the following:

$$k(T) = A \exp\left(\frac{-E}{\mathrm{RT}}\right) \tag{4}$$

in which $A, E, R,$ and T, respectively, represent pre-exponential factors, activation energy, the universal gas constant, and absolute temperature. With the heating rate in Eq. (3) written as ($\beta = dT/dt$) and assuming a reaction according to the $f(\alpha) = (1-\alpha)^n$ model, Eq. (3) could be rewritten into:

$$\frac{d\alpha}{(1-\alpha)^{\mathrm{n}}} = \frac{A}{\beta} e^{\frac{-E}{\mathrm{RT}}} dT \tag{5}$$

Equation (5) being the basis, where the kinetic parameters would be evaluated by using the method of Coats-Redfern.

2.4 Thermodynamic Parameters Evaluation

Besides calculating the activation energy, it is also essential to know the thermodynamic parameter quantity of the peat combustion; those are the enthalpy change ΔH, the Gibbs free energy change ΔG, and entropy change ΔS [14, 15].

$$\Delta H = E - R \cdot T \tag{6}$$

$$\Delta G = E + R \cdot \text{Tm} \cdot \ln\left(\frac{\text{Kb} \cdot \text{Tm}}{h \cdot A}\right) \tag{7}$$

$$\Delta S = \frac{\Delta H - \Delta G}{\text{Tm}} \tag{8}$$

wherein K_b represents the Boltzmann constant (1.381×10^{-23} J/K), h the Planck constant (6.626×10^{-34} J.s), and T_m the temperature peak on the DTG graph.

3 Result and Discussion

3.1 Thermal Behavior of Peat Combustion

Figure 1 shows the peat combustion curve from the data documented by thermogram measurements, figured out as mass loss (TG) and mass-loss rate (DTG) graphs.

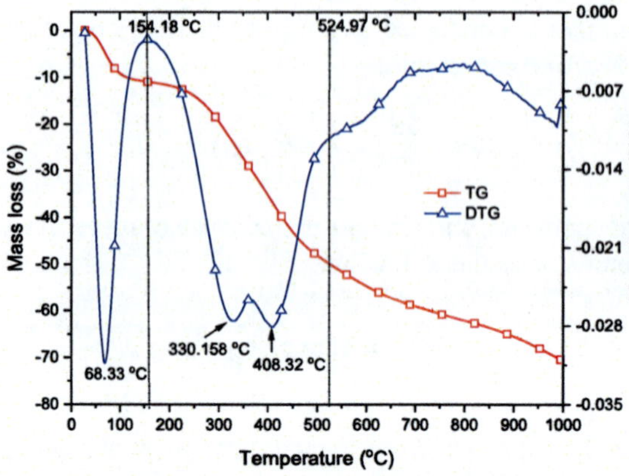

Fig. 1 TG and DTG graphs of peat combustion

According to the TG and DTG graph, it is clear that there are three decomposition phases of combustion throughout the escalation temperature from the room temperature up to 1000 °C. The first phase occurred at the temperature range of 27–154.18 °C, with its peak of mass-loss rate being 0.0313404%/s (DTG graph) at the temperature of 68.33 °C. Said phase shows that a dehydration process took place [16], and a high amount of water evaporates, in which around 10.90% of the mass was lost from the own sample's total mass.

The second phase occurred at the temperature range of 154.18–524.97 °C, indicating the decomposition process of the organic materials of the peat [16], followed by the combustion of volatile matter and fixed carbon. This is signified by the two peaks of the mass-loss rate of the DTG graph at the temperatures of 330.16 and 408.32 °C, respectively. At this stage, the first peak is related to the decomposition of hemicellulose components [17] and the second one has characterized the severe degradation of cellulose components, both of which caused a mass loss percentage of 49.93% out of the total mass.

The third phase, observed from the temperature range of 524.97–994.24 °C, indicates the lignin decomposition process [17] occurring up to the temperature of 803 °C, followed by decomposition of the rest of the solid residual, and the result of which is the loss of 70.40% of the total mass. These thermal characteristics are strong closely related to the material composition of the peat [18]. The temperature characteristics of combustion are detailed in Table 1.

Table 1 Characteristics temperature of peat combustion

Stage	T_o	T_e	T_p	M_p	Mass loss
	°C			%/s	%
I	27.5248	154.18	68.336	−0.0313404	10.90
II	154.18	524.97	408.32	−0.0280975	49.93
III	524.97	994.24	n.d	n.d	70.40

T_o: onset temperature; T_e: temperature at the end of the stage; T_p: temperature at peak; M_p: mass-loss rate peak of stage; n.d: not detected

3.2 Kinetic Analysis of Peat Combustion

Kinetic analysis was conducted according to Eq. (5), which is simplified as the following:

$$\ln g(\alpha) = -\frac{E}{\text{RT}} + \ln\left(\frac{\text{AR}}{\beta E}\right) \tag{9}$$

In Eq. (9), the $g(\alpha)$ value is calculated using the following equations:

$$\begin{aligned} g(\alpha) &= -\left(\ln (1-\alpha)/T^2\right), \text{ if } n = 1, \text{ and} \\ g(\alpha) &= \{1-(1-\alpha)\cdot(1-\text{n})\}/\left\{(1-\text{n})T^2\right\}, \ n \neq 1 \end{aligned} \tag{10}$$

Based on Eq. (9), it could draw a plot between $\ln g(\alpha)$ versus $1/T$ at the predetermined value of n, at the alpha range (α) of 0.05–0.95, so that a straight line with coefficient correlation R^2 is achieved. Then, the $R^2 - n$ plot is made. The correlation coefficient (R^2) value that is closest to 1 is regarded as the best fitting, which is related to the most appropriate reaction order (n). In this experiment, the highest R^2 value is 0.9954, which corresponds to the reaction order (n) value of 1.22. The $R^2 - n$ plot in the active combustion zone of the peat (stage II) is shown in Fig. 2.

After acquiring the value of n according to the highest R^2 value, then the value of $\ln g(\alpha)$ at the determined α range is calculated. The final plot of $\ln g(\alpha)$ versus $1/T$ at T corresponding to the α resulted in a slope and intercept, wherein the slope corresponds to the value of $(-E/R)$ and the intercept to the value of $(\ln((\text{AR})/(\beta E)))$. In this experiment, the last plot of $\ln g(\alpha)$ versus $1/T$ is presented in Fig. 3.

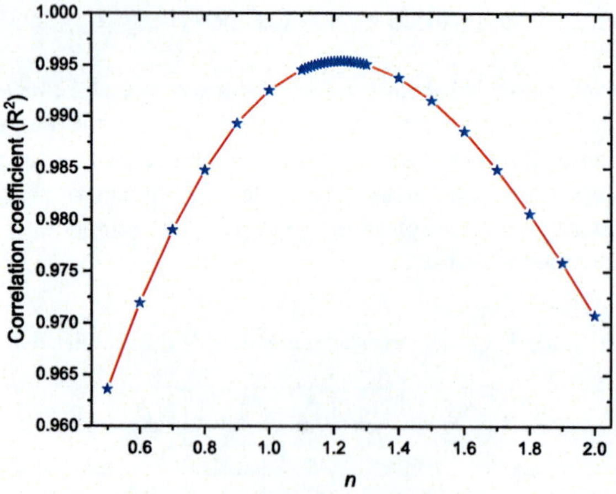

Fig. 2 $R^2 - n$ plot of peat combustion in stage II

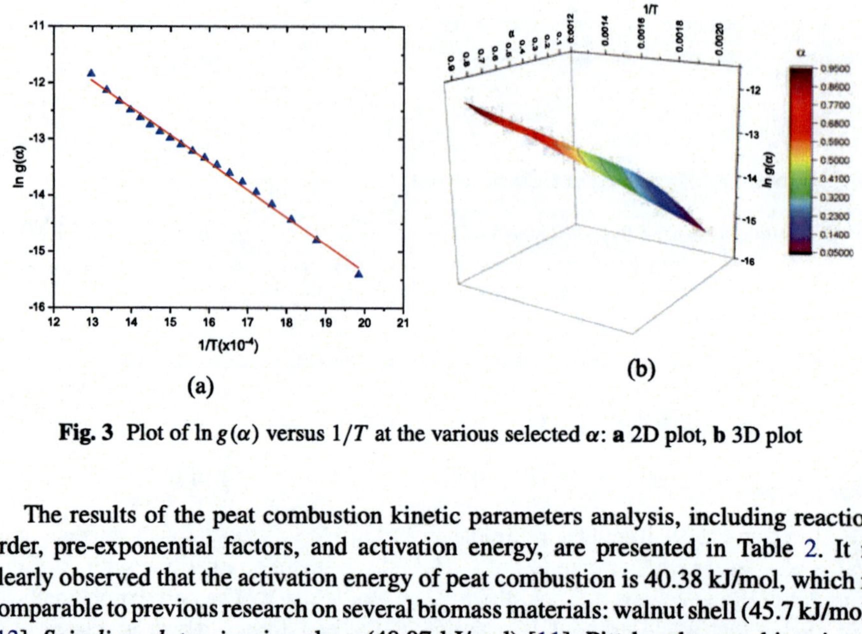

Fig. 3 Plot of $\ln g(\alpha)$ versus $1/T$ at the various selected α: **a** 2D plot, **b** 3D plot

The results of the peat combustion kinetic parameters analysis, including reaction order, pre-exponential factors, and activation energy, are presented in Table 2. It is clearly observed that the activation energy of peat combustion is 40.38 kJ/mol, which is comparable to previous research on several biomass materials: walnut shell (45.7 kJ/mol) [13], *Spirulina platensis* microalgae (48.97 kJ/mol) [11], Pittsburgh seam bituminous coal (45.18 kJ/mol), and Black Thunder Powder River Basin (PRB) sub-bituminous coal (39.87 kJ/mol) [19]. This low activation energy indicated a good oxidation reactivity of the peat sample, which means the energy required to start a reaction would be relatively low [10, 19].

Table 2 Kinetic parameters of peat combustion

(°C/min)	Trendline equation	Kinetic parameters			
		E (kJ/mol)	log A	*n*	R^2
10	$y = -4856.4x - 5.652$	40.38	2.23	1.22	0.9954

3.3 Thermodynamic Parameters of Peat Combustion

The thermodynamic parameters evaluation of peat combustion was performed at the main decomposition phase. These parameters comprised of ΔH, ΔG, and ΔS, each of which was given in Table 3.

Table 3 Thermodynamic parameters of peat combustion

Stage	ΔH	ΔG	ΔS
	(kJ/mol)		
II	35.05	182.81	−0.217

Table 3 presented a positive value of ΔH that was 35.05 kJ/mol. This value indicated the amount of energy required by the reagent during the transition state in the main decomposition phase; in other words, it confirmed that the combustion of peat in this state was an endothermic reaction. The ΔH represented the energy discrepancy between the reagent and the activated complex [15, 20]. The positive value of Gibbs free energy change (ΔG) points out the reaction of activated complex formation that takes place in non-spontaneously mode [20]. As observed in Table 3, it is suggested that energy of around 182.81 kJ/mol was required in the transition state of the sample to form an activated complex. Finally, the negative value of ΔS implied the sample's bond dissociation tends to be more stable during the transition state of combustion. This means that the peat sample had low reactivity, and much more time was needed for the activated complex formation [14, 20, 21]. The very low value of entropy change indicated that the sample is very close to its own thermodynamic equilibrium state, and the material reacts slower to result in the activated complex.

4 Conclusion

The thermal behavior of peat combustion has been characterized, and the results reveal that there are three stages of decomposition: (1) dehydration process, in which the moisture content evaporates in the temperature range of 27–154.18 °C; (2) decomposition and combustion process of organic material in the temperature range of 154.18–524.97 °C; and (3) lignin and solid residue decomposition process in the temperature range of 524.97–994.24 °C. The activation energy, pre-exponential factors, and the reaction order in the main decomposition phase are 40.38 kJ/mol, 170.47 /min, and 1.22, respectively, showing a comparable required energy for combustion with other researched biomass. The thermodynamic parameters include Gibbs free energy, changes in entropy, and changes in enthalpy shows that the formation of an activated complex at the transition processes of the combustion requires some energy.

Acknowledgements. The authors would like to acknowledge the Universitas Negeri Malang for supporting this research.

References

1. World Energy Council 2013 World energy resources: peat presented at the 2013
2. Kim JK, Lee HD, Kim HS, Park HY, Kim SC (2014) Combustion possibility of low rank Russian peat as a blended fuel of pulverized coal fired power plant. J Ind Eng Chem 20:1752–1760. https://doi.org/10.1016/j.jiec.2013.08.027
3. Olsson M (2006) Wheat straw and peat for fuel pellets-organic compounds from combustion. Biomass Bioenerg 30:555–564. https://doi.org/10.1016/j.biombioe.2006.01.005
4. Huang X, Restuccia F, Gramola M, Rein G (2016) Experimental study of the formation and collapse of an overhang in the lateral spread of smouldering peat fires. Combust Flame 168:393–402. https://doi.org/10.1016/j.combustflame.2016.01.017
5. Warren M, Hergoualc'h K, Kauffman JB, Murdiyarso D, Kolka R (2017) An appraisal of Indonesia's immense peat carbon stock using national peatland maps: uncertainties and potential losses from conversion. Carbon Balance Manag 12, https://doi.org/10.1186/s13021-017-0080-2
6. FAO: 9. Energy use of peat http://www.fao.org/3/x5872e/x5872e0b.htm, Accessed 28 July 2021
7. Sippula O, Lamberg H, Leskinen J, Tissari J, Jokiniemi J (2017) Emissions and ash behavior in a 500 kW pellet boiler operated with various blends of woody biomass and peat. Fuel 202:144–153. https://doi.org/10.1016/j.fuel.2017.04.009
8. Sommersacher P, Brunner T, Obernberger I, Kienzl N, Kanzian W (2015) Combustion related characterisation of Miscanthus peat blends applying novel fuel characterisation tools. Fuel 158:253–262. https://doi.org/10.1016/j.fuel.2015.05.037
9. Vershinina KY, Dorokhov VV, Nyashina GS, Romanov DS (2019) Environmental aspects and energy characteristics of the combustion of composite fuels based on peat, oil, and water. Solid Fuel Chem 53:294–302. https://doi.org/10.3103/S0361521919050100
10. Sukarni S, Hamidi N, Yanuhar U, Wardana ING (2015) Thermogravimetric kinetic analysis of nannochloropsis oculata combustion in air atmosphere. Front Energy. 9:125–133. https://doi.org/10.1007/s11708-015-0346-x
11. Sukarni S, Sumarli S, Nauri IM, Prasetiyo A, Puspitasari P (2019) Thermogravimetric analysis on combustion behavior of marine microalgae spirulina platensis induced by $MgCO_3$ and Al_2O_3 additives. Int J Technol 10:1174–1183 https://doi.org/10.14716/ijtech.v10i6.3611
12. Sukarni S (2020) Thermogravimetric analysis of the combustion of marine microalgae spirulina platensis and its blend with synthetic waste. Heliyon 6:e04902. https://doi.org/10.1016/j.heliyon.2020.e04902
13. Açıkalın K (2011) Thermogravimetric analysis of walnut shell as pyrolysis feedstock. J Therm Anal Calorim 105:145–150. https://doi.org/10.1007/s10973-010-1267-x
14. Kim YS, Kim YS, Kim SH (2010) Investigation of thermodynamic parameters in the thermal decomposition of plastic waste-waste lube oil compounds. Environ Sci Technol 44:5313–5317. https://doi.org/10.1021/es101163e

15. Prasetiyo A, Sukarni S, Wulandar R, Puspitasari P (2020) A kinetic study on tetraselmis chuii combustion: the catalytic impact of nanoparticle titanium dioxide (TiO_2) additive. J Adv Res Fluid Mech Therm Sci 71:39–49 https://doi.org/10.37934/arfmts.71.1.3949
16. Yang J, Chen H, Zhao W, Zhou J (2016) Combustion kinetics and emission characteristics of peat by using TG-FTIR technique. J Therm Anal Calorim 124:519–528. https://doi.org/10.1007/s10973-015-5168-x
17. Yang J, Chen H, Zhao W, Zhou J (2016) TG-FTIR-MS study of pyrolysis products evolving from peat. J Anal Appl Pyrolysis 117:296–309. https://doi.org/10.1016/j.jaap.2015.11.002
18. Cancellieri D, Leroy-cancellieri V, Leoni E, Simeoni A, Ya A, Filkov AI, Rein G (2012) Kinetic investigation on the smouldering combustion of boreal peat 93:479–485. https://doi.org/10.1016/j.fuel.2011.09.052
19. Miller BG, Tillman DA (2008) Coal characteristics. Combust Eng Issues Solid Fuel Syst 33–81
20. Parthasarathy P, Fernandez A, Al-ansari T, Mackey HR, Rodriguez R, Mckay G (2021) Thermal degradation characteristics and gasification kinetics of camel manure using thermogravimetric analysis. J Environ Manage 287:112345. https://doi.org/10.1016/j.jenvman.2021.112345
21. Huang L, Liu J, He Y, Sun S, Chen J, Sun J, Chang KL, Kuo J, Ning X (2016) Thermodynamics and kinetics parameters of co-combustion between sewage sludge and water hyacinth in CO_2/O_2 atmosphere as biomass to solid biofuel. Bioresour Technol 218:631–642. https://doi.org/10.1016/j.biortech.2016.06.133

Co-Combustion of Water Hyacinth (*Eichhornia crassipes*) and Coal: Thermal Behavior and Kinetics Analysis Under the Coats-Redfern Method

Sukarni Sukarni[1,2](✉), Nandang Mufti[2,3], Avita Ayu Permanasari[1,2], Ardianto Prasetiyo[1], Poppy Puspitasari[1,2], and Anwar Johari[4]

[1] Center for Renewable and Sustainable Energy Engineering (CRSEE), Department of Mechanical Engineering, Universitas Negeri Malang, Jl.Semarang No 5, Malang 65145, Indonesia
sukarni.ft@um.ac.id

[2] Center of Advanced Materials for Renewable Energy (CAMRY), Universitas Negeri Malang, Jl.Semarang No 5, Malang 65145, Indonesia

[3] Department of Physics, Universitas Negeri Malang, Jl.Semarang No 5, Malang 65145, Indonesia

[4] Centre of Hydrogen Energy, Institute of Future Energy, Universiti Teknologi Malaysia, Johor Bahru, Malaysia

1 Introduction

Two main issues related to the power generation from conventional fossil fuels are the gradual depletion of reserves and adverse emissions effect on the environment. They have been fostered global concerns the renewables and environmentally friendly resources. Among these, the most considered alternative for viable replacement is biomass [1, 2] due to its availability worldwide and the ability to fix CO_2 through photosynthesis. From the abundance point of view, water hyacinth (WH) is a promising feedstock because of its fast and invasive growth ability [3, 4]; therefore, its sustainability is guaranteed. These plants were doubling their biomass in seven days [5], and around 320-ton dry matter can be harvested for every hectare per year [6].

The most common technologies relating to biomass conversion to be energy are biochemical and thermochemical conversion modes [7]. Among various thermochemical conversion techniques, direct combustion of biomass is the most promising choice due to its suitability with the currently available power plant technology with minor modifications [8, 9], and it is the short-term alternative to gaining useful energy by shortening the long chain of the conversion process. However, combustion biomass alone has a drawback related to its nature as low-quality fuel because biomass has a relatively lower heating value compared with conventional solid fuel as coal [7].

Co-combustion of biomass with coal is strongly thought it can provide extensive support for the thermochemical performance of coal-based systems, especially for enhancing the reactivity during the combustion as a result of more product gases from biomass

M. Kolhe et al. (eds.), *Recent Advances in Renewable Energy Systems*,
Lecture Notes in Electrical Engineering 876,
https://doi.org/10.1007/978-981-19-1581-9_36

[10]. The flame stability can be maintained in the coal-based furnace by providing the volatile matter content greater than 35% [9], and therefore, biomass-coal blended fuel finds its place because of biomass rich in the volatile matter. The problems that are frequently faced during biomass combustion, such as the ash deposition and the pipe fouling on a hot surface, can be reduced or eliminated by biomass-coal co-combustion mode [11]. The utilization of biomass-coal blended fuel will extend the duration of fossil fuel depletion and therefore ensure the sustainabilities of future life [4, 12, 13]. Moreover, the co-combustion of biomass on the current existing coal-based power plants will provide several advantages in terms of technical, economic, and environmental [14].

Recently, investigation on the possibilities for the WH biomass to be co-solid fuel has gained more attention by scholars. Liu et al. [15] have studied the effect of five additives on the performance and gas evolved during co-combustion of WH with sewage sludge (SS), and the results revealed that the comprehensive combustibility index (CCI) increased by 0.31-fold in the co-combustion mode and the K_2CO_3 as a potentially an optimal option for decreasing CO_2, NO_2, SO_2, HCN, and NH_3 emissions. The lowest weighted mean values of activation energy (Em) encountered in the WH-SS blend with the addition of the 5% Na_2CO_3 have been reported by Huang et al. [16]. The WH-SS blend maximum mass loss percentage of 92.4% can be achieved at co-combustion parameters that were a temperature of 630.9 °C, a blend ratio of 60.1% SS, and a heating rate of 29.9 °C/min [17]. Implementation of the oxy-fuel atmosphere in the WH-SS–blended fuel combustion showed that the lowest activation energy (Ea) was achieved in $CO2/O2 = 7/3$ atmosphere, and addition 10–40 (wt%) WH has improved 1–1.97 times the combustion performance of SS [18]. To the best of our knowledge, the co-combustion of WH with coal has not been found in the literature. To extend the possibilities of utilization of WH as a solid fuel feedstock in the existing coal-based power plant system and decrease the dependence on fossil-based fuel, in-depth understanding of the thermal behavior and interaction effect between WH and coal during their co-combustion is the critical task.

This paper presented the thermal characteristic of the WH and coal during their co-combustion in the thermogravimetric analyzer. The synergistic effect resulting from the interaction between the materials has been evaluated through both thermogravimetric (TG) and derivative thermogravimetric (DTG) data. In addition, kinetic parameters have been studied according to the fitting method of Coats-Redfern.

2 Methods

2.1 Materials

The details of material preparation and their properties have been presented in the previous papers [3, 4, 19]. The WH samples were collected from both Dams of Selorejo and Sengguruh, district of Malang, Indonesia. For ensuring the WH samples were free from other impurities such as silt, they were washed with distilled water prior to cutting and drying. The cleaned samples were then cut and subsequently subjected to the oven at around 85 °C for 6 h. After that, the dried WH samples were ground and strained into the desired particle size of less than 0.250 mm and put in storage in the tight vacuum bottle for subsequent analysis. The coal samples were gained from Gunung Panjang, Berau,

East Kalimantan, Indonesia. The coal samples were crushed to be less than 0.250 mm, and then, this powder was stored in the tightly sealed bottle. Their ultimate, proximate, and calorific analysis were depicted in Table 1, which was adopted from our previous study [19].

Table 1 Physicochemical properties of the parent sample [19]

Samples	Ultimate analysis^{ar} (wt, %)					Proximate analysis (wt, %)				HHV^{db} (MJ/kg)
	C	O	H	S	N	M^{ar}	A^{db}	VM^{db}	FC^{db}	
WH	39.69	28.91	4.20	0.24	2.70	4.90	21.10	64.40	14.50	15.20
Coal	55.94	13.84	3.91	0.63	1.38	21.80	3.20	46.50	50.30	28.64

M moisture, *VM* volatile matter, *FC* fixed carbon, *A* ash, *ar* as received, *db* dry basis

The decline of the coal reserves has forced the efforts to seek renewable materials for substitute partly or the replacement entirely coal's key role. Therefore, an in-depth understanding of the co-combustion behavior of coal and renewable materials in the higher ratio of renewable materials is critical to investigate the possibility of replacing most of the coal as fuel so that existing coal reserves can be saved. Therefore, this has been being a motivation for making the higher WH content in the blended fuel compared to the coal. For this purpose, the mechanical mixing of WH and coal was performed by means of mortar to achieve the blended samples with a proportion of WH to the coal of 60:40.

2.2 Thermal Experiment

The METTLER TOLEDO TGA/DSC1 apparatus was employed for investigating the thermal combustion characteristic of the samples. A very small sample mass of around 10 mg was used to ensure the absence of the influence of mass and heat transfer restrictions. Each sample of WH, coal, and their blend was placed into the crucible alternately and then exposed into the furnace that was heated up at the rate of 40 °C/min. The combustion process was performed in the non-isothermal mode started from room temperature and terminated at 1000 °C. To ensure sufficient oxidative air during the combustion, a 100 ml/min air atmosphere has continuously flowed into the chamber throughout the experiment. The sample mass loss that represented the thermogravimetric (TG) data, and all together with the derivative thermogravimetric (DTG) data, would be used for characterizing the sample's behavior during the thermal processes. According to the temperature characteristics that were specified from both TG and DTG curves, subsequently, the kinetic parameters were evaluated.

3 Results and Discussion

3.1 Combustion Profile of the Samples

The respective Fig. 1a and b showed the TG and DTG curves of the blend that was compared with WH and coal during combustion at 40 °C/min heating program.

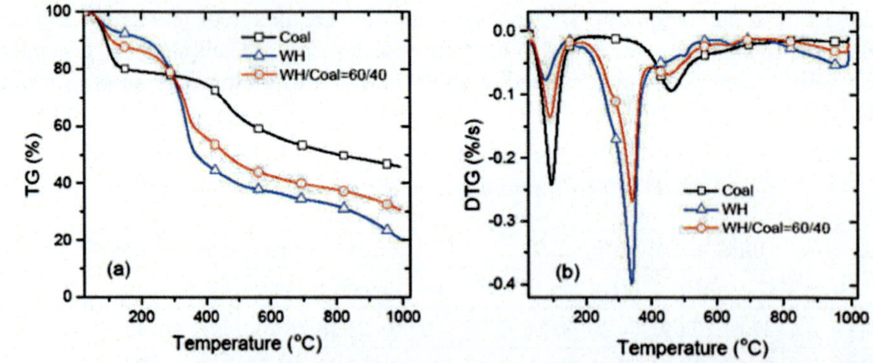

Fig. 1 Combustion profile of the blend compared with WH and coal: **a** TG and **b** DTG curves

From both Fig. 1a and b, several characteristic parameters of the samples could be determined, as depicted in Table 2. The temperature at which volatile matter started to release was the initial reaction temperature (T_i), specified by beginning loss of mass as the TG shows. The temperature at a state where the mass loss curve underwent stabilization in the stage that was being the concern was the terminated temperature (T_t). The maximum rate temperature ($T_{\max}$) was represented as the temperature of mass loss rate at the maximum position (DTG_{max}). The M_R represented the mass leftover at the finalized process (1000 °C).

Table 2 Characteristic parameters of the blend compared with WH and coal

Samples	T_i(°C)	T_t(°C)	$T_{\max}$(°C)	$-DTG_{\max}$(%/s)	M_R(%)
Coal	335.718	563.231	459.949	0.09456	45.7471
WH/coal = 60/40	212.699	429.554	340.656	0.26964	30.6147
WH	192.948	407.671	337.639	0.39941	20.1807

By considering Fig. 1a and b, it was encountered easily that three stages of thermal degradation at a diverse temperature range were experienced on each sample. It was also could be seen that the curve of the blended fuel combustion lay between those of the WH and coal. Similar results have been reported by Wang et al. [20] related to the co-combustion of pinewood and anthracite.

From Fig. 1a, the TG curve, it was identified that the mass loss profile of the blend was identical with the WH and yet dissimilar from that of coal. By looking at the DTG curve (Fig. 1b), it was obviously observed that either blend or WH has a similar profile, and however, both differ from the coal. This phenomenon was highly thought because the WH content placed a higher portion in the blend compared to coal promoted its thermal behavior was similar to the WH.

The combustion process of the blend was initiated by releasing moisture and light volatile in the first stage that stretched from ambient temperature to 212.7 °C. The second stage, correlated with the main degradation of the organic compound that was volatilized and together with carbonaceous species were burned, was occurred at the temperature range of 212.7–429.6 °C. From the TG curve, it was well understood that a huge amount of volatile contents in the WH affected the significant mass loss of the fuel blend. From the DTG point of view, it was observed that the mass loss of the blend combustion was carried out in the earlier temperature compared to the coal due to the earlier release and combustion of the volatiles from WH. This region was associated with the main combustion process. The third stage was taken place from the end of the second stage (429.6 °C) until the completed process at 1000 °C, in which carbonaceous residue was degraded and burned in prolonged mode. As given in Table 2, blending WH with coal has impacted the decreasing residue by 33.08%.

Looking carefully at Table 2, it was found that the blend has initial temperature decomposition (T_i) higher than WH and lower than coal. It means the WH material that contained more volatile and also more oxygen and hydrogen compared to coal (see Table 1) has promoted the ignition process toward a lower temperature than coal. This result was in agreement with Guo et al. [21] for co-combustion of biomass pellets and coal, which revealed that the biomass pellets contained four times higher in the volatile matter than fixed carbon and played a key role in the more reactive combustion process. Furthermore, biomass pellets that are higher in oxygen and hydrogen contents significantly determined the raising of the thermal reactivity. Muthuraman et al. [22] have studied the co-combustion of coal and wood, in which increasing wood in the blend led to the temperature of weight loss shifted to the lower value and reducing the temperature of volatile release. Yanfen [23] found that the co-combustion of coal and paper mill sludge indicated that the more oxygen concentration in the fuels, the more effective the combustion reaction was.

Table 2 also indicated that terminated temperature (T_t) of coal was highest compared to either WH or the blend materials. This is because of the very high fixed carbon content in the coal sample, in which fixed carbon was thermally more stable, and it combusted harder; on the contrary, volatile content was burned very easily. This result was linear with the result of Kastanaki and Vamvuka [14]. From the parameters of (T_i) and (T_t), these were clearly understood that a higher amount of volatile matter in WH have a significant contribution to the easiness initiation and shorter duration of the combustion process of the blend. Table 1 showed that the volatile matter content of WH was around 1.4 times higher compared to coal. Therefore, mixing WH could improve the burning characteristic of coal that has a low volatile matter, indicating that co-combustion of WH and coal was practicable.

From Table 2, it also could be seen that the maximum mass loss rate ($-DTG_{max}$) of the blend that was representing the fuel reactivity was higher than coal. The reactivity of the solid fuel is also often measured by maximum rate temperature (T_{max}), where a lower T_{max} indicated more reactive the materials and vice versa [24]. According to Table 2, it was detected that theT_{max} of blend was lower than that of coal. From parameters of $-DTG_{max}$ and T_{max}, either indicated that the WH in the blend led to fuel more reactive than coal. The lower T_{max} of the blend correlated to the volatile combustion originated from WH, while higher T_{max} of the coal in accordance with the fixed carbon combustion. Among the three samples, the $-DTG_{max}$ of the WH was maximum one, revealed that combustion of WH was taken place very fast due to its high volatiles and oxygen content. This discrepancy was primarily due to the disparity of their chemical composition and structure [25]. The WH was principally composed of cellulose, hemicellulose, and lignin, where their macromolecular structures were linked together with relatively weak ether bonds of R-O-R (bond energy of 380–420 kJ/mol). These ether bonds do not resist heating, not even to low temperatures. On the contrary, coal was composed of dense polycyclic aromatic hydrocarbons that were linked by C=C bonds. These chemical structures were more stable thermally because they had high bond energy of 1000 kJ/mol [26, 27]. Moreover, the increased oxygen concentration in the fuel also promoted the release of the volatile materials quickly, yielding the maximum mass loss rate of the samples being increased [23].

3.2 Kinetic Analysis of Blend Combustion

The kinetic analysis of the WH, coal, and their blend was carried out by using the Coats-Redfern method. The conversion rate during the non-isothermal experiments was expressed as:

$$\frac{d\alpha}{dt} = k(T)f(\alpha) \tag{1}$$

where the respective α, t, and T were conversion degree, time, and absolute temperature. $k(T)$ represented the constant of the temperature-dependent rate or simply referred to as the reaction rate and $f(\alpha)$ was the temperature-independent function of conversion that represented the hypothetical model of the reaction mechanism. According to the Arrhenius law, k was presented as:

$$k(T) = Ae^{\frac{-E}{RT}} \tag{2}$$

in which A was the frequency factor, E was the activation energy, and R was the universal gas constant. The function $f(\alpha)$ was denoted as:

$$f(\alpha) = (1-\alpha)^n \tag{3}$$

and

$$\alpha = (m_i - m_t)/\left(m_i - m_f\right) \tag{4}$$

where n was the reaction order, m_i was the initial mass, m_t was the actual mass at time t, and m_f was the residual mass at the end of the event of the concern.

By combining, Eqs. (4)–(7) could achieve the expression as follow:

$$\frac{d\alpha}{dt} = Ae^{\frac{-E}{RT}}(1-\alpha)^n \tag{5}$$

At the non-isothermal condition, temperature escalated at a heating program of $\beta = dT/dt$, and therefore rearrangement of Eq. (8) gave:

$$\frac{d\alpha}{(1-\alpha)^n} = \frac{A}{\beta}e^{\frac{-E}{RT}}\,dT \tag{6}$$

Integrating Eq. (9) formed the expression as follow:

$$g(\alpha) = \int_0^\alpha \frac{d\alpha}{(1-\alpha)^n} = \frac{A}{\beta}\int_{T_o}^{T} e^{\frac{-E}{RT}}\, dT \tag{7}$$

where $g(\alpha)$ was the integral function of conversion.

Equation (10) was integrated by using the Coats-Redfern method, resulted in:

$$\ln g(\alpha) = \ln\frac{AR}{\beta E} - \frac{E}{RT} \tag{8}$$

where:

if $n = 1$ then $g(\alpha) = -(\ln(1-\alpha))/T^2$.
if $n \neq 1$ then $g(\alpha) = \left(1-(1-\alpha)^{(1-n)}\right)/\left((1-n)T^2\right)$.

Equation (11) was applicable in the condition where combustion occurred as a single-step reaction of n th order at a specific temperature region. It could be verified that for most values of activation energy, E, the expression of $ln(AR/\beta E)$ in Eq. (11) is principally constant [28]. The overall kinetic parameters, included reaction order, activation energy, and frequency factor, were evaluated by the method proposed by Acikalin [29, 30] and were evaluated in detail in various literature [4, 31, 32].

The kinetics evaluation was performed in accordance with the characteristic temperature parameters that were tabulated in Table 2. Started with selecting any n values, then $\ln g(\alpha)$ was calculated at α that was stretched in the range of 0.05–0.95. This step then was followed by plotting the $\ln g(\alpha)$ versus $1/T$ at various n values in order to a linear regression process could be done for gaining the slope with a correlation coefficient of R^2. The best linear fitting was indicated by the closest value of R^2 to the 1.0 value. Hereafter, various resulted R^2 were plotted to the several selected n to determine the most proper n value (Fig. 2a). Based on this final n value, the last ln g(α) was calculated at various chosen α, and then the final plot of $\ln g(\alpha)$ versus $1/T$ was performed to generate the final slope and intercept (Fig. 2b). Finally, the respective activation energy and frequency factor were calculated from the final slope and intercept. The kinetic parameters of WH, coal, and the blend were tabulated in Table 3.

From Table 3 as well as Fig. 2b, it was known that the fitting processes had generated a very high correlation coefficient (R^2) of 0.999, 0.994, and 0.995; therefore, the results of the kinetic analysis were acceptable. The activation energy of the blend was lower than coal. The addition of 60% (wt%) WH to the coal had been reducing the activation energy by 22.67%. This result was in agreement with Lu et al. [33] related to the co-combustion of pine sawdust and anthracite and also in accordance with the result of Gil et al. [10] in terms of pine sawdust-coal co-combustion. Decreasing activation energy in the biomass-coal blend indicated that there was an improvement in the reaction activity during the blend combustion. The result of this study confirmed that the presence of WH in the blended fuel led to shift both the temperature of the releasing volatile and the temperature of the gaseous phase combustion toward the lesser values because of the fact that thermochemical processes of this fuel blend could take place at such as lower activation energy, which was in line with the decreasing of the initial decomposition temperature (T_i) and terminated temperature (T_t) given in Table 2. Reaction with low

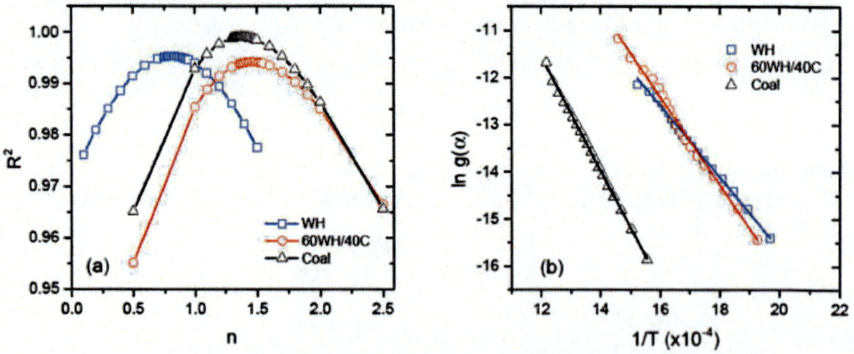

Fig. 2 **a** The $R^2 - n$ curve and **b** linear regression of a plot $\ln g(\alpha)$ with $1/T$

Table 3 The combustion kinetic parameters of the sample

Samples	Trendline equation	R^2	Kinetic parameters		
			E (kJ/mol)	$\log A$ (min^{-1})	n
Coal	$y = -12037x + 2.8834$	0.999	100.079	6.332	1.38
60WH/40coal	$y = -9308.1x + 2.4829$	0.994	77.387	6.047	1.46
WH	$y = -7589.4x - 0.4439$	0.995	63.098	4.687	0.81

activation energy required low temperature or shorter reaction time and vice versa [4, 34].

Pay attention to the frequency factor (A); it was observed that the blended fuel had a lower value compared to the coal one. The frequency factor was a measure of the frequency of molecular collisions of a reaction regardless of their energy level [35, 36]. A smaller frequency factor indicated reducing the effective collision during the reaction of reactant molecules led to a decrease in the probability of secondary reaction [37], which was in agreement with TG-DTG curves Fig. 1a, b.

4 Conclusions

The thermal behavior and kinetics of the blended fuel composed of WH and coal have been studied. The parent fuels of WH and coal were also examined as the control sets. The blended sample experienced three combustion stages during the escalation temperature from ambient temperature to 1000 °C. The thermal profile of the blend was closed to the WH and differed from the coal. The blending of 60% WH to coal impacted the combustion process taken place at a lower temperature. The activation energy of the blend combustion decreased significantly by 22.67% compared to coal. These overall results justified the existence of a synergistic effect during co-combustion of the WH and coal, in which the presence of the WH in the blend promoted the fuel more reactive than individual coal.

Acknowledgements. The support from Universitas Negeri Malang through the PNBP Research Grant 2019 (20.3.117/UN32.14.1/ LT/2019) was acknowledged.

References

1. Furutani Y, Kudo S, Norinaga K, Hayashi J-I, Watanabe T, Asia G (2017) Current situation and future scope of biomass gasification in Japan. Evergreen 04:24–29. https://doi.org/10.5109/1929681
2. Kusrini E, Supramono D, Alhamid MI, Pranata S, Wilson LD, Usman A (2019) Effect of polypropylene plastic waste as co-feeding for production of pyrolysis oil from palm empty fruit bunches. Evergreen 6:92–97. https://doi.org/10.5109/2328410
3. Sukarni S, Zakaria Y, Sumarli S, Wulandari R, Permanasari AA, Suhermanto M (2019) Physical and chemical properties of water hyacinth (Eichhornia crassipes) as a sustainable biofuel feedstock. In: IOP conference series: materials science and engineering. p 012070. IOP Publishing. https://doi.org/10.1088/1757-899X/515/1/012070
4. Sukarni S, Widiono AE, Sumarli S, Wulandari R, Nauri IM, Permanasari AA (2018) Thermal decomposition behavior of water hyacinth (Eichhornia crassipes) under an inert atmosphere. In: MATEC web of conferences, pp 00010. EDP sciences https://doi.org/10.1051/matecconf/201820400010.
5. Gunnarsson CC, Petersen CM (2007) Water hyacinths as a resource in agriculture and energy production: a literature review. Waste Manag 27:117–129. https://doi.org/10.1016/j.wasman.2005.12.011
6. Thomas TH, Eden RD (1990) Water hyacinth—a major neglected resource. In: Energy and environment: into the 1990's—proceedings of the 1st world renewable energy congress, pp 2096–2096. Pergamon
7. Vhathvarothai N, Ness J, Yu QJ (2014) An investigation of thermal behaviour of biomass and coal during copyrolysis using thermogravimetric analysis. Int J Energy Res 38:1145–1154. https://doi.org/10.1002/er.3120
8. Magalhães D, Panahi A, Kazanç F, Levendis YA (2019) Comparison of single particle combustion behaviours of raw and torrefied biomass with Turkish lignites. Fuel 241:1085–1094. https://doi.org/10.1016/j.fuel.2018.12.124
9. Biagini E, Lippi F, Petarca L, Tognotti L (2002) Devolatilization rate of biomasses and coal-biomass blends: an experimental investigation. Fuel 81:1041–1050. https://doi.org/10.1016/S0016-2361(01)00204-6
10. Gil MV, Casal D, Pevida C, Pis JJ, Rubiera F (2010) Thermal behaviour and kinetics of coal/biomass blends during co-combustion. Bioresour Technol 101:5601–5608. https://doi.org/10.1016/j.biortech.2010.02.008
11. Haykiri-Acma H, Yaman S (2008) Effect of co-combustion on the burnout of lignite/biomass blends: a Turkish case study. Waste Manag 28:2077–2084. https://doi.org/10.1016/j.wasman.2007.08.028
12. Sukarni S, Sumarli S, Nauri IM, Purnami P, Al Mufid A, Yanuhar U (2018) Exploring the prospect of marine microalgae Isochrysis galbana as sustainable solid biofuel feedstock. J Appl Res Technol 16:53–66. https://doi.org/10.22201/icat.16656423.2018.16.1.703
13. Sukarni S, Yanuhar U, Wardana IN, Sudjito S, Hamidi N, Wijayanti W, Wibisono Y, Sumarli S, Nauri IM, Suryanto H (2018) Combustion of microalgae nannochloropsis oculata biomass: cellular macromolecular and mineralogical content changes during thermal decomposition. Songklanakarin J Sci Technol 40:1456–1463. https://doi.org/10.14456/sjst-psu.2018.178
14. Kastanaki E, Vamvuka D (2006) A comparative reactivity and kinetic study on the combustion of coal-biomass char blends. Fuel 85:1186–1193. https://doi.org/10.1016/j.fuel.2005.11.004

15. Liu J, Huang L, Sun G, Chen J, Zhuang S, Chang K, Xie W, Kuo J, He Y, Sun S, Buyukada M, Evrendilek F (2018) (Co-)combustion of additives, water hyacinth and sewage sludge: thermogravimetric, kinetic, gas and thermodynamic modeling analyses. Waste Manag 81:211–219. https://doi.org/10.1016/j.wasman.2018.09.030
16. Huang L, Xie C, Liu J, Zhang X, Chang KL, Kuo J, Sun J, Xie W, Zheng L, Sun S, Buyukada M, Evrendilek F (2018) Influence of catalysts on co-combustion of sewage sludge and water hyacinth blends as determined by TG-MS analysis. Bioresour Technol 247:217–225. https://doi.org/10.1016/j.biortech.2017.09.039
17. Liu J, Huang L, Buyukada M, Evrendilek F (2017) Response surface optimization, modeling and uncertainty analysis of mass loss response of co-combustion of sewage sludge and water hyacinth. Appl Therm Eng 125:328–335. https://doi.org/10.1016/j.applthermaleng.2017.07.008
18. Huang L, Liu J, He Y, Sun S, Chen J, Sun J, Chang KL, Kuo J, Ning X (2016) Thermodynamics and kinetics parameters of co-combustion between sewage sludge and water hyacinth in CO2/O2atmosphere as biomass to solid biofuel. Bioresour Technol 218:631–642. https://doi.org/10.1016/j.biortech.2016.06.133
19. Sukarni S, Mufti N, Permanasari AA, Eka R, Johari A (2019) The fitting kinetic evaluation during co-pyrolysis of coal and water hyacinth (Eichhornia crassipes) to explore its potential for energy. In: 2019 the 4th international tropical renewable energy conference (i-TREC)
20. Wang G, Zhang J, Shao J, Ren S (2014) Characterisation and model fitting kinetic analysis of coal/biomass co-combustion. Thermochim Acta 591:68–74. https://doi.org/10.1016/j.tca.2014.07.019
21. Guo F, He Y, Hassanpour A, Gardy J, Zhong Z (2020) Thermogravimetric analysis on the co-combustion of biomass pellets with lignite and bituminous coal. Energy 197 https://doi.org/10.1016/j.energy.2020.117147
22. Muthuraman M, Namioka T, Yoshikawa K (2010) A comparison of co-combustion characteristics of coal with wood and hydrothermally treated municipal solid waste. Bioresour Technol 101:2477–2482. https://doi.org/10.1016/j.biortech.2009.11.060
23. Yanfen L, Xiaoqian M (2010) Thermogravimetric analysis of the co-combustion of coal and paper mill sludge. Appl Energy 87:3526–3532. https://doi.org/10.1016/j.apenergy.2010.05.008
24. Sahu SG, Sarkar P, Chakraborty N, Adak AK (2010) Thermogravimetric assessment of combustion characteristics of blends of a coal with different biomass chars. Fuel Process Technol 91:369–378. https://doi.org/10.1016/j.fuproc.2009.12.001
25. Chen X, Liu L, Zhang L, Zhao Y, Zhang Z, Xie X, Qiu P, Chen G, Pei J (2018) Thermogravimetric analysis and kinetics of the co-pyrolysis of coal blends with corn stalks. Thermochim Acta 659:59–65. https://doi.org/10.1016/j.tca.2017.11.005
26. Ulloa CA, Gordon AL, García XA (2009) Thermogravimetric study of interactions in the pyrolysis of blends of coal with radiata pine sawdust. Fuel Process Technol 90:583–590. https://doi.org/10.1016/j.fuproc.2008.12.015
27. Sadhukhan AK, Gupta P, Goyal T, Saha RK (2008) Modelling of pyrolysis of coal-biomass blends using thermogravimetric analysis. Bioresour Technol 99:8022–8026. https://doi.org/10.1016/j.biortech.2008.03.047
28. Zhou L, Wang Y, Huang Q, Cai J (2006) Thermogravimetric characteristics and kinetic of plastic and biomass blends co-pyrolysis. Fuel Process Technol 87:963–969. https://doi.org/10.1016/j.fuproc.2006.07.002
29. Açikalin K (2011) Thermogravimetric analysis of walnut shell as pyrolysis feedstock. J Therm Anal Calorim 105:145–150. https://doi.org/10.1007/s10973-010-1267-x
30. Açikalin K (2012) Pyrolytic characteristics and kinetics of pistachio shell by thermogravimetric analysis. J Therm Anal Calorim 109:227–235. https://doi.org/10.1007/s10973-011-1714-3

31. Sukarni S, Prasetiyo A, Sumarli S, Nauri IM, Permanasari AA (2018) Kinetic analysis of co-combustion of microalgae Spirulina platensis and synthetic waste through the fitting model. In: MATEC web of conferences. pp 00009. EDP Sciences. https://doi.org/10.1051/mateccconf/201820400009.
32. Sumarli S, Himawan CU, Wulandari R, Sukarni S (2019) Thermogravimetric analysis and the fitting model kinetic evaluation of corn silk thermal decomposition under an inert atmosphere. In: IOP conference series: materials science and engineering 494, pp 012098. https://doi.org/10.1088/1757-899X/494/1/012098
33. Lu G, Zhang K, Cheng F (2017) The combustion characteristics of anthracite and pine sawdust blends. Energy Sources, Part A Recover Util Environ Eff 39, 1131–1139 (2017). https://doi.org/10.1080/15567036.2017.1299258
34. Hamidi N, Yanuhar U, Wardana IN (2015) Thermogravimetric kinetic analysis of Nannochloropsis oculata combustion in air atmosphere. Front Energy 9:125–133. https://doi.org/10.1007/s11708-015-0346-x
35. White JE, Catallo WJ, Legendre BL (2011) Biomass pyrolysis kinetics: a comparative critical review with relevant agricultural residue case studies. J Anal Appl Pyrolysis 91:1–33. https://doi.org/10.1016/j.jaap.2011.01.004
36. Galwey AK, Brown ME (2002) Application of the arrhenius equation to solid state kinetics: can this be justified? Thermochim Acta 386:91–98. https://doi.org/10.1016/S0040-6031(01)00769-9
37. Jeong HM, Seo MW, Jeong SM, Na BK, Yoon SJ, Lee JG, Lee WJ (2014) Pyrolysis kinetics of coking coal mixed with biomass under non-isothermal and isothermal conditions. Bioresour Technol 155:442–445. https://doi.org/10.1016/j.biortech.2014.01.005

Effect of Stored Dexlite and Palm Oil Biodiesel on Fuel Properties, Performance, and Emission of Single-Cylinder Diesel Engines

Atok Setiyawan(✉), Kuntang Winangun, and Vernanda Sania

Mechanical Engineering Department, Institut Teknologi Sepuluh Nopember, Surabaya, Indonesia
atok.setiyawan@gmail.com

1 Introduction

According to data from Indonesia's 2020 energy outlook, the projected energy demand during 2018–2050 is expected to increase by an average of 3.9% per year, and the type of fuel oil (BBM) will continue to dominate. However, dependence on energy imports continues to increase in line with the depletion of energy reserves due to the condition of oil production wells that are getting older and the limitations of existing technology. The use of diesel engines has also increased due to machines that are still being used massively and continuously [1], so that renewable and low-emission fuel solutions are needed, one of which is biodiesel crude palm oil (CPO).

Crude palm oil (CPO) biodiesel was chosen because it has physical and chemical properties similar to petroleum diesel, which has a high oxygen content, and low CO, UHC, and PM emissions. However, CPO biodiesel has hygroscopic properties that affect changes in fuel properties and combustion characteristics.

Research on changes in biodiesel properties during the storage process has been carried out by several researchers. Research on the effect of temperature and storage duration on the properties and characteristics of crude palm oil biodiesel as a mixed fuel was carried out by [2]. The effect of storage duration of biodiesel on water content, density, kinematic viscosity, flash point, and acid number, using biodiesel fuel from crude palm oil stored for 2016 hours and tested regularly every one week and stored at three different temperature conditions. All variations showed that the water content, density, kinematic viscosity, and the acid number increased, while the flashpoint decreased with the length of storage duration. [3] conducted experiments with palm oil biodiesel which was named palm oil diesel (POD). POD has a lower calorific value, resulting in higher fuel consumption than petro-diesel. On average, PODs have a 10% higher BSFC than diesel engines. [4] conducted a study using Dexlite fuel (B0), B30 (30% biodiesel blend with 70% Dexlite), and CPO biodiesel (B100) stored in a tank for 12 weeks. Fuel testing was carried out at weeks 0, 6, and 12 including testing properties and testing performance and emissions from fuel. From the results of the study, it was found that the highest density, viscosity, and water content values were owned by B100 fuel. In performance testing, thermal efficiency and smoke opacity decreased with increasing

M. Kolhe et al. (eds.), *Recent Advances in Renewable Energy Systems*,
Lecture Notes in Electrical Engineering 876,
https://doi.org/10.1007/978-981-19-1581-9_37

fuel storage duration, while for BSFC, it increased. Decrease in thermal efficiency on Dexlite, B30, and B100 fuels, respectively, by 14%, 6%, and 8%.

Based on the description above, the longer the duration of storage of biodiesel fuel, it will have an impact on increasing the value of viscosity, water content, density, and decreasing calorific value. The problem in this study is that the longer storage of Dexlite fuel and palm oil biodiesel will cause degradation of the fuel. The degradation that occurs can affect the physical properties and properties of a fuel. The purpose of this study was to determine the effect of storage duration of Dexlite fuel and palm oil biodiesel on fuel properties, performance, and emissions of single-cylinder diesel engines.

2 Materials and Methods

This study uses an experimental method of sample testing in the laboratory. Testing of fuel properties is carried out at the LPPM-ITS energy laboratory. Testing of fuel properties based on storage time is carried out on a period of 0 months and 3 months. The fuel used is Dexlite fuel and biodiesel from palm oil. Tests of fuel properties include water content, density, viscosity, and calorific value. Engine performance tests include thermal efficiency, power, and fuel consumption. Engine performance testing was carried out at the Thermal Engineering and Energy System Laboratory of Mechanical Engineering FT-IRS ITS.

2.1 Fuel Storage

The fuel used is Dexlite and biodiesel from palm oil, and Dexlite is obtained from PT. Pertamina is sold in general at gas stations, and biodiesel is obtained from PT. Wilmar Nabati Indonesia. The fuel is stored in a tank made of PVC with a capacity of 200 liters, and storage is carried out separately between Dexlite fuel and biodiesel. The storage process is carried out in a room with ambient temperature (Fig. 1).

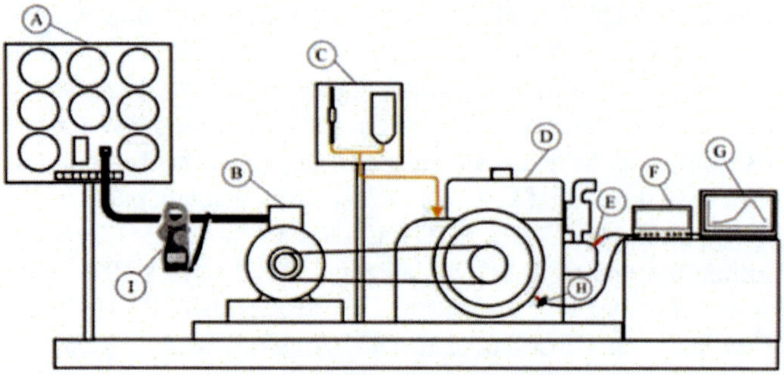

Fig. 1 Schematic of testing equipment

Table 1 Specification of testing machine

Description	Specification
Model	DI 800 H
Type	One cylinder, four stroke
Combustion system	Direct injection
Bore × stroke	82 × 78 mm
Maximum power	7 HP (5.22 kW)/2200 rpm
Continuous power	6 HP (4.47 kW)/2000 rpm
Compression ratio	18:1
Pilot injection timing	13° BTDC
Cooling system	Hopper

2.2 Test Machine

The machine used for testing is Diamond DI 800, direct injection, single-cylinder, water cooling, and four stroke. In detail, the engine specifications are in Table 1.

Information: (A) loading lamp, (B) generator, (C) fuel gage pipette, (D) diesel engine, (E) pressure transducer sensor, (F) combustion analyzer, (G) computer, (H) crank angle sensor, (I) multimeter.

3 Results and Discussion

The results of the test and discussion presented are the effect of storage of Dexlite fuel and biodiesel (B100) on testing fuel properties and performance of single-cylinder diesel engines. Tests of fuel properties include water content, calorific value, density, and viscosity. Engine performance testing includes thermal efficiency, power, and fuel consumption.

3.1 Properties of Fuel

The water content in the fuel affects the combustion quality of a diesel engine. Figure 2a shows the results of testing the water content of Dexlite and (B100) fuel. B100 has a higher water content compared to Dexlite during storage. After being stored for 3 months, the water content increased to 0.064% by volume and 0.129% by volume, respectively, to Dexlite and B100. The increase in water content is due to the hygroscopic glycerol content, namely the ability to absorb water content from the surrounding environment [5]. The increase in water content in Dexlite and B100 fuels is 83% and 18%, respectively.

The calorific value is identified as a very important parameter in fuel because it affects engine power [6]. In general, the calorific value of biodiesel has a lower value than diesel fuel [7] as shown in Fig 2b; the calorific value of Dexlite and B100 is 10,436 cal/g and 9486 cal/g, respectively. However, after 3 months of storage, the calorific value of both

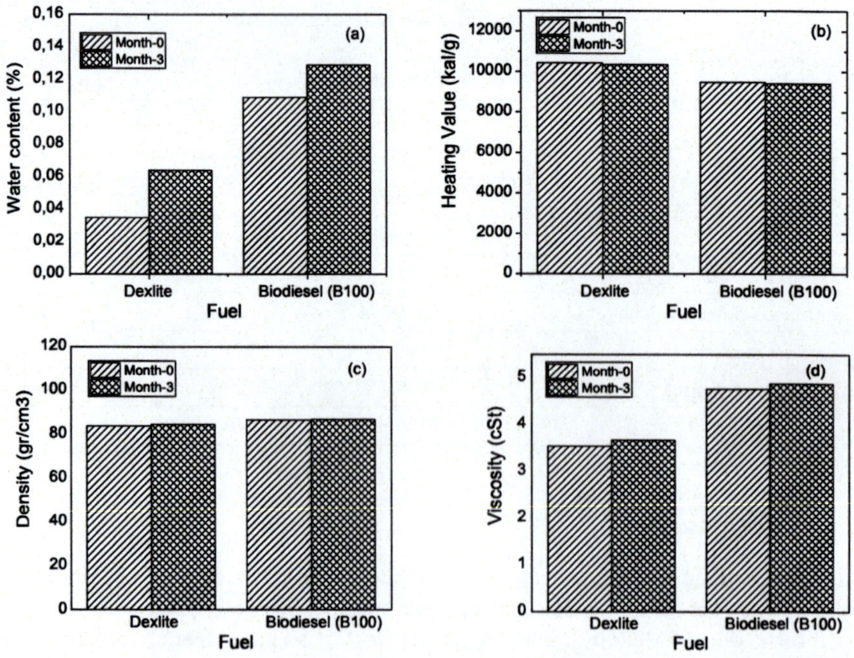

Fig. 2 Water content **a**, calorific value **b**, density **c**, and viscosity **d** to fuel and storage duration

fuels decreased by 0.72 and 0.59%. The calorific value of methyl ester has a lower value than petro-diesel, and biodiesel has a lower heating value than diesel by about 10% [8].

The density of fuel can affect the flow and process of fuel injection into the combustion chamber so that it will affect engine power and exhaust emissions [9]. The higher density causes the viscosity value to increase so that it affects the fuel droplet size [10]. The increased interaction of biodiesel molecules that are degraded due to peroxide causes the density to increase, in addition to the glycerol content in the biodiesel which can absorb water [2]. In Fig 2c, it can be seen that B100 fuel has a higher density than Dexlite during the storage period. The density increase in Dexlite and B100 fuels is 0.98% and 0.29%, respectively.

Viscosity can affect the atomization process when fuel is injected into the combustion chamber, so it will affect the quality of combustion [11]. High viscosity and density affect droplet distribution into the combustion chamber, atomization formation, and heavier pump work [8]. Based on the results of the fuel viscosity test in Fig 2d, it can be seen that B100 fuel has a higher viscosity than Dexlite during a 3 month storage period. After being stored for 3 months, the viscosity increased to 3.76 and 4.87 cSt on Dexlite and B100 fuel.

3.2 Machine Performance Testing

In Fig 3a, it can be seen that the thermal efficiency of all fuels and storage time shows an increasing trend with increasing engine load [12]. As long as fuel storage affects the engine's thermal efficiency, it decreases. Dexlite fuel decreased by 26.42%, and biodiesel fuel decreased by 5.70%. The decrease in engine thermal efficiency is influenced by the heating value, density, and viscosity of the fuel. The higher density and viscosity make it difficult for the fuel to flow, thus causing poor atomization at the time of fuel spraying [13, 14].

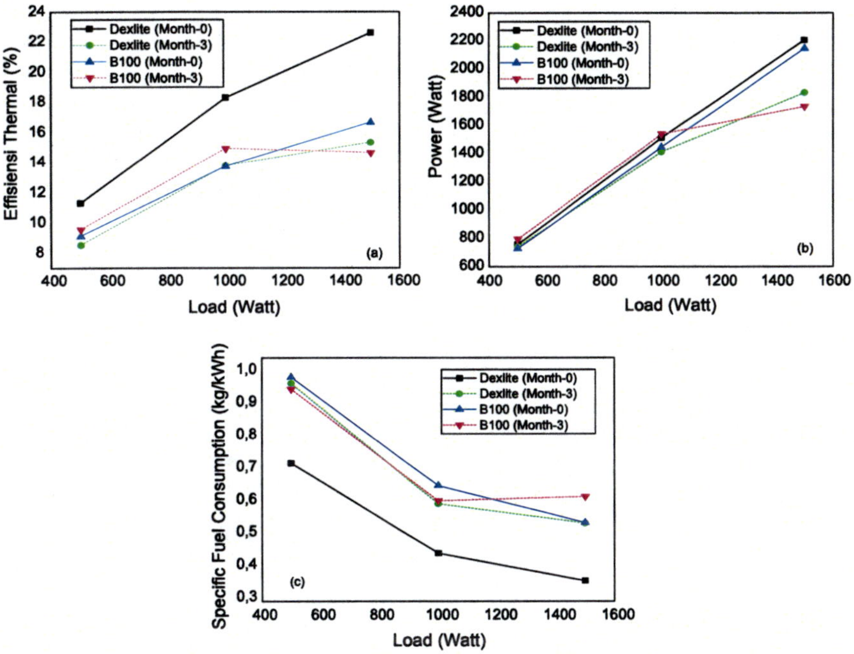

Fig. 3 **a** Thermal efficiency, **b** engine power, and **c** fuel consumption

The variation of the load on the power output of the engine can be seen in Fig 3b. Storage of fuel affects the decrease in engine power by 1833 W and 1733 W on Dexlite and B100 fuel. Decrease in engine power because the density and viscosity values have increased. This is also confirmed by [15].

Figure 3c shows a graph of the SFC against the load at different storage durations and fuel types. In Figure, it can be seen that the SFC in all fuels and storage times shows a decreasing trend with increasing the applied load. The decrease was caused by an increase in cylinder temperature which caused the injected fuel to be more flammable and converted into power produced. So, to produce the same amount of energy requires less fuel [16]. SFC increases Dexlite and B100 fuels by 34.45% and 5.70%, respectively.

3.3 Exhaust Emissions

Smoke emissions are often identified as thick smoke that occurs during the combustion process in diesel engines [17]. Figure 4 presents the increase in smoke emission as the engine load increases. A significant increase occurred in Dexlite fuel and was followed by B100 fuel. Smoke emission decreased significantly due to the long storage time of the fuel, decreasing by 3.8% and 2.3%, respectively, at 1500 W load. The decrease in smoke emissions is caused by the high oxygen content in B100 and the low sulfur content in biodiesel compared to fossil fuels. The oxygen content can help a better combustion process [18, 19].

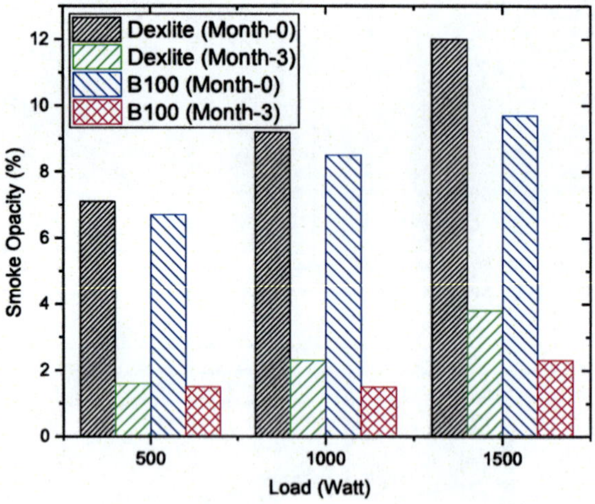

Fig. 4 Variation of load on smoke opacity emission

4 Conclusion

It can be concluded from research on the impact of fuel storage for 3 months on fuel properties, engine performance, and exhaust emissions of single-cylinder diesel engines at low load (500, 1000, 1500 W) are:

1. Storage of fuel results in degradation of physical and chemical properties, and this leads to changes in fuel properties (Dexlite and biodiesel). The Dexlite properties of water content, density, and viscosity increased by 82.86%, 0.98%, and 3.67%, respectively, and biodiesel (B100) increased by 18.35%, 0.29%, and 2.31%, respectively. On the other hand, the calorific value of Dexlite and biodiesel decreased by 0.72% and 0.59%, respectively.
2. Thermal efficiency and power engine using Dexlite decreased by 26.62% and 3.72%, respectively, and biodiesel decreased by 5.69% and 0.85%, respectively. However, the fuel consumption of Dexlite and biodiesel increased by 34.45% and 5.70%, respectively.
3. Smoke emissions of Dexlite and biodiesel decreased significantly by 21.4% and 17.6%, respectively.

Acknowledgments. On this occasion, the author would like to give appreciation to the Department of Mechanical Engineering, Sepuluh Nopember Institute of Technology. The author would also like to thank PT. Wilmar Nabati Indonesia has supported the implementation of this research.

References

1. BPPT, Indonesia Energy Outlook 2020—Special Edition Dampak Pandemi COVID-19 terhadap Sektor Energi di Indonesia. (2020).
2. Zakaria H, Khalid A, Sies MF, Mustaffa N, Manshoor B (2014) Effect of storage temperature and storage duration on biodiesel properties and characteristics. Appl Mech Mater 465–466:316–321
3. Bari S, Hossain SN (2019) Performance and emission analysis of a diesel engine running on palm oil diesel (POD). Energy Procedia 160(2018):92–99
4. Panjaitan SH (2021) Studi Eksperimen Pengaruh Penyimpanan Bahan Bakar Campuran Biodiesel Terhadap Sifat Bahan Bakar, Perfoma, Uji Ketahanan, Dan Emisi Mesin Diesel. [Online]. Available: https://repository.its.ac.id/id/eprint/83987.
5. Jalil RSN, Yusoff Z, Amat HHC (2020) Investigation on performance and emissions of a single cylinder marine diesel engine fuelled with diesel and crude palm biodiesel oil
6. Sanjid A, Masjuki HH, Kalam MA, Rahman SA, Abedin MJ, Palash SM (2014) Production of palm and jatropha based biodiesel and investigation of palm-jatropha combined blend properties, performance, exhaust emission and noise in an unmodified diesel engine. J Clean Prod 65:295–303
7. Silitonga AS, Masjuki HH, Mahlia TM, Ong HC, Chong WT, Boosroh MH (2013) Overview properties of biodiesel diesel blends from edible and non-edible feedstock. Renew Sustain Energy Rev 22:346–360
8. Ali OM, Mamat R, Abdullah NR, Abdullah AA (2015) Analysis of blended fuel properties and engine performance with palm biodiesel-diesel blended fuel. Renew Energy 86:59–67
9. Lee D (2014) Effects of soybean and canola oil-based biodiesel blends on spray, combustion, and emission characteristics in a diesel engine. J Energy Eng 140(3)
10. Akhabue CE, Osa-Benedict EO, Oyedoh EA, Otoikhian SK (2020) Development of a bio-based bifunctional catalyst for simultaneous esterification and transesterification of neem seed oil: modeling and optimization studies. Renew Energy 152:724–735
11. Mofijur M, Masjuki HH, Kalam MA, Atabani AE (2013) Evaluation of biodiesel blending, engine performance and emissions characteristics of jatropha curcas methyl ester: Malaysian perspective. Energy 55:879–887
12. Jaliliantabar F et al (2018) Comparative evaluation of physical and chemical properties, emission and combustion characteristics of brassica, cardoon, and coffee-based biodiesels as fuel in a compression-ignition engine. Fuel 222:156–174
13. Ge JC, Kim HY, Yoon SK, Choi NJ (2020) Optimization of palm oil biodiesel blends and engine operating parameters to improve performance and PM morphology in a common rail direct injection diesel engine. Fuel 260:116326
14. Gad MS, El-Shafay AS, Hashish HA (2021) Assessment of diesel engine performance, emissions and combustion characteristics burning biodiesel blends from jatropha seeds. Process Saf Environ Prot 147:518–526
15. Kalam MA, Masjuki HH, Jayed MH, Liaquat AM (2011) Emission and performance characteristics of an indirect ignition diesel engine fuelled with waste cooking oil. Energy 36(1):397–402

16. Jiaqiang E, Liu T, Yang WM, Li J, Gong J, Deng Y (2016) Effects of fatty acid methyl esters proportion on combustion and emission characteristics of a biodiesel fueled diesel engine. Energy Convers Manag 117:410–419
17. Nguyen HH (2020) The effect of biodiesel on combustion, performance and emission characteristics of Di diesel engine. Adv Mil Technol 15(2):221–230
18. Crabbe E, Nolasco-Hipolito C, Kobayashi G, Sonomoto K, Ishizaki A (2001) Biodiesel production from crude palm oil and evaluation of butanol extraction and fuel properties. Process Biochem 37(1):65–71
19. Ahmed M, Dincer I (2019) A review on photoelectrochemical hydrogen production systems: challenges and future directions. Int J Hydrogen Energy 44(5):2474–2507

Production of Bioethanol from Corn Straw by Co-immobilization of *Saccharomyces cerevisiae* and *Aspergillus niger* in Na-Alginate: Review and Potential Study

Karenina Anisya Pratiwi[1], Petra Arde Septia Graha[1], Shinta Dewi Surya Pertiwi[1], Yuliana Dewi Puspitasari[1], Muhammad Dimas Hafani[1], Afan Hamzah[1](✉), Arief Widjaja[2], and Soeprijanto[1]

[1] Departement of Industrial Chemical Engineering, Faculty of Vocational Studies, Institut Teknologi Sepuluh Nopember, Surabaya, Indonesia
afan@its.ac.id

[2] Departement of Chemical Engineering. Industrial and System Technology Faculty, Institut Teknologi Sepuluh Nopember, Surabaya, Indonesia

1 Introduction

Indonesia has a high dependence on fossil energy. The primary energy is still largely dominated by fossil fuels, oil (33.9%), coal (25.4%), and natural gas (15%), while renewable energy only covers 25.7% [1]. The continuous exploitation of fossil energy will burden the oil supply and increase CO2 emissions and environmental pollution [2]. Meanwhile, Indonesia's government policy No. 79 of 2014 states that the target for renewable energy in 2025 and 2050 will cover at least 23% and 31%, respectively. Bioethanol is considered the main source of renewable energy in the future due to its economic and environmental benefits [3]. The first generation of bioethanol did not preferable because it requires a lot of foods materials. Therefore, the manufacture of bioethanol from edible biomass was replaced with non-edible biomass, lignocellulose (bioethanol 2nd generation) [4]. Lignocellulosic biomass is the most abundant material in nature, and its cellulose content (35–50%) has the potential to be converted to low-cost biofuel [5]. Lignocellulose can be obtained from agricultural waste such as corn straw.

The ethanol fermentation process from lignocellulosic materials usually involves two main processes [2]. The first stage is starch liquefaction by amylase and enzymatic saccharification, and the next stage is glucose fermentation into ethanol. Adelabu (2019) converted lignocellulosic to bioethanol using cell immobilization techniques that have been widely used previously [6]. With the immobilization technique, the cells can be reused after the fermentation is complete by separating the cells from the product so that the production of bioethanol can run continuously [7]. It also increased the yield and quality of bioethanol. The co-immobilization technique offered a simpler process since it can be carried out in one stage where the microorganisms for hydrolysis and fermentation are simultaneously entrapped in one material support. This method can

M. Kolhe et al. (eds.), *Recent Advances in Renewable Energy Systems*,
Lecture Notes in Electrical Engineering 876,
https://doi.org/10.1007/978-981-19-1581-9_38

reduce energy input and increase substrate utilization efficiency [2]. Furthermore, the reusability of immobilized cells made this process more efficient. In this study, the co-immobilization technique was employed in the production of bioethanol from corn straw with less time, less energy, and efficient costs.

2 Material and Method

2.1 Material

Corn straw was obtained from Wlingi Market; potato dextrose agar (PDA), *A. niger*, *S. cerevisiae*, yeast extract, $(NH_4)_2SO_4$, KH_2PO_4, $CaCl_2.2H_2O$, $MgSO_4.7H_2O$, $ZnSO_4.7H_2O$, $MnSO_4.H_2O$, $FeSO_4.7H_2O$, NaCl, NaOH, Na-Alginate, $CaCl_2$, H_2SO_4, and acetate buffer were purchased from Sigma-Aldrich.

2.2 Microorganism Culture

A. niger or *S. cerevisiae* was cultured on PDA media aseptically and then incubated at 30 °C for 72 h. The microorganism then moved to liquid media. It was composed of yeast extract 5 g/L, $(NH_4)_2SO_4$ 0.7 g/L, KH_2PO_4 1 g/L, $CaCl_2.2H_2O$ 0.2 g/L, $MgSO_4.7H_2O$ 0.15 g/L, $ZnSO_4.7H_2O$ 2.5 g/L, $MnSO_4.H_2O$ 0.8 g/L, and $FeSO_4.7H_2O$ 0.7 g/L and corn straw. They were incubated at 120 rpm 30 °C for 72 h. The yeast was filtered using filter paper to remove corn straw. The yeast cells were then separated by centrifugation at 10,000 rpm for 15 min. The results were then precipitated in sterilized 0.9% NaCl [8].

2.3 Corn Straw Pre-treatment

Corn straw was washed, cut, dried in an oven at 100 °C for 24 h, and crushed into 60 mesh. Corn straw (50 g) is mixed with 800 mL 1% NaOH. The mixture was stirred and heated at 80 °C for 2.5 h. It was cooled and filtered; the solids were washed with hot water until neutral (pH 7). The solids obtained were oven at 100 °C to constant weight. The cellulose, hemicellulose, and lignin content in corn straw were determined using a method adapted from Datta et al. [9].

2.4 Co-immobilization

Na-Alginate 2% (w/v) and yeast 2% (w/v) with certain composition were mixed in 150 mL water. It was added into the 3% (w/v) $CaCl_2$ solution to form a bead gel with an average diameter of 3–4 mm. The bead gel granules formed were separated from the $CaCl_2$ solution using filter paper. Bead gel is stored at 4 °C for 24 h before use [10]. The bead gel morphology analysis was carried out using SEM [2].

2.5 Fermentation

In this stage, a review study was done to investigate the potential of this novelty and the fermentation process was planned to process as follow. Corn straw 10 (%, w/v) and co-immobilized bead gel (5% of corn straw) were mixed and incubated at pH 4 at 30 °C, and shaker speed was 150 rpm. Fermentation was carried out for 96 h [2].

3 Result and Discussion

3.1 Lignocellulose Composition of Corn Straw

Pretreatment was carried out to remove lignin content in corn straw because lignin can inhibit the hydrolysis of cellulose and hemicellulose. In this study, mechanical and chemical pretreatment were carried out. Mechanical pretreatment had the objective to reduce the size, while the chemical part diminish the lignin. After pretreatment, the lignin content decreased from 7.4 to 3.5%, as given in Table 1. Before pretreatment, the surface of the corn straw was intact and smooth with a compact structure and neatly arranged, and the cellulose and hemicellulose were tightly wrapped by lignin. After pretreatment, the surface becomes rough with many voids, and the wrapping structure is damaged. So removing lignin can significantly increase the access of microorganisms to the cellulose and hemicellulose so that the conversion of biomass into fuel becomes more effective [11].

Table 1 Lignocellulose composition of corn straw

No.	Corn straw content	Before pretreatment (%)	After pretreatment (%)
1	Cellulose	21.9	35.6
2	Hemicellulose	31.6	31.4
3	Lignin	7.4	3.5

3.2 Co-Immobilization of *A. niger* and *S. cerevisae*

The *A. niger* and *S. cerevisiae* were co-immobilized in Na-Alginate with a cell ratio of 1:1, 1:2, 2:1, 2:3, and 3:2 to produce bead gel containing them. The bead gel is white and gray in the largest ratio, and the average bead gel diameter is 2–4 mm. Figure 2 shows SEM micrograph of Na-Alginate which contains cells while the bead gel that does not contain cells is shown in Fig. 1. The significant difference between Figs. 1 and 2 is the white mark that indicates the cells had been immobilized. The selection of Na-Alginate as a supporting material for co-immobilization is due to its good biocompatibility, low cost, availability, and ease of preparation [10]. Bead gel will be employed in the production of bioethanol and has the potential to yield higher bioethanol yields. This is because Na-Alginate protects cells from possible contamination during the hydrolysis and fermentation processes so that cells are able to effectively perform [12]. In addition, the use of immobilized cells has a high substrate absorption rate, long-term cell activity, and stability, high tolerance to high substrate concentrations, reduced end-product inhibition, and cell protection from inhibitors, and easy product recovery [13].

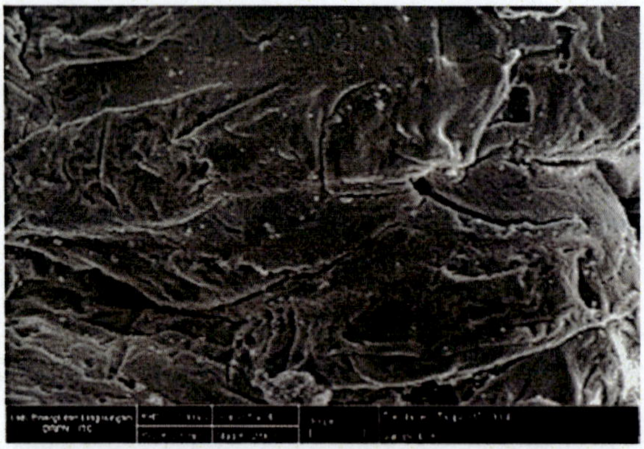

Fig. 1 SEM (1000x) of the gel bead does not contain cells

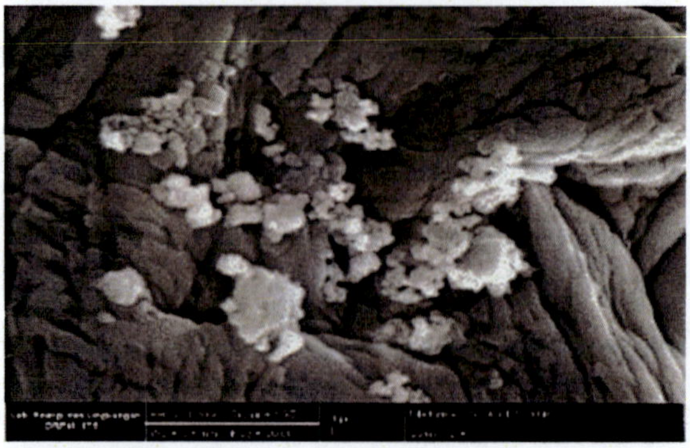

Fig. 2 SEM (1000x) of bead gel with A. Niger: S. cerevisiae = 1:1

3.3 Fermentation Potential Study

The co-immobilized cells will be employed for the fermentation of corn straw. The potential study was carried out before the fermentation process was done. Some studies state that the yields of products are heavily influenced by the ratio of the cells. Table 2 displays several studies about co-immobilized cell application in the hydrolysis. The studies indicate that hydrolysis was able to be implemented using co-immobilized cells and enzymes. Table 2 also shows the best ratio of cells to obtain maximum reducing sugar yield. The fermentation process of the studies still uses conservative ways and yields decent ethanol content. It shows that the yield of reducing sugar and bioethanol from cell immobilized is less compared to the immobilized enzymes. This is due to the ability of enzymes to produce more sugar due to their high selectivity compared to microorganisms [14].

The next studies shown in Table 3 relate to cells co-immobilization for fermentation and hydrolysis was in various technique such as enzymatic, acid, and recombinant cell hydrolysis. These studies indicate that the yeast cell co-immobilization was able to

Table 2 Cells and enzymes co-immobilized on gel bead application on hydrolysis

No	Substrate	Cells or enzymes	Ratio	Reducing sugar	Yield bioethanol	Reference
1	Carboxymethylcellulose (CMC)	*Trichoderma reesei* and *Aspergillus niger*	1:1	0.129%	5.6%	[15]
2	*Ipomoea carnea*	Laccase, cellulase and β-glucosidase	1:1:1	20.5 ± 0.373%	63.43 ± 9.35%	[16]
3	Sweet potato	*Aspergillus oryzae* and *Monascus purpureus*	1:2	57%	41%	[2]

produce decent bioethanol content from reducing sugar produced by various techniques of hydrolysis.

Table 4 shows several studies about the application of Cells and enzymes co-immobilized on gel beads on simultaneous saccharification and fermentation (SSF). The study indicates that hydrolysis and fermentation were able to coincide in one pot and still yield the decent number of bioethanol (11.48–81%).

However, the studies in Table 4 still utilize enzymes in the hydrolysis. The use of enzymes requires high cost; furthermore, the inactivated and denaturized enzymes may inhibit the entire process. Therefore, in this study, *A. niger* cells were applied for hydrolysis and *S. cerevisiae* cells for fermentation as exploitation of the entire cell will significantly reduce the cost. The optimum ratio of those two cells needs to figure out. The optimum ratio mainly will be affected by hydrolysis and fermentation rate as they run sequentially in the normal condition. The rate of sugar production must be faster than the rate of consumption; on the contrary, hydrolysis ought to run first and mostly rather than fermentation [2]. The utilization of cell co-immobilization for hydrolysis and fermentation in this study will yield bioethanol with more efficient time, cost, and energy.

Further investigation was carried out to compare SSF methods and Sequential Hydrolysis and Fermentation (SHF) employing cell and/or enzyme immobilization. Table 5 shows some research related to SSF and SHF implementation on bioethanol production. From Table 5, most of the SSF processes showed higher yields of bioethanol than the SHF process. In addition, SSF is more advantageous than SHF because of its fast ethanol production and higher ethanol concentration, as well as being more energy efficient [26]. Moreover, in the SHF, the conversion to ethanol is faster because sugar is available from the start, but the previous hydrolysis process took about 12 h. Maximum ethanol production occurs in about 12 h and then begins to decline. In contrast, the slow release of sugar from the SSF process makes the conversion to ethanol slower, but it is produced constantly so that it can achieve higher bioethanol yields [27]. Table 5 also shows some data that have higher yields in the SHF. It was due to its sequential process and longer reaction time [30].

The next investigation is about the reusability capability of co-immobilized cell on sodium alginate since it will decide the long-term stability of the cells. Several previous researches in Table 6 show reusability of immobilized yeast with a range of 5–7 reusability. The decrease of the bioethanol yields after some cycles is unavoidable due to cell leakage, and the activity of the cells will decline. Besides, the beads will also become brittle and deformed. The reuse of immobilized yeast has the potential to increase yield and lower operating costs [34].

4 Conclusion

Corn straw is biomass that consists of cellulose, hemicellulose, and lignin and may be used as a substrate to produce ethanol. Cells of *A. niger* and *S. cerevisiae* were co-immobilized in the sodium alginate beads. It was then employed in the SSF process to yield bioethanol. The review and potential study indicate that the overall process with an optimum ratio of cell immobilization is possible to be carried out and hold the potential to reduce the time, cost, and energy required to generate bioethanol.

Table 3 Cells and enzymes co-immobilized on gel beads application on fermentation

No	Substrate	Hydrolysis type	Fermentation	Ratio	Yield	References
1	*Cheese whey*	Enzymatic hydrolysis	*Kluyveromyces marxianus* dan *Saccharomyces cerevisiae*	3:1	40%	[17]
2	Corn Straw	Acid hydrolysis	*Saccharomyces cerevisiae* dan *Pachysolen tannophilus*	1:1	40% and 20.04 g/L	[18]
3	Wheat straw	Enzymatic hydrolysis	*Saccharomyces cerevisiae* dan *Scheffersomyces stipitis*	1:9	17.29%	[19]
4	Cotton stalk	Enzymatic hydrolysis	*Saccharomyces cerevisiae* dan *Pachysolen tannophilus*	1:1	56.4%	[13]
5	Wheat straw	Cell recombinant	*Saccharomyces cerevisiae* dan *Pichia pastoris*	1:1	42% 32.6 g/L	[20]

Table 4 Cells and enzymes co-immobilized on gel beads application on simultaneous saccharification and fermentation (SSF)

No	Substrate	Hydrolysis	Fermentation	Ratio	Yield (%)	References
1	(Sorghum bicolor)	Glucoamylase	*Yeast*	35:9	11.48	[21]
2	Sago starch	Amyloglucosidase (AMG)	*Zymomonas mobilis*	1: 1,2 (77,2 dan 93,2 g dry weight/l beads)	36.87	[22]
3	Starch	Amyloglucosidase (AMG)	*Zymomonas mobilis*	1:3,9 (7,5 and 29,2 g)	45	[23]
4	Starch	Glucoamylase	*Saccharomyces cerevisiae*	1:13 33 (1,5 and 20 g)	81	[24]

Table 5 Bioethanol yield from SSF and SHF processes

No	Substrate	Hydrolysis	Fermentation	*Yield*		References
				SSF	SHF	
1	Sugarcane bagasse	H_2SO_4	*Zymomonas mobilis* dan *Pichia stipitis*	0.414 g *ethanol/g cellulose* 81,17%	0.36 g ethanol/g cellulose 70.65%	[25]
2	Wheat straw	Nov-zyme and cell clast	*Saccharomyces cerevisiae* dan *Scheffersomyces stipitis*	17,29%	18.79%	[19]
3	Empty fruit bunch (EFB)	Cellic® CTec2 and Cellic® HTec2	*Saccharomyces cereviceae*	97%	76%	[26]
4	Rice husk	Cellic CTec2	*Saccharomyces cerevisiae*	32.6 ± 2.9%	30.2 ± 0.7%	[27]
5	Corn starch	SAN extra L	*Saccharomyces cerevisiae var. Ellipsoideus*	0.51 ± 0.007 g/g 90.32 ± 1.52%	0.43 ± 0.009 g/ 76.79 ± 1.54%	[28]
6	Sugar bagasse	Cellulosic enzymes	*Zymomonas mobilis*	70,09%	79.09%	[29]

Table 6 Reusability of immobilized yeast in bioethanol production

No	Substrate	Cycle	Yield	References
1	Cellobiose	7	60%	[31]
2	Corn straw	5	12%	[4]
3	Sugar beet pulp	7	0.446 ± 0.017 g/g 87.30 ± 0.03%	[8]
4	Batang kapas	5	25,1%	[13]
5	Lignocellulose biomass	6	35,1%	[32]
6	Rice straw	5	84.7%	[33]

Acknowledgements. The authors gratefully acknowledge the financial support by Research and Higher Education Directorate of the Ministry of Education and Culture of Research.

References

1. Khatiwada D, Silveira S (2017) Scenarios for bioethanol production in Indonesia: how can we meet mandatory blending targets. Energy 119:351–361. https://doi.org/10.1016/j.energy.2016.12.073
2. Lee WS, Chen IC, Chang CH, Yang SS (2012) Bioethanol production from sweet potato by co-immobilization of saccharolytic molds and Saccharomyces cerevisiae. Renew Energy 39:216–222. https://doi.org/10.1016/j.renene.2011.08.024
3. Mingxiong H, Feng H (2009) Direct production of ethanol from raw sweet potato starch using genetically engineered Zymomonas mobilis. African J Microbiol Res 3:721–726
4. Adelabu BA, Kareem SO, Oluwafemi F, Abideen Adeogun I (2019) Bioconversion of corn straw to ethanol by cellulolytic yeasts immobilized in mucuna urens matrix. J King Saud Univ Sci 31:136–141. https://doi.org/10.1016/j.jksus.2017.07.005
5. Pang J, Liu ZY, Hao M, Zhang YF, Qi QS (2017) An isolated cellulolytic Escherichia coli from bovine rumen produces ethanol and hydrogen from corn straw. Biotechnol Biofuels 10:1–10. https://doi.org/10.1186/s13068-017-0852-7
6. Marlinda Ramli Ardis (2019) PENGARUH KONSENTRASI IMOBILISASI SEL SACCHAROMYCES CEREVISIAE PADA PEMBUATAN BIOETANOL DARI NIRA NIPAH. In: Prosiding Seminar Nasional Penelitian & Pengabdian Kepada Masyarakat. pp 135–139
7. Awaltanova E, Bahri S (2015) Fermentasi nira nipah menjadi bioetanol menggunakan teknik immobilisasi Sel Saccharomyces cerevisiae. Jom fteknik 2
8. Vučurović VM, Razmovski RN (2012) Sugar beet pulp as support for Saccharomyces cerivisiae immobilization in bioethanol production. Ind Crops Prod 39:128–134. https://doi.org/10.1016/j.indcrop.2012.02.002
9. Datta R (1981) Acidogenic fermentation of lignocellulose–acid yield and conversion of components. Biotechnol Bioeng 23:2167–2170. https://doi.org/10.1002/bit.260230921
10. Pathania S, Sharma N, Handa S (2017) Immobilization of co-culture of Saccharomyces cerevisiae and Scheffersomyces stipitis in sodium alginate for bioethanol production using hydrolysate of apple pomace under separate hydrolysis and fermentation. Biocatal Biotransform 35:450–459. https://doi.org/10.1080/10242422.2017.1368497

11. Liu Z, Li L, Liu C, Xu A (2018) Pretreatment of corn straw using the alkaline solution of ionic liquids. Bioresour Technol 260:417–420. https://doi.org/10.1016/j.biortech.2018.03.117
12. Rattanapan A, Limtong S, Phisalaphong M (2011) Ethanol production by repeated batch and continuous fermentations of blackstrap molasses using immobilized yeast cells on thin-shell silk cocoons. Appl Energy 88:4400–4404. https://doi.org/10.1016/j.apenergy.2011.05.020
13. Malik K, Salama ES, El-Dalatony MM, Jalalah M, Harraz FA, Al-Assiri MS, Zheng Y, Sharma P, Li X (2021) Co-fermentation of immobilized yeasts boosted bioethanol production from pretreated cotton stalk lignocellulosic biomass: Long-term investigation. Ind Crops Prod 159:113122. https://doi.org/10.1016/j.indcrop.2020.113122
14. Binod P, Janu KU, Sindhu R, Pandey A (2011) Hydrolysis of lignocellulosic biomass for bioethanol production. Elsevier Inc
15. Liu YK, Yang CA, Chen WC, Wei YH (2012) Producing bioethanol from cellulosic hydrolyzate via co-immobilized cultivation strategy. J Biosci Bioeng 114:198–203. https://doi.org/10.1016/j.jbiosc.2012.03.005
16. Kirupa Sankar M, Ravikumar R, Naresh Kumar M, Sivakumar U (2018) Development of co-immobilized tri-enzyme biocatalytic system for one-pot pretreatment of four different perennial lignocellulosic biomass and evaluation of their bioethanol production potential. Bioresour Technol 269:227–236. https://doi.org/10.1016/j.biortech.2018.08.091
17. Beniwal A, Saini P, Kokkiligadda A, Vij S (2018) Use of silicon dioxide nanoparticles for β-galactosidase immobilization and modulated ethanol production by co-immobilized K. marxianus and S. cerevisiae in deproteinized cheese whey. Lwt 87:553–561. https://doi.org/10.1016/j.lwt.2017.09.028
18. Zhang W, Bai A, Chen X, Wei G (2012) Ethanol production from lignocelluloses hydrolyzates with immobilized multi-microorganisms. Energy Sources Part A Recover Util Environ Eff 34:1206–1212. https://doi.org/10.1080/15567031003681960
19. Karagöz P, Özkan M (2014) Ethanol production from wheat straw by Saccharomyces cerevisiae and Scheffersomyces stipitis co-culture in batch and continuous system. Bioresour Technol 158:286–293. https://doi.org/10.1016/j.biortech.2014.02.022
20. Zhang Y, Wang C, Wang L, Yang R, Hou P, Liu J (2017) Direct bioethanol production from wheat straw using xylose/glucose co-fermentation by co-culture of two recombinant yeasts. J Ind Microbiol Biotechnol 44:453–464. https://doi.org/10.1007/s10295-016-1893-9
21. Dyartanti ER, Pranolo SH, Setiani B, Nurhayati A (2015) Bioethanol from sorghum grain (Sorghum bicolor) with SSF reaction using biocatalyst co-immobilization method of glucoamylase and yeast. Energy Procedia 68:132–137. https://doi.org/10.1016/j.egypro.2015.03.241
22. Bandaru VVR, Somalanka SR, Mendu DR, Madicherla NR, Chityala A (2006) Optimization of fermentation conditions for the production of ethanol from sago starch by co-immobilized amyloglucosidase and cells of Zymomonas mobilis using response surface methodology. Enzyme Microb Technol 38:209–214. https://doi.org/10.1016/j.enzmictec.2005.06.002
23. Altuntaş EG, Ö; zçelik F (2014) Ethanol production from starch by co-immobilized amyloglucosidase—zymomonas mobilis cells in a continuously-stirred bioreactor. Biotechnol Biotechnol Equip
24. Giordano RLC, Trovati J, Schmidell W (2008) Continuous production of ethanol from starch using glucoamylase and yeast co-immobilized in pectin gel. Appl Biochem Biotechnol 147:47–61. https://doi.org/10.1007/s12010-007-8067-1
25. Wirawan F, Cheng CL, Lo YC, Chen CY, Chang JS, Leu SY, Lee DJ (2020) Continuous cellulosic bioethanol co-fermentation by immobilized Zymomonas mobilis and suspended Pichia stipitis in a two-stage process. Appl Energy 266:114871. https://doi.org/10.1016/j.apenergy.2020.114871

26. Dahnum D, Tasum SO, Triwahyuni E, Nurdin M, Abimanyu H (2015) Comparison of SHF and SSF processes using enzyme and dry yeast for optimization of bioethanol production from empty fruit bunch. Energy Procedia 68:107–116. https://doi.org/10.1016/j.egypro.2015.03.238
27. Arismendy Pabón AM, Felissia FE, Mendieta CM, Chamorro ER, Area MC (2020) Improvement of bioethanol production from rice husks. Cellul Chem Technol 54:689–698. https://doi.org/10.35812/CelluloseChemTechnol.2020.54.68
28. Nikolić S, Mojović L, Rakin M, Pejin D (2009) Bioethanol production from corn meal by simultaneous enzymatic saccharification and fermentation with immobilized cells of Saccharomyces cerevisiae var. ellipsoideus. Fuel 88:1602–1607. https://doi.org/10.1016/j.fuel.2008.12.019
29. Wirawan F, Cheng CL, Kao WC, Lee DJ, Chang JS (2012) Cellulosic ethanol production performance with SSF and SHF processes using immobilized Zymomonas mobilis. Appl Energy 100:19–26. https://doi.org/10.1016/j.apenergy.2012.04.032
30. Kim JH, Lee JC, Pak D (2011) Feasibility of producing ethanol from food waste. Waste Manag 31:2121–2125. https://doi.org/10.1016/j.wasman.2011.04.011
31. He B, Zhu X, Zhao C, Ma Y, Yang W (2018) Sequential co-immobilization of β-glucosidase and yeast cells on single polymer support for bioethanol production. Sci China Chem 61:1600–1608. https://doi.org/10.1007/s11426-018-9319-1
32. Ashoor S, Comitini F, Ciani M (2015) Cell-recycle batch process of Scheffersomyces stipitis and Saccharomyces cerevisiae co-culture for second generation bioethanol production. Biotechnol Lett 37:2213–2218. https://doi.org/10.1007/s10529-015-1919-9
33. Watanabe I, Miyata N, Ando A, Shiroma R, Tokuyasu K, Nakamura T (2012) Ethanol production by repeated-batch simultaneous saccharification and fermentation (SSF) of alkali-treated rice straw using immobilized Saccharomyces cerevisiae cells. Bioresour Technol 123:695–698. https://doi.org/10.1016/j.biortech.2012.07.052
34. Amutha R, Gunasekaran P (2001) Production of ethanol from liquefied cassava starch using co-immobilized cells of Zymomonas mobilis and Saccharomyces diastaticus. J Biosci Bioeng 92:560–564. https://doi.org/10.1016/S1389-1723(01)80316-9

Analysis of the Characteristic and Performance Development of Coconut Biodiesel

Tri Vicca Kusumadewi[1](✉) and Digdo Listyadi Setyawan[2]

[1] Department of Mechanical Engineering, Sepuluh Nopember Institute of Technology (ITS), Surabaya, Indonesia
vicca@its.ac.id
[2] Department of Mechanical Engineering, University of Jember, Jember, Indonesia

1 Introduction

Indonesia is a tropical country with many potentials of energy scattered across various islands. As the way to meet zero-emission, Indonesia developed energy conservation such as the development of renewable energy. Biodiesel from coconut is one of renewable energy development as a biofuel. Coconut plants in Indonesia can be found in the tropical area and produced coconut by 18.30 million tons within a year [1]. Fuel consumption in Indonesia is increasing due to population and economic growth. The availability of fuel is very important in realizing more advanced economic development. Thus, research on the provision and consumption of fuel in Indonesia is very important and interesting to do. The energy conservation development to supply fuel in Indonesia continues to be developed to meet the needs of clean fuel energy. The development of biodiesel in Indonesia has been carried out since Indonesia declare that renewable energy shared in 2030 will be increase to 26%. The biodiesel from japatra, candlenut, cooking waste, and coconut has been developed. These materials can be used as biodiesel because of the content of vegetable oil. Coconut is also the basic ingredient of making biodiesel. In coconut seeds, it contains 50% fatty acids and 7% capric acid, both of these can be converted into very beneficial energy [2]. Making biodiesel from natural ingredients or vegetable oils has better combustion properties and is environmentally friendly. In addition, extensive plantation land in Indonesia is a positive point for the development of biodiesel.

Nowadays, biodiesel has been produced much better quality than before. Various examinations, research, and improvements are carried out by biodiesel producers to improve the quality of biodiesel to be able to meet user demand and meet biodiesel specifications set by the government [3]. As fuel, biodiesel must have standard characteristics that are following Indonesian national standards such as flame, flash point, density, viscosity, and others. In addition to examining the standard of the characteristics of biodiesel, the combustion characteristics examination must be examined to determine the characteristics of fire in biodiesel, namely, the temperature, the height cone, and the color of the flame. Types of flame flow there is laminar and turbulent flow, which are distinguished by the form of streamlined flow with regular or random movements. The

M. Kolhe et al. (eds.), *Recent Advances in Renewable Energy Systems*,
Lecture Notes in Electrical Engineering 876,
https://doi.org/10.1007/978-981-19-1581-9_39

height of the flame cone shows the structure of the flame where this height depends on the flame flow, the ratio of fuel and air, temperature, and combustion zone [4]. The color of the fire can be affected by the fuel content and the mixture of air that goes on the flame. The color of the flame which tends to be red is caused by the lack of oxygen in the combustion process thereby reducing the heating value, while the color of the flame tends to be blue due to the amount of oxygen in the combustion process [5]. Also, the reduction happens because the biodiesel fuel has a high density so that the fuel is difficult to mix and burn. Digdo [6] examined the candlenut biodiesel mixed by diesel oil to find the characteristic of the flame in the various air–fuel ratio, percentage of biodiesel with the premix combustion examination. The result estimated that the temperature will be decreased with the addition of biodiesel due to the calorific value of biodiesel lower than diesel oil.

The purpose of this study was to determine the effect of biodiesel and Pertamina dex mixtures on the flame temperature distribution in the combustion process of the fuel and the performance of the biodiesel as well. The result and discussion will be presented in the paper.

2 Methodology

This research includes two main activities, namely, the production of coconut biodiesel and experimental of the flame characteristics and performance examination. In the process of making biodiesel, coconut has been sliced or shredded and dried in the sunshine within 2–3 days. After that, the oil is distilled using a pressing machine. Then, the pure coconut oil reacted with KOH as a catalyst to become biodiesel and boiled at a temperature of 65 °C. This process called transesterification, which is the process of exchanging the organic group of an ester with the organic group of an alcohol, these reactions are often catalyzed by the addition of an acid or base catalyst.

The combustion process to examine the characteristic of the flame. The combustion is a chemical reaction between a fuel and an oxidizer accompanied by the production of heat and light in the form of phosphorescence or fire. The biodiesel flame characteristics were tested such as flame color, temperature, and flame cone height. Pertamina dex must be prepared for the flame characteristics experimental mixed by coconut biodiesel and divided the percentage of the coconut biodiesel B0 (Pertamina dex 100%), B5 (biodiesel mixed by 5%), and B10 (biodiesel mixed by 10%). The equivalent ratio (ϕ) used is 1 with the discharge fuel flow of 2 ml/min. Based on stoichiometric calculations, for an ideal combustion reaction in biodiesel fuel, 14.86 mol of air and 1 mol of biodiesel fuel are needed. The equivalent ratio is the ratio between the stoichiometric air–fuel ratio and the actual air–fuel ratio (AFR). Between the variation of AFR $\phi = 1.5$, $\phi = 1$ and $\phi = 0.5$, the decreasing of the flame color will be steadily than $\phi = 1.5$ [6]. The equipment for the combustion examination can be shown in Fig. 1.

The premixed combustion examination was carried out to get the flame data [7]. The examination carried out to determine the flame temperature at 3 points by using thermocouple type *K*, the position of thermocouple shows in Fig. 1. Thermocouple type *K* used to examine the flame temperature at the 3 points namely TC1, TC2, and TC3. The camera Image-J used to examine the flame color. Then the performance of the biodiesel

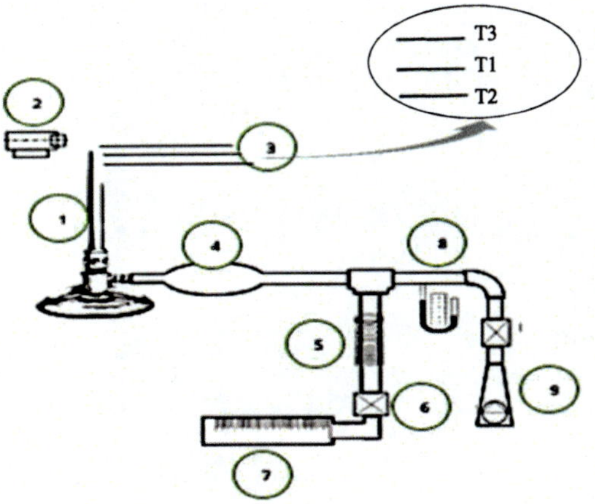

Fig. 1 Equipment for combustion examination

B0, B5, and B10 examined by Diesel Nissan SD 22 series with the different revolution start from 1600 to 2500 rpm.

3 Result and Discussion

The results show the characteristic of coconut biodiesel meets the National Standard of Indonesia (SNI). The data can be seen in Table 1.

Table 1 The characteristic of coconut biodiesel

Examination	Coconut biodiesel	SNI standard (7182:2015)
Heat (calorie/gr)	9319.55	9536.8
Density, D1298/D4052 (gr/ml)	0.881	0.85–0.89
Viscosity, D445 (mm^2/s (cSt))	5.191	2.3–6.0
Flashpoint, D93 (^oC, min)	123	100

The flame characteristics of the coconut biodiesel mixed with Pertamina dex carried out the flame color data collection to determine the height of the flame cone from each component of the fuel mixture with an equivalent ratio (ϕ) 1, which can be seen in Fig. 2.

It can be seen that B0, B5, and B10 has blue flame color even with the increase of 5% biodiesel composition in the fuel mixture. The blue color indicates that the combustion is near stoichiometry condition. The blue color will be decrease steadily along with the increasing the coconut biodiesel content. The increase in the percentage coconut biodiesel will be caused density of the fuel to increase. This condition will make the fuel difficult to burn and the temperature also will be decreased.

In Fig. 3, from the results of the characteristics of the flame premixed combustions on biodiesel B0, B5, and B10 with the equivalent ratio $\phi = 1$, the flame temperature data

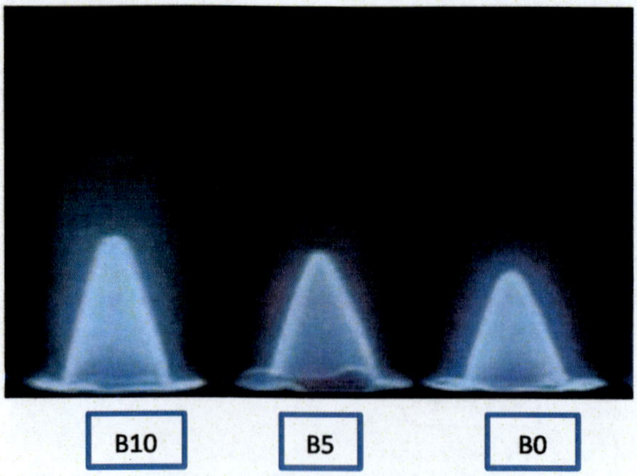

Fig. 2 The flame color with equivalent ratio (ϕ) 1

measurement for each thermocouple point that the highest flame temperature was found in the biodiesel 100% (B0) at the thermocouple 1 (TC1) which are placed near inner flame by 1142 °C and the lowest temperature is B10 at the thermocouple 3 (TC3) which are placed above TC1 and TC2 by 1067 °C. For the trend line of the flame temperature to the percentage of biodiesel mixing, it can be seen that the more percentage biodiesel mixture, the temperature value will be decreased. The position of the thermocouple also affects the temperature values. T1 in all biodiesel mixtures has the highest value because of its position in the middle of the fire point.

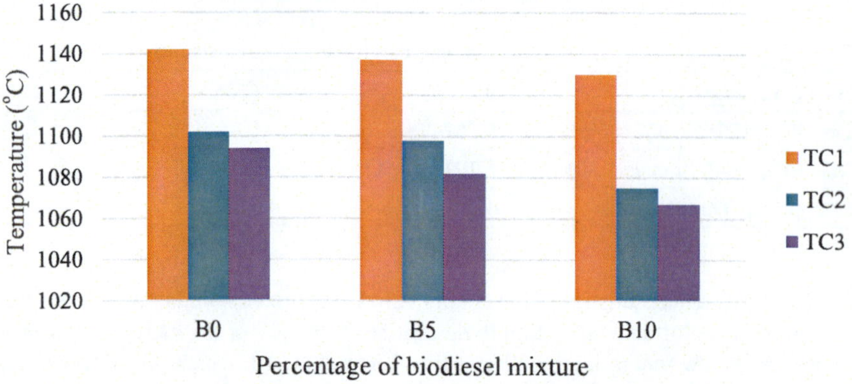

Fig. 3 Fuel flame temperature at the T1, T2, and T3

Besides knowing the temperature of the flame, the height of the flame cone will also be known. The height of the flame cone is divided into two parts, namely, the height of the outer and inner flame cone. The chart of flame inner cone with the equivalent ratio $\phi = 1$ can be shown in Fig. 4.

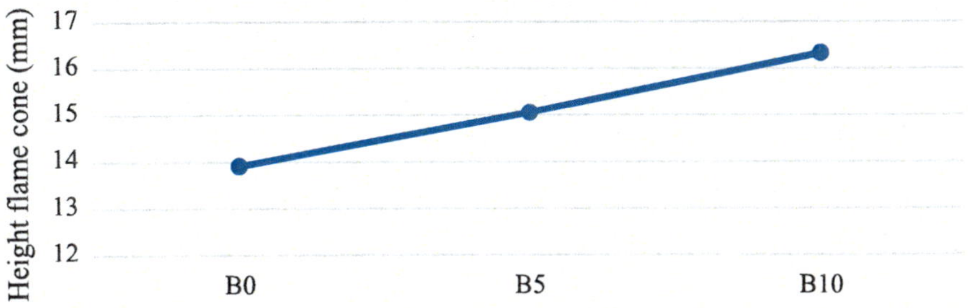

Fig. 4 Inner flame cone height

From the trend line above, it shown that the highest flame cone height is B10 by 16.33 mm. The lowest flame cone height is B0 by 13.91 mm. B0 has short cone due to the density of Pertamina dex. The lack of an optimal combustion process due to high fuel density so that the combustion results tend to form a diffuse flame that has a reddish color. This causes the combustion process to be not optimal because the high density can make fuel more difficult to evaporate and burn so that the energy produced in combustion will be reduced.

Biodiesel performance depends on the biodiesel mixtures and the petroleum diesel characteristic. The performance of biodiesel can show by the torque, power, and specific fuel consumption (SFC) biodiesel in each rotation per minute, which are given. The result will show in Table 2 and Fig. 5.

The result show the torque will be on the maximum at 1800 rpm for all biodiesel mixtures. The increased engine revolution will lead to an increasing in power generation because the load provided is also increased. In general, increased engine revolutions (rpm) lead to an increased fuel consumption because at high revolutions the combustion process occurs so fast that the mixture of air with fuel cannot burn perfectly as shown in Fig. 6. The coconut biodiesel performance at the 2500 rpm, the fuel consumption B0 is the highest in each rotation (rpm) but at the 2300 rpm the consumption will be decreased at the point 0.30 kg/kWh. B10 is the lowest in each rotation (rpm) especially at the 1600 rpm by 0.27 kg/kWh. B0 consumed 0.341 kg/kWh, and the power of biodiesel was 30.33 kW. B5 and B10 consumed 0.29 kg/kWh with the power of biodiesel by 30.56 kW and 30.33 kW, respectively. The average specific fuel consumption B0, B5, and B10 as follows 0.318 kg/KWh 0.284 kg/KWh, and 0.28 kg/KWh, respectively. The specific fuel consumption will be decreased due to increasing the biodiesel percentage mixtures. From the average of specific fuel consumption B10 could save fuel by 11.95%.

4 Conclusion

Here is the following conclusion can be shown from this research:

1. The temperature of biodiesel will be decreased due to the increase of the biodiesel percentage mixture.

Table 2 Torque, power, and SFC at the different revolution (rpm)

Fuel	rpm	Torque (Nm)	Power (KW)	SFC (kg/kWh)
B0	2500	115.90	30.33	0.34
	2300	117.65	28.32	0.30
	2000	119.41	25.00	0.33
	1800	119.41	22.50	0.31
	1600	117.65	19.70	0.31
B5	2500	116.77	30.56	0.29
	2300	118.53	28.53	0.29
	2000	121.16	25.36	0.28
	1800	122.92	23.16	0.28
	1600	121.16	20.29	0.28
B10	2500	115.90	30.33	0.29
	2300	117.65	28.32	0.28
	2000	120.29	25.18	0.28
	1800	122.04	22.99	0.28
	1600	120.29	20.14	0.27

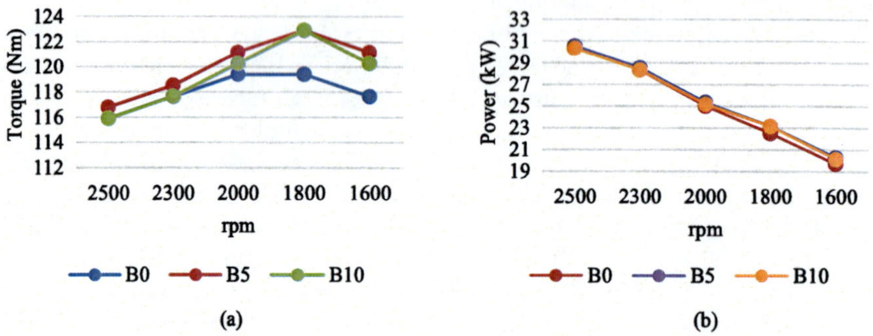

Fig. 5 **a** Torque versus revolution, **b** power versus revolution

2. The opposite of the temperature, the flame height cone will be increased due to the increasing of the biodiesel percentage mixtures due to biodiesel evaporation temperatures higher than Pertamina dex.
3. The performance of B10 is better than B0 and B5 because the specific fuel consumption (SFC) is the lowest one. So, the more we increase the biodiesel content, the SFC will be reduced.

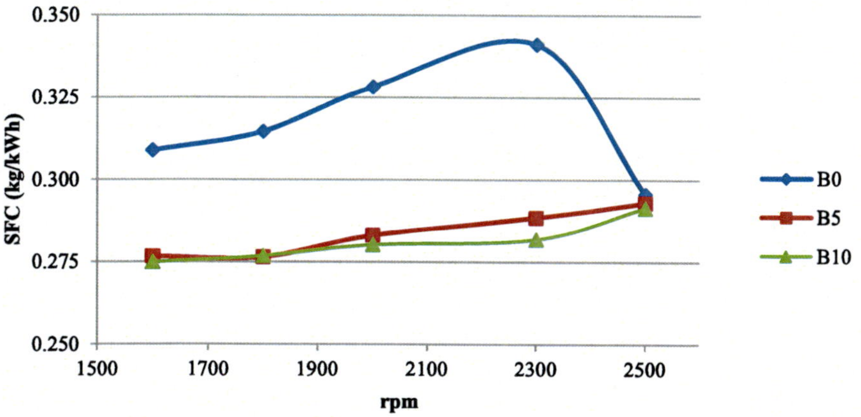

Fig. 6 Specific fuel consumption versus rpm

Acknowledgements. The authors would like to thank the University of Jember for the foundation supports on this research

References

1. Mardiatmoko G, Ariyanti M (2018) Production of coconut plants (Cocos nucifera L.), Faculty of agriculture, Ambon
2. Riwu DB (2016) Pengaruh penambahan LPG (liquified petroleum gas) pada proses pembakaran premixed uap minyak jarak pagar terhadap warna dan temperatur api, Jurnal Lontar 3:55–60
3. Khoiria FO, Dedi ZS, Fauzia L, Tiara HS, Anggono TI (2021) Gusti Ngurah Agung Surya Pradipta Negara, Biodiesel, Jejak panjanh sebuah perjuangan, Badan Litbang ESDM
4. Hu S, Gao J, Gong C, Zhou Y, Bai DX (2017) Assessment of uncertainties of the laminar flame speed of premixed flames as determined using a bunsen burner at varying pressures. J Appl Energy 9:100–110
5. Agarwal AK (2006) Biofuels applications as fuels for internal combustion engines, progress in energy and combustion science. J Energy Fuels 8:1–38
6. Mahandari CP (2012) Fenomena Flame Lift—Up Pada Pembakaran premixed Gas Propana, Disertasi Program Pascasarjana Fakultas Teknik Universitas Indonesia, Indonesia
7. Digdo LS, Nasrul I, Hary S, Tri Vicca K, Raka R (2020) Analysis of mixed premixed combustion characteristics of biodiesel candlenut oil (Aleurites Moluccana) with biodiesel fuel. In: AIP conference proceedings, 2278, 1, article id 020027

The Analysis of Biogas Production from Mixed Cow Dung with Watermelon Rind as an Alternative Fuel

Faisal Manta(✉), Doddy Suanggana, Putra Dilto Tondok, and Firman Ali Nuryanto

Department of Mechanical Engineering, Kalimantan Institute of Technology, Balikpapan, Indonesia
faisal86@lecturer.itk.ac.id

1 Introduction

In 2007, the government of the Republic of Indonesia announced a policy to change the use of kerosene to liquefied petroleum gas (LPG) due to the depletion of Indonesia's oil reserves and the increasing price of crude oil in the international market. According to 2006 ESDM data, Indonesia's oil reserves are only nine billion barrels left, and it is estimated that they will run out in the next two decades if they continue to be exploited without finding alternative energy [1]. An alternative energy that is quite easy to apply in everyday life is biogas. Biogas is a technology that produces alternative energy that can be renewed and comes from anaerobic degradation [2]. Biogas is a gas fuel that is feasible to develop because it is renewable and can be produced with simpler technology. In addition to producing biogas, the manure in the holding tank can also be used as fertilizer. From an economic point of view, the cost of biogas technology is highly dependent on the raw materials and raw materials used to make a biogas digester. In general, biogas technology is very cost effective if organic raw materials are available at low prices, and biogas digester tanks can be made by using local materials [3].

1.1 Biogas

Biogas is a renewable energy produced through the decomposition of organic matter by bacteria. Biogas can be made from household waste, agricultural waste, and industrial waste in anaerobic digestion [3]. The basic principle of biogas technology is the process of decomposition of organic materials by microorganisms in conditions without air (anaerobic) to produce a mixture of several gases, including methane and CO_2 [4]. The function of an anaerobic digester is to be a place to carry out biogas processing [5].

1.2 Biogas Process

Basically, almost all types of organic substances can be processed into biogas. For simple biogas production, the most commonly used organic materials in Indonesia are animal dung and urine. The choice of biogas material can be determined from the ratio of

M. Kolhe et al. (eds.), *Recent Advances in Renewable Energy Systems*,
Lecture Notes in Electrical Engineering 876,
https://doi.org/10.1007/978-981-19-1581-9_40

Table 1 Biogas composition

Composition	Percentage (%)
CH_4	81.1
CO_2	14.0
H_2S	2.2
$N_2 + O_2$	2.7

the C (carbon) and N (nitrogen) content in the material [6]. Table 1 gives the biogas composition [7].

Several parameters influence the formation of biogas, including digester temperature, pH (acid level), pressure, and humidity. The optimal temperature of the digester is in the range of 30–35 °C. This temperature is the best condition for bacterial growth and methane formation in the digester. It is very important to keep the temperature stable so that the formation of biogas is maximal [8]. The level of acidity is very influential on anaerobic decomposition because at an inappropriate pH, microbes cannot grow optimally and can cause microbial death. This condition can inhibit the formation of methane gas [9]. The availability of essential nutrients as nutrients for anaerobic bacteria, lack of nutrients will inhibit the growth of bacteria in the digester. Additional nutrients such as glucose can encourage bacterial growth [6].

Stirring is useful for mixing evenly the material contained in the digester and to maintain an even temperature in the digester. Stirring can also minimize the occurrence of the material deposition so that all materials undergo a perfect fermentation process [10]. The digester has five parts, including the inlet as a place for the entry of biogas raw materials, a reactor that functions as a place for anaerobic processes, a gas reservoir that functions as a gas reservoir, an outlet that functions as a separator and finally, the slurry serves as a drain hole. Cow dung is a waste product of cow digestion. Cow dung can be defined as solid waste from livestock which is usually mixed with urine or gas. Nutrient content depends on the amount produced, feed, and type of livestock [11].

The anaerobic decomposition process in the digester is assisted by several microorganisms. At a temperature of 30–55 °C, microorganisms can work optimally to remodel organic materials in the digester. The anaerobic process in the digester consists of three stages, among others [12].

Hydrolysis (Dissolution), in this stage, the easily soluble materials are decomposed with the help of water. This decomposed compound is cellulose with the help of exoenzymes from anaerobic bacteria converted into monomers [13].

The reaction of cellulose to glucose is as follows:

$$(C_6H_{10}O_5)n + n\,H_2O \rightarrow n\,C_6H_{12}O_6$$

Acetogenesis (Acidification), the sugar produced by the hydrolysis process is eaten by anaerobic bacteria and then forms acid. The decomposition of this monomer produces acetic acid, propionic acid, formic acid, lactic, alcohol, carbon dioxide, hydrogen, and ammonia [13].

The reactions that occur at this stage are:

$$n\, C_6H_{12}O_6 \rightarrow 2n\left(C_2H^5OH\right) + 2n\, CO_2(g) + \text{Heat}$$

$$2n\,(C_2H_5OH) + n(g)\, CO_2 \rightarrow 2n\,(CH_3COOH)\,(aq) + n\, CH_4(g)$$

Methanogenesis, in this stage, methane gas is formed, assisted by the bacteria Methanococcus, Methanobacillus, and Methanobacterium. This gas is created from the reaction of acetate decarboxylation and CO_2 reduction [13].

The reaction at this stage is

$$2n\,(CH_3COOH)(aq) \rightarrow 2n\, CH_4(g) + 2n\, CO_2(g)$$

These three stages form methane gas with a small amount of carbon dioxide into biogas.

2 Methods

There are several variables used for biogas analysis from a mixture of cow dung and watermelon rind. The variables used to obtain experimental data in this study are listed in Table 2.

Table 2 Variable experimental

Variable	Cow dung (%)	Water melon (%)	Water (%)
d1	40	30	30
d2	30	40	30
d3	35	35	30

The process of making biogas begins with the preparation of tools to be used, including digesters, PVC pipes, and measuring instruments (manometer and thermometer). After all the tools are available, then assemble them all into a digester. Before being put into the digester, mix cow dung first with the watermelon skin that has been chopped or mashed within a ratio of 40%:30% (d1), 30%:40% (d2), and 35%:35% (d3) with the addition of 30% water in each variation. After the variations of the mixture are ready, the next step is to put the substrate into the digester and close the digester until it becomes airtight (anaerobic). Then, place the digester in the exact position that has been provided. Data were collected twice a day at 10.00 WITA and 17.00 WITA for the whole 30 days. After all the data is obtained, an analysis will be carried out to determine the best variation among those three. The data collected at those hours consisted of Table 3.

Table 3 Data collected

Variable	Point	Periods
Digester temperature	°C	30 day
Environment temperature	°C	
Digester pressure	Bar	
Biogas mass	gram	

3 Results

3.1 Ambient and Digester Temperature

It is a graph of ambient temperature and digester temperature at 10.00 WITA and 17.00 WITA. In Fig. 2, it is known that the highest environmental temperature data is on the fourth and 24th day simultaneously, which is 35 °C, whereas the lowest is on the 25th day, which is 25° and has 31.6 °C as the average temperature. Then at d1, the highest temperature was found on the fourth day which is 36.3 °C, the lowest was observed at 26.7 °C for the 25th day and has 32.28 °C as the average temperature. Meanwhile, at d2, the highest temperature was found on the 24th day which is 36 °C, the lowest is observed at 26.3 °C on the 25th day and the average temperature is 32.27 °C. The last one, d3, has the highest temperature on the 24th day of 35.3, the lowest on the 25th day of 26 °C and the average temperature is 32.45 °C. Figure 3, it is known that the highest ambient temperature is on the 29th day which is 32 °C, and the lowest on the 25th and 26th day which is 26 °C and has an average of 29.2 °C. At d1, the highest temperature was found on the 29th day which is 32.6 °C, the lowest was found on the 26th day which is 26.4 °C and has 29.94 °C as the average temperature. Meanwhile, at d2, the highest temperature is found on the 29th day which is 33.5 °C, the lowest is found on the 25th day which is 26.9 °C and has an average temperature of 29.94 °C. On the last variation d3, the highest temperature was found on the 29th day which is 33.6, whereas the lowest was found on the 26th day which is 26.9 °C and has an average temperature of 30.04 °C.

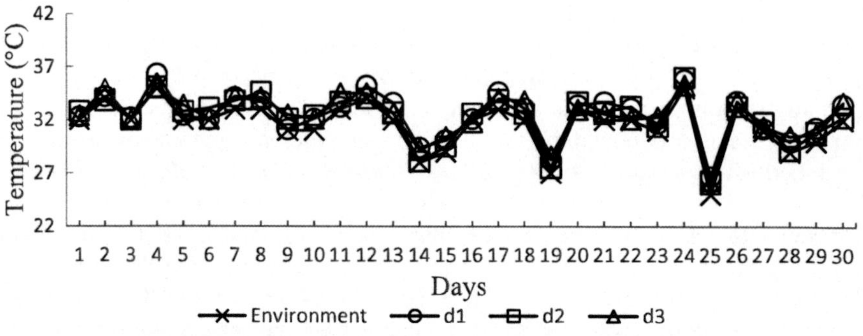

Fig. 2 Environmental and digester temperature at 10.00 WITA

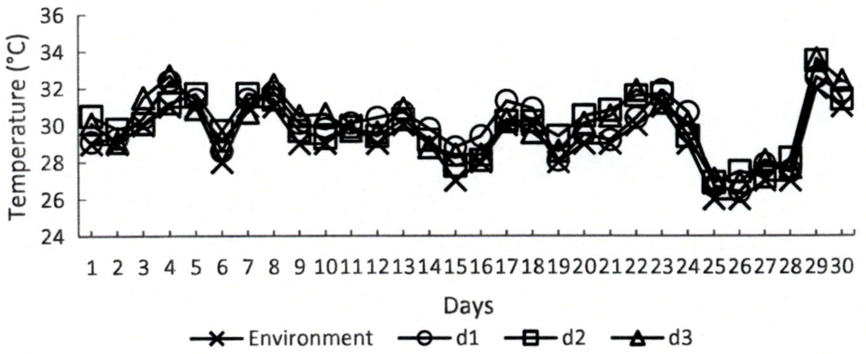

Fig. 3 Environmental and digester temperature at 17.00 WITA

3.2 Biogas Pressure

Figure 4 shows that the pressure at d1 peaked on day four of 0.06292 bar, whereas the lowest point was observed on days 23 and 26 simultaneously of approximately 0.00225 bar, and the average pressure of biogas produced in digester d1 was 0.018 bar. At d2, the highest pressure was found on the third day which is 0.05272 bar and the lowest pressure on the 20th day which was 0.00235 bar, and the average pressure of the biogas digester d2 was 0.012. At d3, the pressure peaked on the third day at an exact 0.06145 bar and the lowest on the 24th day at 0.00176 bar with an average pressure of 0.014 bar. In Fig. 5, the results showed that d1 experienced the greatest pressure on day two of 0.0292 bar and the lowest on day 21 was 0.0009 bar, and the average pressure of biogas produced in digester d1 was 0.0085 bar. Whereat d2 the highest pressure was found on day 1 of 0.02 bar, and the lowest pressure on day 16 was 0.0007 bar, and the average pressure of biogas digester d2 was 0.0049. Lastly, at d3, the biggest pressure is on day one which is 0.0243 bar and the lowest is on day 19 which is 0.0004 bar with an average pressure of 0.0054 bar.

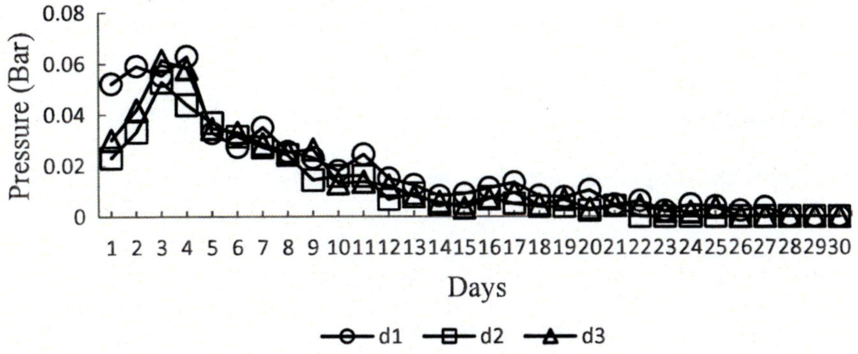

Fig. 4 Biogas pressure at 10.00 WITA

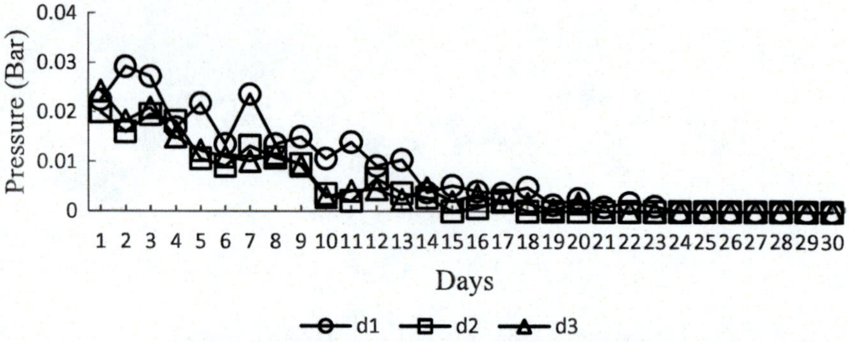

Fig. 5 Biogas pressure at 17.00 WITA

3.3 Biogas Mass

In Figs. 6 and 7, the results show that the biogas mass fluctuates. The total masses obtained when combined are 8.9 g for d1, 5.7 g for d2, and 7.1 g for d3. From the mass data obtained, digester d1 has the highest mass.

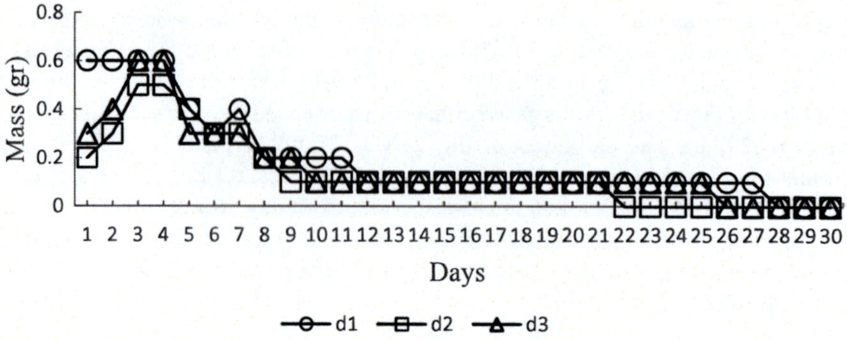

Fig. 6 Mass of biogas at 10.00 WITA

4 Conclusion

In the measured temperature obtained, the average daily temperatures of digesters are as follows: 31.21 °C for d1, 31.10 °C for d2, and 31.24 °C for d3, then the average daily ambient temperature is 30.4 °C. In the mass measurement obtained with three different comparison variations with the results that have been totaled, the digester results d1 of 6.8 g, d2 of 4.0 g, and d3 of 5.6 g. The highest mass was found in digester d1 with a substrate ratio of 40%:30%:30%. The measured pressures obtained with three different comparison variations which consisted of the daily pressure of each digester are 0.0133 bar within d1, 0.0087 bar within d2, and 0.0101 bar within d3. Total production of d1 lasts for 27 days, d2 lasts for 21 days, and d3 for 25 days.

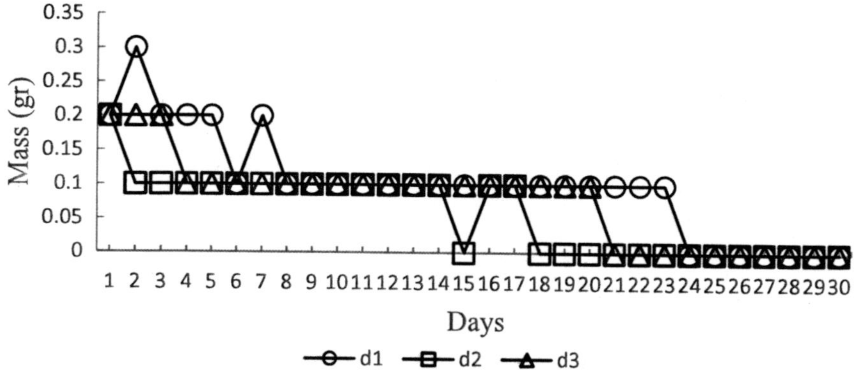

Fig. 7 Mass of biogas at 17.00 WITA

References

1. Sulistiyanto Y, Sustiyah S, Zubaidah S, Satata B (2016) Pemanfaatan Kotoran Sapi Sebagai Sumber Biogas. J Udayana Mengabdi 15(2):150–158 (2016)..
2. Siswanto JE, Susanto A (2018) Analisa Biogas Berbahan Baku Enceng Gondok dan Kotoran Sapi. Chempublish J. 3(1):11–20
3. Haryanto A, Okfrianas R, Rahmawati W (2019) Pengaruh Komposisi Subtrat dari Campuran Kotoran Sapi dan Rumput Gajah (Pennisetum purpureum) terhadap Produktivitas Biogas pada Digester Semi Kontinu. J Rekayasa Proses 13(1):47–56
4. Darnegsih D (2016) Pengaruh Perbandingan Bahan Baku Terhadap Konsentrasi Biogas Dari Eceng Gondok Dengan Menggunakan Starter Kotoran Sapi. J Chem Process Eng 1(1):9
5. Wulandari C, Labiba Q (2017) Pembuatan Biogas Dari Campuran Kulit Pisang Dan Kotoran Sapi Menggunakan Bioreaktor Anaerobik. Inst Teknol Sepuluh Nop, pp 6–15
6. Fitri MA, DhaniswaraTK (2018) Pemanfaatana Kotoran Sapi dan Samoah Sayur Pada Pembuatan Biogas dengan Fermentasi Sampah Sayuran. J Res Technol
7. Horikawa MS, Rossi F, Gimenes ML, Costa CM, Da Silva MG (2004) Chemical absorption of H_2S for biogas purification. Braz J Chem Eng21(3):415–422
8. Putri DA, Tsani ST (2015) Pengaruh Suhu dan Konsentrasi Rumen Sapi Terhadap Produksi Biogas Dari Vinasse. Jurnal Bahan Alam Terbarukan 4(1):1–5
9. Kamal N (2019) Kajian Pengaruh Media Penambat pada Reator biogas fluidized bed. J Teknol 1(33):12–33
10. Indrawati R (2017) Penurunan Bod Pada Biogas Kotoran Sapi Campuran Limbah Cair Industri Penyamakan. J Res Technol 3(2)
11. Melsasail L, Kamagi YEB (2019) Analisis Kandungan Unsur Hara Pada Kotoran Sapi Di Daerah Dataran Tinggi Dan Dataran Rendah. Cocos 2(6)
12. Yahya Y, Tamrin T, Triyono S (2018) Produksi Biogas Dari Campuran Kotoran Ayam, Kotoran Sapi, Dan Rumput Gajah Mini (Pennisetum Purpureum cv. Mott) Dengan Sistem Batch. J Tek Pertan Lampung (J Agric Eng) 6(3):151
13. Novita E, Wahyuningsih S, Pradana HA (2018) Variasi Komposisi Input Proses Anaerobik Untuk Produksi Biogas Pada Penanganan Limbah Cair Kopi. J Agroteknologi 12(01):43

Tar Reduction in a Three-Stage Refuse-Derived Fuel Gasification System by Adjusting Intake Air Ratio

Bambang Sudarmanta(✉), Atok Setiawan, Is Bunyamin Suryo, Harsono, and Sigit Mujiarto

Institut Teknologi Sepuluh Nopember, Sukolilo, Surabaya, Indonesia
sudarmanta@me.its.ac.id

1 Introduction

1.1 Background of Research

In its development, energy demand continues to increase while its reserves are decreasing. In recent decades, the only major source of energy supply has been fossil fuels. On the other hand, all human activities will always leave waste. And as much as 69% of that waste is landfilled, and 10% of the waste is just dumped into disposal site without further processing [1]. Gasification technology is able to convert this waste into renewable energy sources [2, 3].

One of the potential alternative energy sources to be developed is refuse-derived fuel (RDF) obtained from municipal solid waste (MSW). RDF is a product of MSW processing which is flammable and separated from non-combustible parts. To produce RDF, MSW must first be chopped and then carefully sorted to remove all non-combustible materials such as glass, metal, and plastic [4, 5]. Lower heating value (LHV) of RDF can reach up to 5.38 kkal/kg [6]. Gasification is a suitable method for the conversion of RDF into energy in form of synthesized gas, or syngas. Gasification-based waste-to-energy technology is characterized by low emissions and high efficiency.

Thermochemical treatment includes the conversion process caused by the application of heat in a chemical reaction to extract energy [2, 8]. In recent years, gasification has become an interesting research topic in the energy and environmental fields [9]. As an alternative to fossil fuels, syngas can be used to generate heat and electricity [10]. There are downsides to gasification, however, such as high amount of contaminants like tar, ash, and char that can make operations difficult and unsuitable for use in common combustion engines, for example, turbines and gas engines [7].

Using multi-stage gasification system, by which means, introducing free air to pyrolysis reactions can improve tar content by 30% [9]. Air ratio (AR) between first, second, and third stage air intake also influences the performance of gasification and tar reduction, which is why AR needs precise adjustment to achieve best performance in each zone [10, 11], in terms of gas content and reduction of solid char products.

Multi-stage gasifier can increase the overall temperature along the gasification zones and decrease more tar content compared to single-stage gasifier [12, 13]. Multi-stage

M. Kolhe et al. (eds.), *Recent Advances in Renewable Energy Systems*,
Lecture Notes in Electrical Engineering 876,
https://doi.org/10.1007/978-981-19-1581-9_41

gasification can also initiate oxidative pyrolysis reaction that produces heat, which is used for devolatilization. In the reduction zone, similar reactions occur between introduced air and char that stimulates heterogeneous oxidation reaction, which can accelerate the rate of endothermic reactions, increasing flammable gas concentrations, especially CO, and capture tar components [14, 15]. The chemical reactions that occur in multi-stage gasification can be seen in Fig. 1.

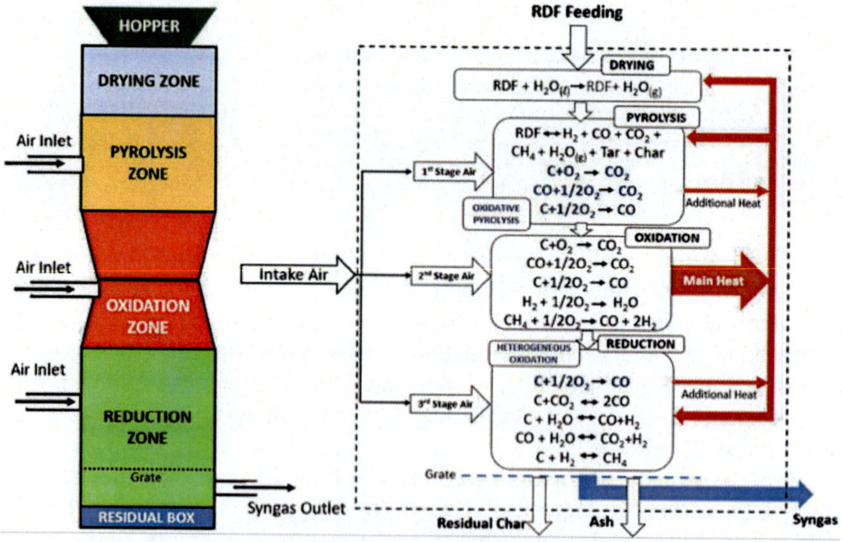

Fig. 1 Multi-stage gasification reactions

Based on multi-stage gasification tar reduction capability, as explained above, the research in this paper was conducted.

1.2 Purpose of Research

The purposes of this research are to find the effects of air ratio adjustments to syngas composition, mainly, in the contents of CO, CH_4, and H_2. Other than that this research also intends to reduce the tar content of syngas, for the purpose of fulfilling the requirement of syngas use in internal combustion engine.

2 Experimental

2.1 Properties of RDF

The raw materials used in this study were RDF in form of pellets or briquettes, obtained from TPSP Bakti Bumi Sidoarjo, East Java. The raw form of this biomass bulk is already in the form of shreds which consists of organic contents (60%) and inorganic contents (40%).

The pelletized fuel used in this study has dimensions of 10 mm in diameter, 50 mm in length, with pellet composition that consists of 60% organic waste and 40% non-organic waste. The characteristics of the RDF are provided in Table 1 after undergoing proximate and ultimate analyses. The analyses were performed in accordance with ASTM standard.

Table 1 Analyses of RDF

Proximate analysis (% mass)		Ultimate analysis (% mass) [4]	
• Moisture	3.68	Carbon	52 ± 2
• Volatile matter	7.70	Hydrogen	8.2 ± 0.7
• Fixed carbon	4.10	Oxygen	28 ± 3
• Ash	14.71	Nitrogen	1.4 ± 0.2
		Sulfur	0.22 ± 0.07
		Lower heating value (KKal/Kg)	5.38

2.2 Experimental Setup

The system diagram, shown in Fig. 2, is a development from previous gasification system [9, 10], which was done with the addition of an automation component that is gasification control unit (GCU).

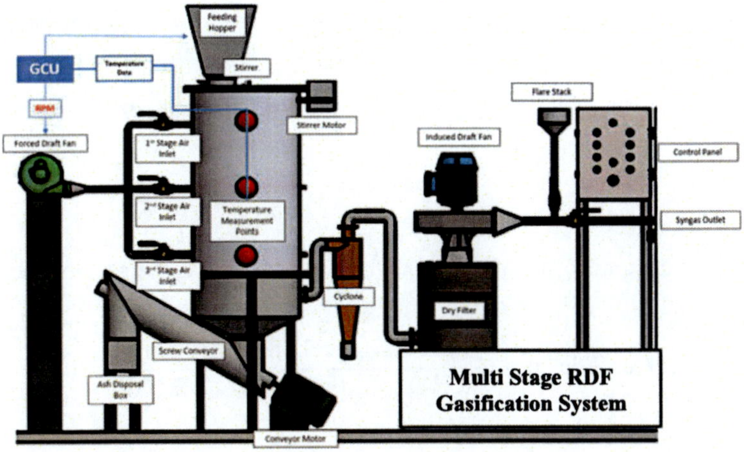

Fig. 2 Diagram of multi-stage RDF gasification system

This gasifier has a height of 1260 mm with internal diameter of 500 mm and ceramic fiber insulation thickness of 40 mm to reduce the rate of heat transfer, while the outer gasifier wall was made using stainless steel material. The air was supplied to the gasifier using a 2-inch forced draft (FD) fan with equivalent ratio (ER) of 0.4, and then distributed to three-stage air intakes. A grate was placed on the bottom of the reactor to suspend RDF pellets above, while allowing ash to pass through. A sweeper is used to assist in ash disposal with a certain sweeping interval. Additionally, a conveyor with an inclination of 30° is employed to dispose of the ash into the ash box for storage. An outlet passage for syngas is located on the lower side of the reactor wall which is connected to the dry filter. For reactor startup, an ignition port is provided on the reactor wall which is used to ignite the fuel.

2.3 Gasification Operation and Procedures

Process stages of multi-stage gasification consist of drying, pyrolysis, oxidation, and reduction. Before RDF was fed into the reactor with screw conveyor, small amount of charcoal is placed on the grate to assist with the starting ignition. Starting ignition is done through the ignition port using a propane torch. ID fan is then operated to prevent pressure buildup inside the gasifier. 9 K-type thermocouples were used to measure temperatures in each zone and recorded using a data logger. Air ratio (AR) is adjusted using intake valves as necessary and measured using a manometer. Oxidation temperature is set to 800 °C. Then do a test for the flammability of syngas in the flare stack. After the temperature reaches steady state, gas and tar samples can be collected.

To calculate equivalent ratio (ER), air ratio (AR), tar content, and cold-gas efficiency, following equations are used:

$$\text{Equivalent Ratio} = \frac{\text{AFR}_{\text{actual}}}{\text{AFR}_{\text{stoichiometric}}} \tag{1}$$

where A/F_{actual} is the amount of actual air that was needed for the gasification reaction and $A/F_{\text{stoichiometric}}$ is the amount air needed for complete combustion.

$$\text{Air Ratio} = \text{Total Air}\left(\text{m}^3/\text{h}\right) \times \text{AR Percentage}(\%) \tag{2}$$

$$\text{Tar Content} = \frac{m_{\text{tar}}}{V_{\text{syngas}}} \tag{3}$$

where m_{tar} is the mass of tar after being weighed and $V\text{gas}$ is the gas volume for each sample

$$\text{Cold Gas Efficiency} = \frac{\text{LHV}_{\text{syngas}}(\text{kj/kg})\dot{m}_{\text{syngas}}(\text{kg/h})}{\text{LHV}_{\text{biomass}}(\text{kj/kg})\dot{m}_{\text{biomass}}(\text{kg/h})} \times 100\% \tag{4}$$

3 Results and Discussion

3.1 Multi-Stage Air Intake Effects in Different Zones

Multi-stage air intakes cause the occurrence of exothermic oxidative reactions, as explained before. The occurrence of these reactions can be seen through temperature

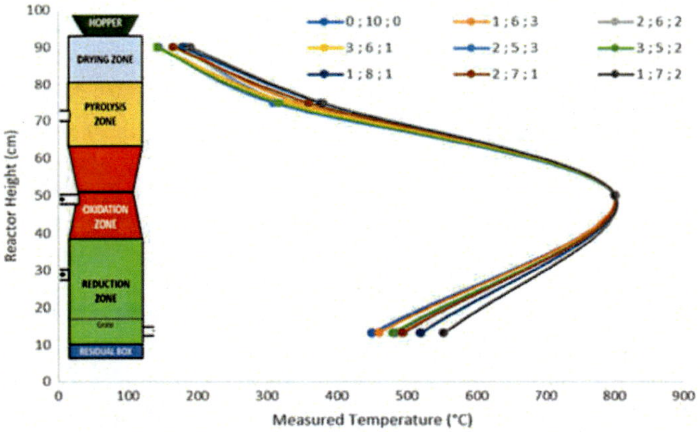

Fig. 3 Temperature distribution along reactor

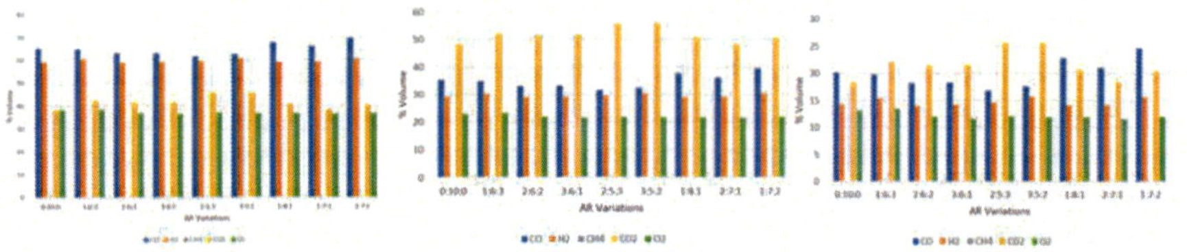

(a) (b) (c)

Fig. 4 Gas composition in the gasifier zones, **a** pyrolysis zone, **b** oxidation zone, and **c** reduction zone

profile and gas composition in the respective gasification zones. For example, inside the pyrolysis zone, compared to single stage AR variation (0:10:0), temperature on different AR variations has relatively higher temperature, which can be seen in the Fig. 3.

In Fig. 4a, it can also be seen that CO and CO_2 composition changes along with variations of AR, compared to single-stage variation (0:10:0). These phenomena shows that oxidative pyrolysis reaction did occur inside the pyrolysis zone. The chart also indicates that the addition of air is not always favorable for the resulting syngas, as shown in Fig. 4 that has fluctuating composition pattern.

While inside the partial oxidation zone, relatively, there was no change in the reactions and temperature, as seen in Fig. 3, since the temperature was set at a fixed point at 800 °C, only that the amount of gas that enter oxidation zone from pyrolysis zone changes due to the occurrence of oxidative pyrolysis in the pyrolysis zone as shown in the following Fig. 4b.

In Fig. 4b, it can be seen that the amount of CO and CO_2 gases in this zone also change along with AR variations. The change in those gases can also be seen to correlate with the previous Fig. 4a in the pyrolysis zone.

On the other hand, the reduction zone also takes effects from AR variations. In multi-stage gasifier, additional air is introduced into this zone which causes the occurrence of heterogeneous oxidation reaction, shown in the previous Fig. 1. In the reduction zone, heterogeneous oxidation occurs between char and intake air that contains O_2. This reaction provides further gas forming reactions, and heat that increases zone temperature.

3.2 Final Syngas Properties

Final syngas product is defined as the syngas that was produced by gasification reactor. Syngas samples were extracted using sample bags which were tested for their flammable (CO, H_2, and CH_4) and non-flammable (O_2 and CO_2) contents. Syngas samples were taken at a set temperature of the oxidation zone with different variations of air ratio.

Figure 5 shows an increase in composition, both flammable and non-flammable gases. The highest composition of CO gas is achieved 19.67% vol. by using a variation of AR 1:7:2, and the highest composition of H_2 using a variation of 3:5:2 was achieved at 10.64% vol. Corresponding to the previous results, the increase in CO and H_2 can be attributed to the increase in overall zone temperature and additional oxidative reactions. While the reduction zone is where the forming process of H_2 gas occurs, when the air input is adjusted correctly, the H_2 gas production is improved, causing further combustion and increasing the temperature in the reduction zone itself. To achieve syngas with most flammable composition, a precise adjustment to AR is needed so that the reactions within the reaction zones can be directed toward flammable gas forming reactions because an improper amount of air in a zone may cause unwanted results, like excessive exothermic reactions or insufficient zone temperature that can hinder flammable gas forming reactions.

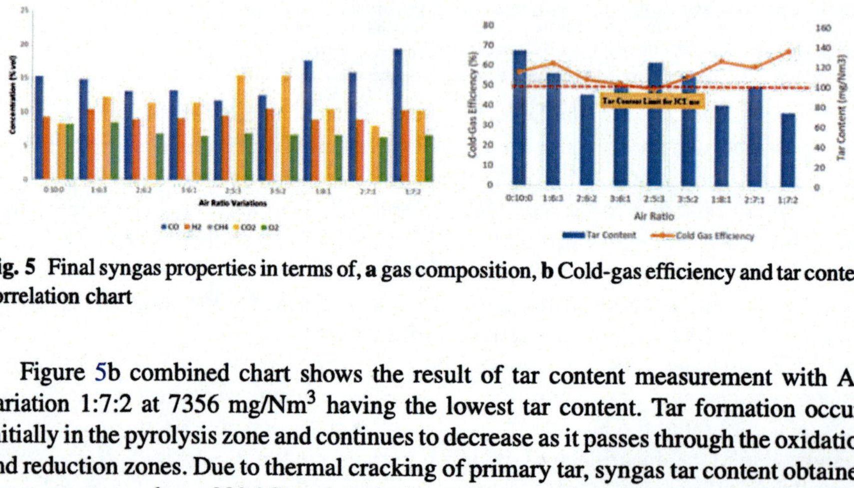

Fig. 5 Final syngas properties in terms of, **a** gas composition, **b** Cold-gas efficiency and tar content correlation chart

Figure 5b combined chart shows the result of tar content measurement with AR variation 1:7:2 at 7356 mg/Nm^3 having the lowest tar content. Tar formation occurs initially in the pyrolysis zone and continues to decrease as it passes through the oxidation and reduction zones. Due to thermal cracking of primary tar, syngas tar content obtained at temperatures above 800 °C tends to be lower because the molecular mass of tertiary tar is lower than primary and secondary tars. This is where the advantage of multi-stage gasifier lies. Due to increased overall zone temperatures from using multi-stage air intake, primary and secondary tars undergo more extensive thermal cracking process, which leads to the formation of lighter tertiary tar.

Meanwhile, in terms of cold-gas efficiency (CGE), the measurement was intended as a parameter for energy conversion efficiency between system input (RDF mass) and output (Syngas mass). The result of the experiment in Fig. 5b shows that varying air ratio had effects on the resulting CGE calculation, with the highest CGE achieved with 1:7:2 air ratio at 67.58%, and the lowest was on AR 2:5:3 at 48.13%.

CGE calculation was directly proportional to syngas and biomass LHV, which explains that if an AR variation has higher syngas LHV compared to others, and with constant RDF LHV, means that said variation has higher CGE. Syngas LHV is directly affected by its flammable composition, whose formation requires a certain temperature range. Therefore, precise AR variations are required to achieve best CGE results since they affect operating temperature, and, subsequently, flammable gas composition.

From the analysis provided above, the AR variation that yields the best results of CGE and tar content are AR 1:7:2, which also means it yields the best syngas quality.

4 Conclusion

The conclusion of this research is that while using a fixed oxidation temperature, variations of AR affect flammable gas composition. In per zone basis, a certain variation of AR can achieve best results. Based on the measurement of gas composition in each zone shows that AR 1;7;2 consistently produced the best composition of flammable gas in all zones.

In terms of final syngas product composition, best result was achieved using AR variation 1;7;2 with gas composition of CO at 19.68% vol, H_2 at 10.55%vol, CO_2 at 10.44%vol, and O_2 at 6.92%vol.

A precise AR variation can minimize the amount of tar content inside the produced syngas and maximize its cold-gas Efficiency, in order to fulfill the criteria for ICE use. In this research, a consistent result could be achieved between CGE and tar content using AR 1;7;2 with its respective values at 67.58% and 73.56 mg/Nm^3.

Therefore, it can be concluded that the best AR variation to use in three-stage downdraft gasifier using RDF fuel is 1;7;2.

Acknowledgements. The authors would like to express their utmost gratitude to the Ministry of Education and Culture—National Research and Innovation Agency of Republic of Indonesia (Kemendikbud—BRIN). Also, to the Gasification Team of Thermal Engineering and Energy System Laboratory, Department of Mechanical Engineering, Institut Teknologi Sepuluh Nopember, which have provided supports to make this important research viable and useful.

References

1. Lokahita B, Samudro G, Huboyo HS, Aziz M, Takahashi F (2019) Energy recovery potential from excavating municipal solid waste dumpsite in Indonesia. Energy Procedia 158:243–248
2. Basu P (2013) Biomass gasification, pyrolysis, and torrefaction. Academic Press, India
3. Sudarmanta B, Gafur A, Saleh AR, Dwiyantoro BA, Sampurno (2018) The effect of two stage gasifying agent on biomass downdraft gasification to the gasifier performance. In: AIP Conference Proceedings 1983
4. Šuhaj P, Haydary J, Husár J, Steltenpohl P, Šupa I (2019) Catalytic gasification of refuse-derived fuel in a two-stage laboratory scale pyrolysis/gasification unit with catalyst based on clay minerals. Waste Manage 85:1–10

5. Násner AML, Lora EES, Palacio JCE, Rocha MH, Restrepo JC, Venturini OJ, Ratner A (2017) Refuse derived fuel (RDF) production and gasification in a pilot plant integrated with an Otto cycle ICE through Aspen plus modelling: thermodynamic and economic viability. Waste Manage 69:187–201
6. Mujiarto S, Sudarmanta B, Fansuri H, Rahman A (2021) Comparative study of municipal solid waste fuel and refuse derived fuel in the gasification process using multi stage downdraft gasifier. Automot Exp 4(2):97–103
7. Kardani N, Zhou A, Nazem M, Lin X (2021) Modelling of municipal solid waste gasification using an optimised ensemble soft computing model. Fuel 289:119903
8. Sudarmanta B, Setiyawan A, Bachtiar AKP, Hidayatulloh D, Ependi DP, Saleh AR (2019) Numerical study of the effect of oxidation zone inlet air temperature variation on municipal solid waste pellet gasification process on downdraft type reactor characteristics. In: AIP conference proceedings, vol 2187, issue 2019-12-10. Conference Proceeding
9. Saleh A, Sudarmanta B, Fansuri H, Muraza O (2020) Syngas production from municipal solid waste with a reduced tar yield by three-stages of air inlet to a downdraft gasifier, Fuel, vol 263, issue 2020-03-01
10. Sudarmanta B, Sampurno, Dwiyantoro BA, Gemilang SE, Bachtiar AKP (2019) Performance characterization of waste to electric prototype uses a dual fuel diesel engine and a multi-stage downdraft gasification reactor. In: Materials science forum, vol 964 MSF, issue 201-901-01 | Book SeriesDilihat di Kardani
11. Bhattacharya SC, Hla SS, Pham HL (2001) A study on a multi-stage hybrid gasifier-engine system. Biomass Bioenerg 21(6):445–460
12. Li D, Briens C, Berruti F (2015) Oxidative pyrolysis of kraft lignin in a bubbling fluidized bed reactor with air. Biomass Bioenergy 76:96–107
13. Shi H, Si W, Li X (2016) The concept, design and performance of a novel rotary kiln type air-staged biomass gasifier. Energies 9(2):1–18
14. Striugas N, Zakarauskas K, Džiugys A, Navakas R, Paulauskas R. An evaluation of performance of automatically operated multi-fuel downdraft gasifier for energy
15. Aljbour SH, Kawamoto K (2013) Bench-scale gasification of cedar wood—part II: effect of operational conditions on contaminant release. Chemosphere 90(4):1501–1507

The Effect of Corncob-Active Carbon Adsorbent Mass on Methane and Carbon Dioxide Content on Biogas Purification

Slamet Wahyudi(✉) and Lia Fitriya

Brawijaya University, Malang 65141, Indonesia
slamet_w72@ub.ac.id

1 Introduction

The "Waste to Energy" program, where waste that is no longer used by the community can be converted into renewable energy, one of which is biogas. Biogas is produced from a biological process (anaerobic digester) which produces gas consisting of 55–60% methane (CH_4), 35–40% carbon dioxide (CO_2), and several other compounds such as water vapor (H_2O) and hydrogen sulfide (H_2S) [1]. In biogas, some gases can reduce the calorific value which has an impact on the combustion process.

The quality of biogas produced from several types of livestock manure is still low because it contains carbon dioxide. One of the efforts to improve the quality of biogas is the biogas purification process by reducing the carbon dioxide content to increase the calorific value [2, 3].

Adsorption is a binding interaction process of a gas or liquid molecule (adsorbate) on the surface of a solid (adsorbent) and is influenced by the physical and chemical properties of the adsorbent (pore size, surface area, and chemical properties of the adsorbent), the composition, and operational conditions of adsorption [4].

One of the adsorbent materials is activated carbon which contains 87–97% carbon compounds, and the rest is in the form of hydrogen, oxygen, sulfur, nitrogen, and other compounds formed from the carbon activation process. Activated carbon can adsorb gases [5] and certain chemical compounds; its adsorption properties depend on the pore volume and surface area. Corncobs are one of the most widely available lignocellulosic wastes [6] in Indonesia and become a source of porous activated carbon.

Several studies of biogas purification to reduce CO_2 levels have been carried out [7] using natural zeolite and activated carbon as adsorbents that can provide better adsorption performance than without activation, due to the more micropores produced accompanied by crystal agglomeration. the smaller one. Meanwhile, research [8] showed that biogas flow rate, mass, and zeolite concentration affected the CO_2 gas adsorption process. Meanwhile [9] utilizes tiles—zeolite as an adsorbent in the biogas purification process to reduce CO_2 levels by using a 2^3-factorial design.

The purpose of this study was to reduce the concentration of CO_2 gas to increase CH_4 gas through biogas purification by utilizing corncob-activated carbon as an adsorbent. According to [10], with the use of activated carbon on corncobs, it is easier to absorb

M. Kolhe et al. (eds.), *Recent Advances in Renewable Energy Systems*,
Lecture Notes in Electrical Engineering 876,
https://doi.org/10.1007/978-981-19-1581-9_42

CO_2 when compared to synthetic zeolite, and with a slow level of activation of activated carbon saturation, activated carbon can absorb contaminant gases for longer. According to [8], absorption of CO_2 gas by the adsorbent is carried out physically because of the Van Der Waals force where the attraction between carbon dioxide and the adsorbent is greater than the attraction between methane gas and hydrogen sulfide in biogas, so it can absorb CO_2 content.

2 Methodology

2.1 Material

1. Biogas from chicken and cow manure from farms Gondanglegi Malang
2. The adsorbent of corncob carbon activation as shown Fig. 1

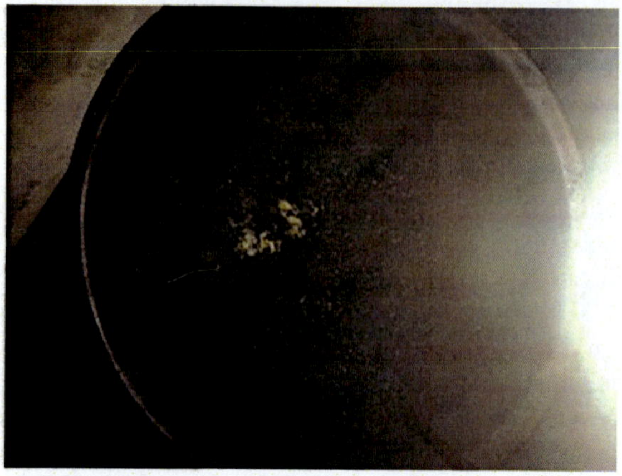

Fig. 1 Active carbon of corncob

3. Purification and compression equipment as shown Fig. 2
4. Gas chromatography–mass spectrometry (GC–MS), Greenhouse Gas Laboratory, Agricultural Environmental Research Institute, Pati, Central Java.

2.2 Method

2.2.1 Making Adsorbent from Corncob Charcoal

Corncobs are washed and then dried in the sun to dry. Furthermore, the corncobs are put into the combustion drum and burned, a few hours until the temperature is ±500 °C so that it becomes carbon charcoal. Corncob charcoal was activated by soaking it in 500 mL 2 M HCl for 36 h. Wash the carbon charcoal with water to pH 7 and then heat it in the oven at 600 °C for 5 h. Cool the corncob carbon charcoal to room temperature.

Fig. 2 Biogas purification equipment

2.2.2 Biogas Purification

The data collection procedure is shown in Fig. 2. In the purification tube installation, there are three parts, measuring = 0.075 m and 0.5 m long. Tube I contains 1 kg of corncob-activated carbon, and the mass is fixed, while tube II contains 50-gr corncob carbon charcoal (can be replaced by 150 gr); in this part of the tube, the adsorbent mass varies (50 or 150 gr), while the third tube contains air which aims as a mixer, then the adsorbent tube is tightly closed so that no gas leaks out.

Sampling was carried out at biogas flow rates of 1 and 2 L/min and time variations of 5 and 15 min. After the sample is taken after biogas purification, repeat the previous step by closing the regulator first and then replacing the activated carbon mass on the corncobs in tube II.

If all samples have been taken, biogas samples without treatment or with certain variations of treatment are tested for the content of CH_4 and CO_2 of biogas using gas chromatography–mass spectroscopy (GC–MS) at the Greenhouse Gas Laboratory of the Pati Environmental Research Institute, Central Java [11]. The results of CO_2 and CH_4 data from the GC–MS test were analyzed using a factorial design to determine the main parameters that affect biogas quality.

3 Result and Discussion

3.1 Result

3.1.1 CO_2 and CH_4 Concentration from Biogas Purification

Biogas that has been purified using activated carbon from corncobs as an adsorbent produces biogas products consisting of carbon dioxide gas and methane gas in the sample bag. The research data are presented in Table 1.

3.1.2 Analysis of 2^3 Design Factorial

(1) Carbon Dioxide (CO_2)

Table 1 CO_2 and CH_4 biogas concentration from purification

M (gr)	*Q* (L/minutes)	*T* (min)	CO_2 (%)	CO_2 (%)	$\bar{y}$	CH_4 (%)	CH_4 (%)	$\bar{x}$
50	1	5	12.29	2.94	7.62	22.65	6.88	14.77
150	1	5	11.49	1.79	6.64	27.70	3.17	15.44
50	2	5	16.54	13.50	15.02	29.29	25.82	27.55
150	2	5	13.56	13.78	13.67	32.59	29.95	31.27
50	1	10	15.90	3.17	9.54	29.76	6.67	18.21
150	1	10	13.58	14.63	14.11	33.20	31.44	32.31
50	2	10	1.84	16.09	8.97	31.67	29.26	30.46
150	2	10	5.12	13.89	9.51	12.01	30.55	21.28
Without treatment			14.86			19.75		

Through the analysis of 2^3-factorial designs [12], the following equation is obtained:

$$y = 10.632 + 0.347M + 1.158Q - 0.103t - 0.549MQ \\ + 0.93Mt - 2.45Qt - 0.457Qt \quad (1)$$

Equation 1 is the percentage of carbon dioxide which was analyzed to determine the effect of interaction between variables on the concentration of carbon dioxide after purification.

From Eq. (1), the coefficient M is positive, which means that the greater the mass of the adsorbent will increase the concentration of CO_2, due to the lack of homogeneity of the adsorbent so that the contact area between carbon dioxide and the adsorbent is not optimal. Unlike research [9], the mass of the adsorbent increases, the carbon dioxide content in the biogas will decrease.

The *Q* coefficient is positive; increasing the flow rate in purification will increase the concentration of carbon dioxide, according to research [13, 14].

The negative *MQ* coefficient or mass interaction factor with biogas flow rate influences decreasing CO_2 concentration although the value is small.

Equation 2 is the result of factorial design 2^3 which was analyzed to determine the effect of interaction between variables on the results of CH_4 concentration after purification.

$$x = 23.912 + 1.163M + 3.729Q + 1.656t - 2.531MQ + 0.067MT \\ - 3.426Qt - 3.292MQt \quad (2)$$

It can be explained that the M coefficient is positive indicating that an increase in the mass of the adsorbent will increase the percentage of CH_4 in large quantities. With the increase in the mass of the adsorbent, the CH_4 concentration will increase due to a decrease in the value of the CO_2 concentration in the biogas purification result. The larger the mass of the adsorbent, the volume will also increase which will increase the area of the adsorbent; the contact area between CO_2 gas and the adsorbent will increase, and if the CO_2 content decreases, the ratio of CH_4 to CO_2 will be higher.

A positive Q coefficient means that the increase in flow rate will increase the post-purification CH_4 concentration, according to [13, 14]. The addition of flow rate to the activated carbon adsorbent of corncobs can bind CO_2 in the pores of the adsorbent due to van der Waals forces.

The mass interaction coefficient and MQ flow rate are negative, meaning that there is an interaction effect between variables on the decrease in CH_4 although it is relatively small. The increase in mass has the potential to increase the CH_4 level because the CO_2 content of the biogas is reduced [15, 16]. Meanwhile, increasing the flow rate has the potential to reduce the ratio between CO_2 and CH_4; because even though a lot of CH_4 is obtained, a lot of CO_2 is not adsorbed so that the ratio decreases [17].

The interaction coefficient of flow rate and time Qt is negative, which means that there is an interaction effect between variables on the decrease in CH_4. This happens because the increase in flow rate has the potential to reduce the ratio between CO_2 and CH_4; because even though a lot of CH_4 is obtained, a lot of CO_2 has not been adsorbed so that the ratio decreases [17]. While the increase in the purification time will make the CO_2 adsorption increase to the optimum point, then decrease, thereby reducing the ratio of CO_2 and CH_4 levels [18].

3.2 Discussion

3.2.1 Biogas Purification Analysis on Carbon Dioxide Concentration.

Figure 3 is the concentration of CO_2 after biogas purification with various treatments. There are differences in the concentration of CO_2 biogas absorbed by the adsorbent in each variable. In the 50-g corncob-activated carbon mass, the CO_2 content is higher (15.02–7.62%) when compared to the 150 g mass (14.106–6.636%), meaning that if the mass is greater, the bond between CO_2 levels and the adsorbent will be even greater. The absorbed CO_2 is greater and the CH_4 ratio increases [5].

There is a difference in the binding of CO_2 to the mass of 50 g adsorbent for a higher flow rate of 2 L/minute (15.02–8.96%) when compared to a flow rate of 1 L/minute (14.106–6.636%), the higher the flow rate of biogas, the higher the CO_2 flow rate. The larger the biogas, it means that the absorption capacity of the adsorbent is reduced. While the adsorbent mass of 150 g with a flow rate of 1 L/minute, and a purification time of 15 min has a high CO_2 concentration (14.106%) when compared to the same mass and flow rate, but the purification time is 5 min (6.636%). This happens because at minute 5 the adsorbent is not yet in a saturated state so that CO_2 cannot be adsorbed optimally [19]. At a mass of 50g corncob-activated carbon with a flow rate of 2 L/minute and a purification time of 5 min produced higher biogas CO_2 (15.02%) when compared to a purification time of 15 min with the same treatment (8.96%). It means that the longer the purification time, the more CO_2 will be bound by the adsorbent. Similarly, the adsorbent mass of 150 g and the biogas flow rate of two liters/minute with a purification time of five minutes resulted in a higher CO_2 content (13,669%) when compared to a purification time of fifteen minutes (9,507%). However, at the biogas flow rate of 1 L/minute with a mass variation of 50 and 150 g, there are deviations. This is because the longer the

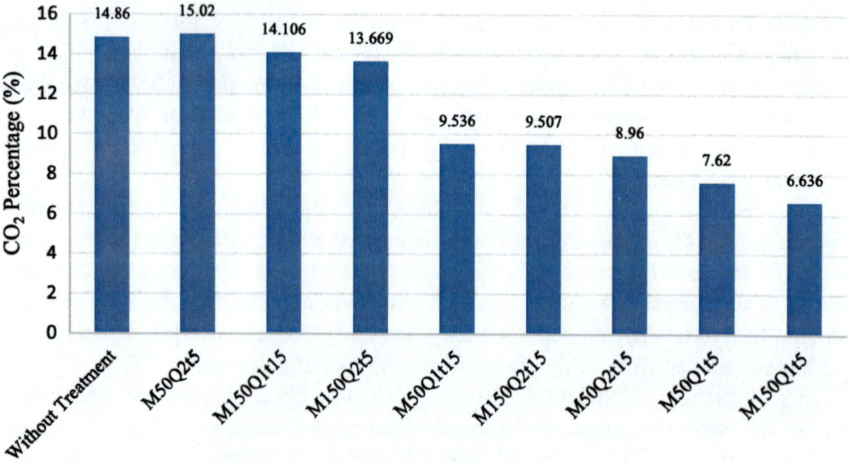

Fig. 3 CO_2 concentration of biogas purification

purification time, most of the pores of the adsorbent have been filled with CO_2 molecules; the adsorption ability decreases significantly. After all, the adsorbent is already at the saturation point [7].

These results are consistent with what was stated [20] that increasing the mass of corncob-activated carbon as an adsorbent will reduce the value of CO_2 concentration from biogas purification, which means the volume increases and increases the contact area between carbon dioxide gas and the adsorbent so that the absorption capacity increases due to van der Walls forces. The decrease in CO_2 concentration in biogas will increase the ratio of CH4 concentration in biogas, thereby increasing its calorific value [21].

3.2.2 Effectiveness Analysis of Biogas Purification

Pareto curves as shown Fig. 4 resulting from biogas purification for reducing carbon dioxide levels, each factor has a response magnitude to carbon dioxide levels. The effectiveness of this binding is how much the concentration of carbon dioxide is bound by the adsorbent after purification and compared with the percentage of carbon dioxide before the purification process so that the effectiveness of carbon dioxide binding in each variation of biogas purification is obtained.

In Fig. 4, the highest effectiveness value occurs in the M150Q1t5 treatment (mass 150 g—flow rate 1 L/minute—purification time 5 min) of 55.340%. The lower the percentage of carbon dioxide concentration, the higher the value of CO2 binding effectiveness. This happens because of variable interactions that make the contact area between the adsorbent mass of 150 gr corncob-activated carbon and carbon dioxide gas more evenly distributed, the binding of carbon dioxide by the adsorbent is more evenly distributed, and the biogas that is wasted is lower so that the binding of this variable interaction makes the effectiveness of carbon dioxide binding by the corncob-activated carbon adsorbent becomes high.

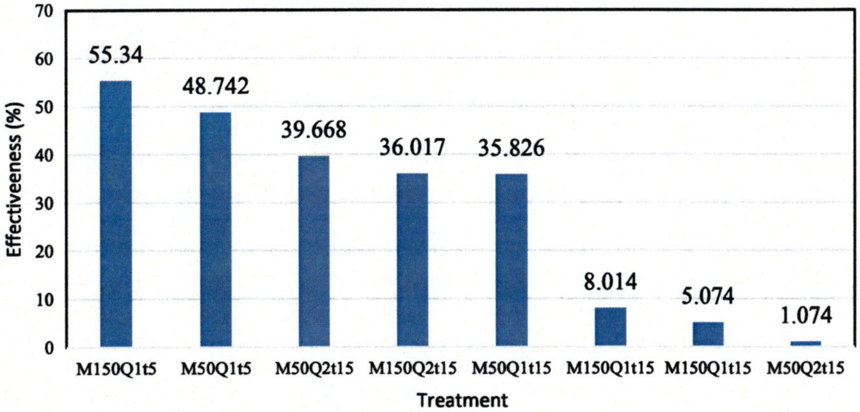

Fig. 4 Pareto curve of the percentage of carbon dioxide binding effectiveness by response factors of treatment

4 Conclusion

Conclusions from the analysis of biogas purification research with red brick powder as an adsorbent include:

1. Corncob-activated carbon can be used as an adsorbent in biogas purification to improve biogas quality.
2. The mass factor of corncob-activated carbon adsorbent has the highest effect (0.347%) on the decrease in carbon dioxide concentration.
3. The increase in biogas flow rate worsens the adsorption capacity (+1.158%) because it increases the carbon dioxide level, while the addition of biogas purification time has a very small effect (−0.103%) on the decrease in carbon dioxide levels.
4. The interaction of adsorbent mass, biogas flow rate, and treatment purification time of M150Q1t5 resulted in the lowest carbon dioxide concentration of 6.636% with the percentage of carbon dioxide absorption effectiveness value of 55.340%.

References

1. Sawyerr N, Trois C, Workneh T, Okudoh V (2019) An overview of biogas production: fundamentals, applications and future research. Int J Energy Econ Policy
2. Hermawan D, Hamidi N, Sasongko MN (2016) Performansi purifikasi biogas dengan koh based absorbent. Jurnal Rekayasa Mesin 7(2):65–73
3. Sari K (2019) Efisiensi proses adsorpsi menggunakan silika gel terhadap kadar bioetanol bonggol jagung (zea mays). Skripsi
4. Al-Jlil SA, Alsewailem FD (2009) Lead uptake by natural clay. J Appl Sci 9 (22):4026–4031
5. Sembiring M, Sinaga T (2003) Arang Aktif (Pengenalan dan Proses Pembuatannya). Jurusan Teknik Industri USU, Sumatera Utara
6. Gupta VK, Carrott PJ, Singh R, Chaudhary M, Kushwaha S (2016) Cellulose: a review as natural, modified and activated carbon adsorbent. Bioresour Technol 216:1066–1076

7. Padang Y, Tira H (2016) Removal of CO_2 and H_2S from raw biogas using active natural zeolite. In: Proceedings of the international mechanical engineering and engineering education conferences (IMEEEC), 030006-1-030005
8. Budinova T, Ekinci E, Yardim F, Grimm A, Björnbom E, Minkova V, Goranova M (2006) Characterization and application of activated carbon produced by H_3PO_4 and water vapor activation. Fuel Process Technol 87:899–905
9. Apriyanti E (2012) Adsorpsi CO_2 Menggunakan Zeolit: Aplikasi Pada Pemurnian Biogas. Majalah Ilmiah 10(22)
10. Wahyudi S, Arif M, Hidayati N (2020) Aplikasi Desain Faktorial 23 pada Kadar CO_2 Proses Purifikasi Biogas dengan Adsorben Genteng—Zeolite. Rekayasa Mesin 11(3):409–414
11. Anggara RD (2015) Meningkatkan Kualitas Biogas dalam Proses Pemurnian Biogas dengan Menggunakan Zeolit Sintetik dan Karbon Aktif dari Tongkol Jagung sebagai Adsorben. skripsi
12. Pavia DL et al (2006) Introduction to spectroscopy, 4th edn. Brooks Cole, United States of America
13. Ritonga AM, Masrukhil M (2017) Optimasi Kandungan Metana (CH_4) Biogas Kotoran Sapi Menggunakan Berbagai Jenis Adsorben. Jurnal Rona Teknik Pertanian
14. Eze JI, Agbo KE (2010) Maximizing the potentials of biogas through upgrading. Am J Sci Ind Res 1(3):604–609
15. Lhanafi S, Anfar Z, Chebli B, Benafqir M, El Haouti R, Azougarh Y, Abbaz M, El Alem N (2018) Factorial experimental design to enhance methane production of dairy wastes co-digestion. Sustain Environ Res 28(6):289. Agadir
16. Saleh A et al (2014) Increasing percentage of methane (CH_4) from biogas with purification by using zeolite membrane. In: The 5th Sriwijaya international seminar on energy and environmental science & technology, Palembang
17. Walozi R et al (2016) Application of low-pressure water scrubbing technique for increasing methane content in biogas. Univ J Agric Res 4(2):60–65. Uganda
18. Widhiyanuriyawan D et al (2014) Purifikasi Biogas dengan Variasi Ukuran dan Massa Zeolit terhadap Kandungan CH_4 dan CO_2. Jurnal Rekayasa Mesin 5(3):27–32
19. Mrosso R et al (2020) Removal of hydrogen sulfide from biogas using red rock. Hindawi J Energy. Arusha
20. Gil MV, Gutiérrez N, Martínez M, Rubiera F, Pevida C, Morán A (2015) Carbon adsorbents for CO_2 capture from bio-hydrogen and biogas streams: breakthrough adsorption study. Chem Eng J 269
21. Kapdi SS, Vijay VK, Rajesh SK, Prasad R (2004) Biogas scrubbing, compression and storage: perspective and prospectus in Indian context. Centre for Rural Development and Technology, Indian Institute of Technology, New Delhi

Author Index

A
AL-bonsrulah, Hussein A. Z., 1
Amarta, Valiant Tirta, 153
Ardita, I Nengah, 53
Ariani, Betty, 289
Arifah, Septia Kurniawati, 231
Arsana, Verry Mardiananta, 197

B
Bambang Arip, D., 145
Budianto, Arif, 145

C
Carolina, Ester, 69

D
Dewi, Nadia Sari, 87
Djanali, Vivien Suphandani, 61, 113
Dwiyantoro, Bambang Arip, 189

F
Farid, Abdul Rochman, 171
Fathallah, Aguk Zuhdi M., 289
Firdaus, Firiana, 45
Firmansyah, Iman, 241
Fitriya, Lia, 377

G
Graha, Petra Arde Septia, 341
Gunawan, Hafiz Rayhan, 79
Guntur, Harus Laksana, 61

H
Hafani, Muhammad Dimas, 341
Halim, 303
Hamzah, Afan, 341
Hantoro, Ridho, 69, 223
Harsono, 369
Hidayat, Yuniawan, 231
Huda, Khoirul, 189

I
Ikhwan, Nur, 249
Izdiharrudin, Mokhammad Fahmi, 223

J
Johari, Anwar, 321

K
Khoirul Effendi, M., 131
Kurnadi, Mohamad, 131
Kurniawan, Aris, 121
Kusumadewi, Tri Vicca, 205, 353

L
La, Deluxe, 205
Lototskyy, Mykhaylo, 19

M
Made Ariana, I., 289
Manta, Faisal, 361
Marbun, Aripin Gandi, 161
Marzuki, 257
Mufti, Nandang, 321
Mujiarto, Sigit, 369

N
Nugroho, Ardi, 281

M. Kolhe et al. (eds.), *Recent Advances in Renewable Energy Systems*,
Lecture Notes in Electrical Engineering 876,
https://doi.org/10.1007/978-981-19-1581-9

Nuryanto, Firman Ali, 361
Nur Yuniarto, M., 105

O
Oktavianto, Denny, 249

P
Pambudi, Puguh, 153
Parsons, Adrian, 19
Pasupathi, Sivakumar, 19
Permanasari, Avita Ayu, 321
Pertiwi, Shinta Dewi Surya, 341
Prabowo, 189, 205, 241
Prasetiyo, Ardianto, 311, 321
Pratama, Rizal Mahendra, 181
Pratiwi, Karenina Anisya, 341
Purwadi, Diki, 113
Puspitasari, Poppy, 321
Puspitasari, Yuliana Dewi, 341
Putra, Arief Laga, 281
Putra, Ary Bachtiar Krishna, 257, 303

R
Rahmawati, Fitria, 231
Ramadhan, Radix Kautsar, 79
Ramadhansyah, Ferdina, 153
Reddy, M. V., 1

S
Sania, Vernanda, 333
Saputro, Galang Adi, 79
Sarwono, 69
Schnell, Uwe, 29
Septyaningrum, Erna, 69, 223
Setiawan, Andri, 171
Setiawan, Atok, 369
Setiyawan, Atok, 333
Setyawan, Digdo Listyadi, 353
Shafi, Muhammad Haekal, 153
Simon Prayoga, Adriska, 271
Sirojuddin, 87, 97
Soeprijanto, 341
Somantri, Wahyu, 213
Suamir, I Nyoman, 53
Suanggana, Doddy, 361
Subekti, 61
Sudarmanta, Bambang, 295, 369
Sukarni, Sukarni, 311, 321
Sukarno, Ragil, 87, 97
Suryo, Is Bunyamin, 29, 369
Susanto, Adi, 37
Sutardi, 121, 197
Sutikno, 131
Suwarno, 113
Syaifudin, Achmad, 61, 171

T
Temaja, I Wayan, 53
Tolj, Ivan, 19
Tondok, Putra Dilto, 361

U
Utomo, Christiono, 257

V
Veeman, Dhinakaran, 1

W
Wahjudi, Arif, 37, 45
Wahyudi, Slamet, 377
Wibawa, Agus, 105
Widjaja, Arief, 341
Widjayanto, Teguh, 189
Widodo, Wawan Aries, 45, 213, 281
Wikarta, Alief, 153
Winangun, Kuntang, 333
Windharto, Agus, 171

Y
Yohanes, 105
Yuansah, 213
Yudisaputro, Hendra, 105
Yulianto, Danan Tri, 295
Yuwono, Tri Yogi, 29, 181

Z
Zahara, Alya Awanis, 97